Small Comforts for Hard Times

HUMANISTS ON PUBLIC POLICY

introduced by Florian Stuber

with a foreword by James Gutmann

·Michael Mooney & Florian Stuber · editors

Small Comforts for Hard Times

HUMANISTS ON PUBLIC POLICY

columbia university press   new york

1977

A slightly longer version of the Nisbet paper appeared as "The New Egalitarians" in *The Columbia Forum*, n.s. 4 (Winter 1975): 2–11. The Gans response to Nisbet appeared as "How Equal, Equal How?" in *The Columbia Forum*, n.s. 4 (Spring 1975): 36–40.

A longer version of the Frankel paper appeared in the *Mississippi Law Journal* 46 (1975): 591–609, as the second of three "Memorial Lectures" given by Mr. Frankel in March 1975 at the University of Mississippi Law School.

Part of the Gaylin paper appeared as "Harvesting the Dead" in *Harper's*, September 1974.

A more fully annotated version of the Mayer paper appeared in the *Journal of Modern History* 47 (1975): 409–36.

Lines of poetry on pages 30–31 and 79–80 are quoted by permission of the publisher from *The Collected Poems of Wallace Stevens*, © 1954 Alfred A. Knopf, Inc.

The novel selection on pages 81–82 is quoted by permission of the publisher from Saul Bellow, *Herzog*, © 1964 The Viking Press, Inc.

Library of Congress Cataloging in Publication Data

Main entry under title:
Small comforts for hard times.

Papers from a conference series conducted under the auspices of Columbia's program of University Seminars.
Includes bibliographical references and index.
1. Equality—Congresses. 2. Civil rights—
Congresses. 3. Technology and civilization—Congresses.
4. War and society—Congresses. 5. Education,
Humanistic—Congresses. I. Mooney, Michael.
II. Stuber, Florian, 1947–   III. Columbia University.
University Seminars.
JC575.S52   309.1   77-5851
ISBN 0-231-04042-3

Columbia University Press
New York   Guildford, Surrey

For James Gutmann

"guide, philosopher, and friend"

Contents

Foreword by James Gutmann   xi

Acknowledgments   xv

Introduction   1
   Florian Stuber

Justice and Human Equality   11

1. Justice—Compensatory and Distributive   13
   Herbert A. Deane

2. Equality, Race, and Preferential Treatment   26
   Paul A. Freund

3. The Costs of Equality   34
   Robert A. Nisbet

4. The Costs of Inequality: In Response to Robert A. Nisbet   50
   Herbert J. Gans

5. Justice, Equality, and the Economic System   59
   William S. Vickrey

6. Equality and Fraternity: A Note on Subjective Realities   72
   Benjamin DeMott

Private Rights and the Public Good   85

7. Private Rights and the Public Good   87
   Charles Frankel

8. Public Rights and Private Interests:
   In Response to Charles Frankel   103
   Hannah Arendt

9. On Privacy and Community   109
   Emile Capouya

10. Do Rocks Have Rights? Thoughts on Environmental Ethics   120
    Roderick Nash

Technology and the Ideal of Human Progress   135

11. Living with Scarcity   137
    Roger L. Shinn

12. The Technology of Life and Death   152
    Willard Gaylin

13. Biomedical Progress and the Limits of Human Health   170
    Daniel Callahan

14. Technology and the Structuring of Cities   182
    David P. Billington

15. The Aesthetics of Technology:
    In Response to David P. Billington   199
    Mario G. Salvadori

War and the Social Order   205

16. On National Frontiers: Ethnic Homogeneity and Pluralism   207
    William H. McNeill

17. The Lower Middle Class as Historical Problem   220
    Arno J. Mayer

18. Reflections on War, Utopias, and Temporary Systems   246
    Elisabeth Hansot

Education and the Good Society   261

19. The University and American Society   263
    George W. Pierson

20. The University, Society, and the Critical Temper:
    In Response to George W. Pierson   277
    Wm. Theodore de Bary

21. Some Questions in General Education Today   281
    Steven Marcus

22. Some Inconsistent Educational Aims   303
    Onora O'Neill

23. The Disestablished Humanities   308
    Rosemary Park

24. A View from the Ivory Tower: In Response to Rosemary Park 321
    Robert W. Hanning

The Humanities and Public Policy   333

   25. A Philosophic Perspective   335
       Abraham Edel

    Notes   385
    Index   393

Foreword

James Gutmann

That humanists are concerned with matters of public policy should surprise no one in this day and age; most humanists have long since emerged from their so-called ivory tower. But it was not always so.

Those of us whose memories go back to the first World War and earlier have witnessed a change in the humanities—literature, the arts, and social studies—not unlike some of the changes which have taken place in the attitudes of natural scientists since the beginning of the atomic age. The explosion of the first atom bomb forced the natural scientists to reconsider their aloofness in secluded laboratories and to face their responsibilities regarding social and political issues. Of course this was not an unprecedented outlook in scientists. The physical chemist Frederick Soddy, for example, refused to continue his services as technical adviser to the British War Office and, after the 1918 armistice, declared that he would not prostitute his science to bellicose nationalism. There could be, however, as Soddy himself knew, no retreat to the confines of the academy to pursue "knowledge for its own sake," and with our advance into a nuclear age, followed by troubling developments in chemical and biological science, researchers came to accept a regard for consequences, human and ecological, as part of their scientific task.

A change similar to such reorientation of natural science took place in humanists' view of their responsibilities as citizens some twenty-five years earlier. Though certain humanists undoubtedly continued, by conviction or habit, to dwell in their ivory tower, many literary and artistic leaders and most social scientists changed their outlook after World War I and emphasized, or at least included in their purview, an interest in the living scene. Recent events and the study of contempo-

rary civilization came to occupy their attention in apparent defiance
of the Hegelian dictum that the owl of Minerva flew only at dusk.

Yet the boundaries of "contemporary civilization" were at first limited to civilization in the Western world: witness the pioneering course
of this title in Columbia College (to speak only of the institution I
know best) which, in 1919, grew out of the study of "Issues of the
War." The same was largely true of Columbia's General Honors
sequence organized by John Erskine, the source of the well-known
seminars on the Great Books at the University of Chicago and St.
John's College and of the successor to Erskine's General Honors, the
Columbia College Colloquim. Even the Columbia Humanities
courses, established in 1937 as a parallel to the courses in Contemporary Civilization, were limited to the great works of occidental cultures. Though there were exceptions here and there, the study of
oriental humanities and civilization as a part of liberal education was
not instituted until after World War II, in part because many Western-
ers, certainly most Americans, could be truly said to have first become aware of the values of Eastern culture through wartime experience. How inadequate this awareness and the education which depended on it was is readily seen in the lack of genuine understanding
manifested by Americans during the Korean and Vietnamese wars.

However, some awareness that liberal education must transcend
the frequently illiberal traditions of the liberal arts of the nineteenth-
century academies might well be said to have accompanied the enlarged world view of the twentieth century. Doubtless there were
some losses of values attained by earlier scholarship. But liberal education, reconsidered as liberalizing the minds of students—and of
their teachers—from narrow conventionalism emphasizes the variety
of human experience available to scholarly inquiry. In this sense, too,
the concept of academic freedom was redefined to go beyond mere
resistance to interference by forces and interests outside the academy.

That such resistance is essential cannot be denied by anyone who
witnessed the incursions of the McCarthy era. Provoked, in part, by
academic and some journalistic opposition to the excesses of the cold
war, McCarthy's "anticommunist" crusade was at long last halted by
those who took the constitutional guarantee of freedom of speech
and of the press seriously—a journalist spokesman like Edward R.

Murrow and an aroused Academy for which academic freedom was by no means academic in the pejorative sense of the term. Twenty-five years of revitalized liberalism in education, despite its admitted defects, made a telling difference in effective resistance to McCarthy.

Another quarter century during which the cold war generated the heated wars in Southeast Asia can now be seen in perspective. It is important to recognize that spokesmen for the humanities and especially the social studies were sharply divided in their attitudes toward the Marshall and Truman doctrines, as they were later toward the wars in Korea and Vietnam. However, the recognition by the vast majority of their involvement as humanists in questions of public policy was part of the stimulus which aroused student reactions in the sixties.

On balance, efforts by humanists in our century to engage and test public policy have been both sobering and emboldening. If educators have gained confidence in their ability to illuminate not merely the patterns of nature and the cosmos but also the ways of society and history, they have also been brought to realize how frequently their focus on events has been too narrow, and how seemingly timeless issues change and enlarge as the humanists' own disciplines develop and renew themselves. The change in language over the last two decades from "Adult Education" to the now more fashionable "Continuing Education" may signal a recognition by educators that there will never be a day when they can cease to rethink the essential functions of their calling.

As Plato envisaged his philosopher-kings not as philosophers become rulers but as governors with philosophic insight into their problems, so humanists pondering public policy are to be seen as life-long students. The function of the humanities in respect to public policy issues is not to supply answers to age-old problems but rather to ask such questions as illuminate and clarify the problems faced by successive generations of mankind in terms of the needs relevant to today and, if the humanists be truly wise, tomorrow.

Acknowledgments

Since this book is the culmination of a three year enterprise, we are hard put to it to single out a few from among the many whose aid and advice proved helpful to us in the various stages of the project. Rightly, we should name all those who participated in the Conferences, for without their humanistic wit and wisdom and concern for public policy, there would have been no reason for this book at all. We can hope only that they will understand the space limitations we labored under and that they will agree that these selected papers offer a just and accurate reflection of what went on during those Fridays at Columbia's School of International Affairs.

There are some, however, whom not to name would be arrogant dismissal indeed. More than thanks are certainly due to Stephen Goodell of the National Endowment for the Humanities, whose idea for this project it originally was, and to Charles Frankel of Columbia, who gave that idea its ultimate shape and structure. George R. Collins, Robert Gorham Davis, Ward H. Dennis, William T. R. Fox, and Walter Gellhorn assisted in planning the project in detail. Later, as the conferences got under way, they could not have gone on so well as they did without Jeffrey Field, who oversaw and encouraged their progress from Washington, and the moderators of the five conference series who oversaw and directed discussions in New York: Wm. Theodore deBary, Ainslie Embree, Louis Henkin, Harvey Picker, I. I. Rabi, and Aaron Warner. Coordinating these sessions was made easier for us thanks to Alice Maier, the associate director of Columbia's University Seminars, whose knowledge of administrative procedures was invaluable.

Editing *Small Comforts for Hard Times* was made possible in part thanks to a grant from Columbia's Program of General Education in the Humanities. The pleasures of preparing the volume were due, in no small measure, to the people we have been privileged to work with: when faced with some painful editorial decisions, we found we could count on the understanding cooperation of our essayists. To each of them, our thanks, especially to Arno Mayer of Princeton, who took time from research in Paris to rework his essay for us at a very critical moment, and to Abraham Edel, whose essay makes our book a book. We also owe large debts of gratitude to John D. Moore, editor-in-chief of Columbia University Press, whose patience and wisdom served us well when they were needed most, and to Arlene Jacobs, our typist, and our friend.

It is useless even to try to say in a simple acknowledgment what we feel about James Gutmann, director of the Conferences and for many years head of the University Seminars. By dedicating this book to him, however, we have the pleasure of knowing we act for many more than ourselves who have often sought a way to honor him.

M.M.—F.S.
New York City
December 1976

Small Comforts for Hard Times

Let us dismiss the Epicureans without any derogatory comment as they are excellent fellows and happy in their belief in their own happiness. Only let us warn them to keep to themselves as a holy secret (though it may be extremely true) their doctrine that it is not the business of a wise man to take part in politics; for if they convince us and all our best men of the truth of this, they themselves will not be able to live the life of leisure which is their ideal.

Cicero
De oratore

Introduction

FLORIAN STUBER

George Pierson once told the following story. Sometime in the early 1950s, a dinner was held at Yale University in honor of President Harry Truman. Arriving late at the dinner because of an evening class, Pierson found himself at the end of a short receiving line and able, as a result, to have a brief conversation with the President. Learning that Pierson taught history, Truman became enthusiastic about his favorite subject: he had always considered himself a student of history, he said, and read it voraciously; he had the greatest admiration for the professional historian; a keen sense of history, he thought, was a most important aid in guiding a Chief Executive to make appropriate decisions regarding public policy. Pierson responded by asking Truman whether, in light of his views, he had ever considered establishing the position of Historian as a permanent part of the President's cabinet. The suggestion struck Truman as a remarkably funny joke.

The exchange could be taken as representative of the attitude many public officials have had toward the role professional humanists should, or rather should not play in the formulation of public policy. It would be representative indeed if admiration for the work of the humanist was more widespread than it is; but most are indifferent and some contemptuous of the "pointy-headed intellectual" and the "egghead vote." Since philosophy bakes no bread, study of the humanities provides at best pleasant distraction from the real business of our daily lives. The humanities themselves are esoteric, arcane, irrelevant, a body of knowledge communicable only in rarified atmospheres.

Ironically enough, this attitude is reinforced by a corresponding attitude taken by many humanists toward their own profession. For them, their pursuits are pure precisely because they are removed from the public arena. They take warning from the example of Socrates, a humanist killed by the State for bringing philosophy into the public forum. As students of human affairs, they will analyze more objectively, more truly, if they remain mere observers of the human scene, above the battle, above the fray. In weighing the respective merits of the active and contemplative lives, a long-standing humanistic tradition self-servingly places the highest value on the latter. The active life involves too many compromises with one's ideals, too many hands wash too many backs, politics being a dirty business that contaminates all involved with it. A grudging admiration is conceded to those staunchest advocates of the active life—an Isocrates in Athens, a Cicero in Rome, a Bruni in Florence—whose lives and works may even be offered to fire the ideals of the young. But, it is felt, the truly mature, self-respecting humanist must, in the end, retire to the Garden of Epicurus where, with dispassionate ease, he can contemplate the truths of nature and history hidden from those too close to the scene.

Thus, on both sides, suspicion matches suspicion, breeding mutual distrust, even mutual contempt.

These are undoubtedly extreme views, and are rarely embodied in their pure state. And indeed, there are minimal, if essential points of contact between the politician who manages the culture of the present and the humanist who sees himself as the custodian of the culture's past. In a democracy such as ours, for example, the humanist as citizen participates in forming public policy when he votes and petitions and, as Watergate has made vividly clear, no public official can expect to remain in office for long without an understanding and respect for the values and traditions which inform the culture he has in his momentary charge. Whether these points of contact can or should be broadened remains the question.

There have been times when contact has become identity. Plato's ideal of the philosopher-king may never have been realized (perhaps never should be), but a number of humanists have had notable careers in public life. Consider, for example, the career of Joseph Addison, who began his adult life as a fellow of Magdalen College at Oxford and served as Secretary of State for the English government for a year before his death in 1719. His life was spent shuttling back and forth between the world of letters and that of state affairs. Whatever were the public policies he helped formulate, his most substantial contribution was a humanistic enterprise: the writing of the *Spectator* papers. These papers, written with Richard Steele and others, appeared daily from March 1, 1711, to December 6, 1712. Their scope is enormous, for their topics range from criticism of literature and the other arts to discourse on the possibilities increased trade held for forming a world community, from speculations on the discoveries of a New Science and the applications of a new psychology to observations on the changing life in town and country and the relations between the sexes. Written for a newly self-conscious middle class, the *Spectator* papers articulated, and formed, the values of that class; during the two centuries following their first publication, they were a text second only to the Bible in their influence on British manners and morals. The essays created a distinct cultural tone, and two generations after they were written, Samuel Johnson paid tribute to Addison's civilizing accomplishment in his *Life of Addison* by saying: "That general knowledge which now circulates in common talk was in his time rarely to be found. Men not professing learning were not ashamed of ignorance, and, in the female world, any acquaintance with books was distinguished only to be censured. His purpose was to infuse literary curiosity by gentle and unsuspected conveyance into the gay, the idle, and the wealthy; he therefore presented knowledge in the most alluring form, not lofty and austere, but accessible and familiar. When he showed them their defects, he showed them likewise that they might be easily supplied. His attempt succeeded; inquiry was awakened, and comprehension expanded. An emulation of intellectual elegance was excited, and from this time to our own, life has been gradually exalted, and conversation purified and enlarged."

Johnson wrote those words in 1777 by which time, on the other side of the Atlantic, two men whose values and prose had been formed in great measure by study of the *Spectator* had incorporated them in a Declaration of Independence. Ten years later, when they themselves were internationally acclaimed humanists and philosophes, Benjamin Franklin and Thomas Jefferson would take leading roles in constituting a government to preserve those values, incorporating them in a Bill of Rights.

☙

It was thus in recognition of the vital role the humanities can play in both forming and reflecting a culture that on the eve of the nation's third century Congress established a National Endowment for the Humanities. And it would have been reasonable to suppose that among the first projects to be subsidized by that Endowment would be one that undertook to investigate the ways in which humanistic learning could help test and formulate public policy. Nor is it surprising that Columbia University should be the place where such a project would originate: long before the student riots of 1968 associated the university with an unfortunate sort of activism in the public mind, its pedagogic traditions were those which tried to investigate connections between the past and the present (the introductory course in history and philosophy required of all freshmen since 1919 still is called, and not irrelevantly, "An Introduction to *Contemporary* Civilization"). The project was organized and conducted under the auspices of Columbia's program of University Seminars, which drew on some thirty years of experience in dealing with such subjects as The State and The Renaissance from a multidisciplinary approach.

The result was a substantial project indeed, a series of sixteen conferences on the humanities and public policy issues, held over the course of three academic semesters and involving some one hundred fifty humanists and scholars from over forty institutions across the country. Taking as a premise that "the humanities give light when used as aids to the understanding of current urgencies," the program proposed to determine ways in which professional humanists could assist in the exploration of various types of public issues.

To this end, five interlocking series of three conferences each were set up to probe currently debated public policy issues embraced by the following five themes: Justice and Human Equality, Private Rights and the Public Good, Technology and the Ideal of Human Progress, War and the Social Order, Education and the Good Society. A conference took place every two or three weeks, on Fridays, and five conferences were held each semester, thus completing in thirteen weeks one round of each series. At both the morning and afternoon session of each conference, a paper was delivered, and a designated respondent would speak to the points raised therein, thus opening and framing the seminar-like discussions which followed. In this way, humanists addressing issues from the perspectives of their specialized disciplines during the first fifteen conferences provided empirical data for the work of a final conference on methods.

As administrative coordinators of the conference series, Michael and I were to make sure that the mechanical aspects of the project went smoothly—that papers were received, typed, xeroxed, and sent to participants in time for study before a conference; that rooms were organized and set up for the conferees; that our $100,000 budget was kept in order and our various payments made as efficiently as bureaucracy would allow. These graceless labors were rewarded when, as graduate students in the humanities, we participated in discussing our country's concerns with some of our most illustrious scholars. Our discussions took place on the top floor of Columbia's School of International Affairs, in glass-enclosed rooms which overlook Harlem on the east, midtown Manhattan on the south, the university campus on the west and north. We often reflected on the irony of the setting: here were the humanists in their tower, all right, and there was the life on the mean streets below. But it was with the quality of that life, as it was and as it might be, that we were concerned. And as we argued, sometimes with passion and often with eloquence, there gradually emerged a clearer understanding not only of the forces which shape the issues confronting our world but also of the present state and purpose of our profession. Indeed, it was with a renewed faith in the significance of humanistic learning in our time that we, as editors, have prepared this book.

All the papers in this volume derive from the work of the conference series. Most of them were originally written for and delivered at one of the conferences while others (namely those with the subtitle "In Response to . . .") are written versions of oral comments made during discussion of the principal papers. Drawn from tape-recordings of the sessions, these essays have been shaped specifically for this book while most of the others have been revised, in some cases substantially, for inclusion here.* (A few of the essays had since their presentation found their way to publication in various scholarly

---

* All the conference papers in their original form as well as the tape recordings of the sessions are held in the Special Collections of the Columbia libraries and are open for study to any interested citizen.

journals and are here returned to the context which generated and
gives edge to their meaning.)

A book, as I. A. Richards once said, is a machine to think with. In
selecting these twenty-five essays from the thirty-eight papers and
eighty hours of talk that made up the conferences, we have tried to
construct such a machine. The first five sections of the book are
named after the five conference themes: in each section, the essays
are arranged so that their juxtaposition gives a semblance of debate
over the issues at hand. The book as a whole has an inner argument
which culminates in the final section containing a single, major essay
by Abraham Edel. In this essay, Edel reads through and analyzes all
the preceding papers in a new way, placing them in a philosophic
perspective which delineates the interface between humanistic learning and the formulation of public policy.

While all the essays in the book remind us that public policy decisions ultimately deal with notions of community, affecting as they
must the very texture of our social fabric, perhaps none brings the
matter so clearly home to us as the papers in the first section of the
book, "Justice and Human Equality." It was Tocqueville who noticed, in the early nineteenth century, the power of the notion of
equality in our culture; for him, it was the point of entry into the
American mind, and its advance in all forms of life and society enjoyed, he thought, a kind of divine warrant such that it could with
greater or lesser ease be accommodated but could not be successfully resisted. Now, and in the name of justice, Americans are demanding not only equality before the law and equality of opportunity
but equal pay for equal work or, more boldly, across-the-board economic equality. Thus, in two sharp debates, a historian of political
thought and a professor of law ask how "just" are notions of compensatory justice while two sociologists speculate on the changes our
social order must experience if programs calling for a redistribution of
wealth are instituted. In the last two essays in this section, the terms
of the discussion are themselves questioned as an economist
wonders whether justice is indeed the best guide to social policy and
a literary critic suggests that concern over equality may distract us
from the challenge of realizing a higher social ideal, that of fraternity.

A declaration has it that the American nation was founded on two
principles: that all men are created equal, and that they are endowed

with inalienable rights to life, liberty, and the pursuit of happiness. While the first principle is the subject of the first series of papers, the second principle is the subject of the next, "Private Rights and the Public Good." It can be argued that the enduring challenge of any society is to care for the public good without violating the rights of individual citizens and that, happily, the two interests often coincide. But the American dream puts particular stress on how an individual may live life as he pleases, and testimony to the power behind this dream can be seen in recent decisions of the Supreme Court reaffirming a "right to privacy" which no law by Congress can restrict. What are these private rights, and what ideal of community do they serve? These questions are debated in four essays written by two philosophers, a literary critic, and a historian.

If the essays in the opening sections of the book are about contemporary attempts to put into practice two of our most cherished values, those in the third section deal with the conflicts in value which arise from the technological achievements that for so much of this century have been a primary source of our national pride. No longer, we are painfully aware, does the advance of technology mean a progress pure and simple; of technology's possible futures, some are sinister and others unpredictable. In "Technology and the Ideal of Human Progress," a theologian ponders the kinds of decisions we must make as we face the "limits to growth," a psychiatrist and a philosopher assess the potential social impact of the "miracles" of modern medicine, and in Addisonian debate, an engineer and an architect remind us that technology, properly imagined, controlled, and directed, may yet serve as handmaiden to beauty.

America's place in world affairs is the general subject of the fourth section of essays, "War and the Social Order." Together, the three essays included here remind us that the humanities deal not only with matters of sweetness and light but that they are also a source for understanding the darker side of the human condition, what Melville called our "little lower layer." The principal subject is viewed first from a world-historical perspective in a paper which traces the changing shape of national frontiers on the world's surface as civilizations and their various social orders have had their rise and fall through war. In this context, the second paper in the section takes on particular point, for here another historian analyzes the role a particular class plays in sustaining and maintaining a social order during times of crisis. The class singled out for attention is one which recent appeals to a "Silent Majority" have made the subject of much American concern—the lower middle class. In a final essay, a political scientist asks whether utopian dreams of international peace, if realized, could truly answer all our human needs, whether indeed any social ordering, short of the disorder of war, can fulfill them.

Essays on "Education and the Good Society" comprise the fifth section of the volume and here, as it were, the book turns in upon itself. In all the preceding essays, humanists were looking out, from the points of view offered by their various disciplines, on the state of our society and the world. Now they look in upon themselves, assessing the state of their own profession both within the academy, which some take to be their bastion, and in our culture. The section opens with an historical overview which maps out the changing public attitudes toward higher, specifically humanistic, education in American society. Then, after a warning about the nature of the critical temper which is the humanist's distinguishing tool, a literary critic and a philosopher consider how our educational institutions currently reflect the tensions of contemporary society: a probing analysis of an apparent lack of confidence in our culture is perhaps the most disturbing essay in the book while that concerning the university's aim to help individuals realize the American dream of both happiness and success is possibly the most radical critique. In the section's closing essays, a professor of education and a professor of English survey humanistic institutions other than the university in our society, and glimpses of a potentially rich relation between humanistic learning and the quality of life in our culture are seen. Thus, the section prepares the way for Abraham Edel's concluding essay which directly treats with this, our major question.

∽

Just before we began editing our book, one of the participants in the conferences, John Lowenthal of Rutger's Law School, suggested that we take a look at Arthur Koestler's recent novel, *The Call Girls*. The novel is a satire on a symposium much like our own, and he thought it would amuse us.

The satire is grim indeed. Amidst signs of an imminent World War III, Nikolai Borissovitch Solovief, an aging physicist of world renown, gathers at the Kongresshaus in Schneedorf twelve of the academy's finest minds to deliberate "Approaches to Survival." Here, in the tranquility of *Höhenluft,* the human condition is to be analyzed, a consensus reached, and a new "Einstein letter" formulated—a message sent to world leaders such as that sent by Einstein to Roosevelt in 1939, to legendary effect. To Solovief, his colleagues are the world's wisest, all Nobel laureates or near-misses, acute, urbane, and caring; but to the more knowing among them, they are mere academic call girls, who for "return fare economy class and a modest honorarium" would be delighted, thank you, to make the conference, and every other on the circuit. By week's end, however, all hope for consensus founders on the very jealousies and aggressive traits for which a solution was to be found: in lieu of a message to the world, the entire conference proceedings are to be published, with Foundation support, as the official report. But during the final night, the tapes go up in flames—the work of an overwrought secretary— and as the bus pulls out next morning, the world mobilizing for war, Solovief is left to ponder the words he mumbled pensively to his wife in bed the first night: "About that Einstein letter. He and his buddies knew what the problem was, and were searching for the solution. We cannot even define the problem. Each one of us has a different solution. And that, precisely, defines our problem."

We may take some comfort from the fact that the notables gathered at Schneedorf were scientists, not humanists, and that the premise from which their discussions proceeded—that there is a "harmony in the cacophony of the universe" which, once discovered, gives clear, unmistakable directions—is not one many humanists would start from. We are saved from such idle confidence by the very tools that distinguish and bedevil us: historical consciousness and a sense for the absurd. Like mimes at court, we trade as much in folly as in dignity and heroism, too richly aware, perhaps, of our own and others' ill-spent efforts and misbegotten hopes, of ideals that stir and lead one generation only to ruin the next.

And yet . . . When we gather at meetings and conduct our symposia, do we not harbor an ill-defined hope that in some unknown manner our talk will produce the Way of Reason, a oneness of spirit

and heart that knows itself to be sound, plausible, even unim-
peachable? Would we undertake our endless rounds of discussion if
we didn't believe that our efforts could produce something firm and
endurable—solutions, not mere options? Disguised in such hope is
no naive expectation of an intellectual *tour de force,* as though our
work might actually result in a humanistic equivalent of the Einstein
letter. What is present in our sentiment, rather, is an extension of that
confidence that quietly builds as we perceive the effect of our work in
the humanities on our own lives—on the richness of our consciousness and on our ability to take possession of our destiny. Might not
similar strengths, we wonder, be gathered by a culture reflecting
humanistically upon itself?

Analogies indeed are possible. Engaged with issues of pressing
public concern, the humanities offer distinct forms of illumination and
hope: they free us from the tyranny of the moment by placing us
within a cultural context and making available to us our inheritance;
they broaden the range of alternatives we are likely to canvass and
indicate that solutions that appear attractive at the moment may have
been tried before and that something can be learned from the experience; they also humble us in the face of the present, reminding us
that threats to received wisdom are often large with promise—and
that some have been able to make that wisdom wiser. Generally,
they sharpen our perception of events, define issues, give voice to
doubts, and make more vivid and comprehensive the values by
which the search for solution is guided.

No, *Small Comforts for Hard Times* is not an Einstein letter to the
nation although some of our essayists do put forth attractive proposals to help alleviate the situations we face. Instead, the essays presented here are in the tradition of the *Spectator* papers—essays of
clarification, and ultimately, inspiration. For if our humanists convince
us that the times are hard, their intelligence and sensitivity are themselves a promise of what will see us through. These may seem small
comforts, but they are nonetheless real.

Justice and Human Equality

Does the difference of knowledge, of means, and of wealth, observable hitherto in all civilized nations, between the classes into which the people constituting those nations are divided; does that inequality, which the earliest progress of society has augmented, or, to speak more properly, produced, belong to civilization itself, or to the imperfections of the social order?

Marie J.A.N. Caritat, marquis de Condorcet
The Progress of the Human Mind, 1795

1

Justice—Compensatory and Distributive

HERBERT A. DEANE

Though the term *compensatory justice* was already in currency, the claims it embodies were brought dramatically to the fore in May 1969 when James Foreman interrupted a service in Riverside Church in New York to deliver the "Black Manifesto," demanding $500 million "in reparation" from white churches and synagogues for their part in the exploitation of American black people. Foreman argued, in brief, that, since the year 1619, blacks in the United States had been "exploited and degraded, brutalized, killed and persecuted" by whites, and that white America—in this case the white religious establishment—owed reparations to the present black community as the heirs of the victims. The monetary compensation demanded was seen as only a fraction of the astronomical wealth exacted through exploitation by white society, and shared in by its religious establishment, and was to be used to fund the development of those institutions, services, and skills for living blacks which their heritage had deprived them of and which were needed to enable them to compete as the socioeconomic equals of whites.[1]

Perhaps because of the revolutionary rhetoric in which the legal and moral argument was couched, the Manifesto attracted little immediate attention from philosophers and political scientists. Social and political commentators were rather more responsive, yet their assessments of the Manifesto's claims were anything but uniform and

Herbert A. Deane is Lieber Professor of Political Philosophy at Columbia University.

even so staunch an advocate of social equality as Michael Harrington
rejected the scheme as "outlandish."[2] In 1972, however, in an effort
to sharpen its argument, Hugo Adam Bedau recast the Manifesto in
sixteen cumulative assertions and endeavored to identify and rebut
some of the principal objections to it.[3] Since that time, a considerable
literature on the subject, including several philosophical defenses of
"compensatory justice," has been published in professional journals,
notably in *Analysis* and in *Social Theory and Practice* and *Philosophy
and Public Affairs,* two journals founded in 1972 as forums for the
philosophical examination of public issues.

At the same time, an analogous body of literature has appeared in
various law reviews. These articles were occasioned in the main by
the Supreme Court's hearing (but eventually declaring moot) the
case of *DeFunis v. Odegaard,* in which a student sued for admission
to the University of Washington Law School on the grounds that minority students with entrance test results clearly inferior to his own
had been admitted to the school because of their race. Indeed, the
DeFunis case rather than the Black Manifesto is more representative
of the current debate over compensatory justice: quixotic claims to
monetary compensation have given way to the actual use of preferential policies that employ racial, ethnic, or sexual classifications, justified on grounds of past discrimination against such groups. And it is
this statement of the issue that I mean to address: granted that certain
members of certain groups have, because of their membership in
these groups, been discriminated against in the past, does justice
require that we now "reverse" such discrimination by giving preferential treatment—in hiring, admissions, and the like—to current
members of these groups? I am aware that some arguments for preferential treatment of minorities and other groups have been mounted
in terms of distributive justice and still others in terms of social utility
or public welfare; but these are not my concern. Here I wish merely
to test the theoretical tenacity and practical wisdom of "compensatory justice," asserted as a basis for giving something more than
equal treatment to members of certain groups in our society.[4]

I

Since I am an historian of political thought, my first impulse when
confronted with an issue such as this one is to turn to the classic texts

in which such problems have been discussed to see if they offer us any help in ordering or clarifying our thoughts. So I turned to Book Five of Aristotle's *Nicomachean Ethics* where he discusses the concept of justice and its various subdivisions. There, it will be recalled, he distinguishes "general justice" or justice as such, which deals with the whole of goodness or virtue insofar as it involves our dealings with our neighbors, from "particular justice," which concerns the fair or equal; he then divides the latter into two kinds: *distributive justice,* which is manifested in the apportioning of offices, honors, money, or other goods that are to be divided among the members of a polity, and *corrective justice,* which plays a rectifying part in transactions between individuals.

In considering distributive justice, Aristotle is careful to distinguish his own view from that of the democrats, who hold that offices and honors should be distributed to all citizens on the basis of arithmetic equality, and that of the oligarchs, who argue for a distribution in proportion to the wealth of different citizens or groups of citizens. Both in the *Ethics* and in the *Politics,* Aristotle criticizes and rejects these views, since both neglect the essential end or purpose of the state—the performance of good or virtuous action. He argues for a conception of distributive justice which holds that offices, honors, and money are to be distributed in proportion, not to wealth, but to worth or merit, i.e., to the contribution a citizen makes to this end of the state, the performance of good actions.

But let us turn to his other major division of particular justice, corrective justice, since at first glance at least this seems to have more to do with our problem of compensatory justice. Here the situation is quite different: in the course of various voluntary transactions between individuals (such as sales, purchases, loans, or deposits) or as a result of certain involuntary transactions between individuals (whether clandestine, like theft, adultery, poisoning, or assassination; or violent, like assault, murder, robbery with violence, or abuse), one individual has suffered loss or injury and the other has inflicted damage or harm, and, in so doing, has achieved gain or profit that is unfair. The law, Aristotle says, looks only to the difference created by the injury and treats the two men as previously equal; the judge seeks to remove the arithmetic inequality that has been introduced into their relationship and to restore the original state of equality by imposing a penalty upon the offender equal to the gain he has illegitimately obtained. "It is," he says, "as though there were a line divided into unequal parts, and he took away that by which the greater segment exceeds the half, and added it to the smaller segment. And when the whole has been equally divided, then they say that they have 'their own'—i.e., when they have got what is equal" (*Ethics* 1132a). And in this same section of the *Ethics*, Aristotle also criticizes the Pythagorean view that justice is reciprocation—that A shall have done to him what he has done to B; this kind of reciprocal treatment is not always the appropriate form of rectification or punishment.

What, then, can we glean from this examination of Aristotle's discussion of justice that may illuminate our problem? First, it is absolutely clear, I think, that for Aristotle distributive justice *always* involves the distribution of common goods such as offices and honors among the citizens of a state, "among those who have a share in the constitution," as he puts it in the *Ethics* (1130b). At one point, it is true, Aristotle speaks briefly about distribution from the common funds of a partnership (1131b), but what he means here, I think, is simply that such distribution will be made according to the ratio which the funds put into the business by the partners bear to one another. It is distribution in accordance with a geometrical proportion, and *in this respect* it is like the distribution of offices and honors in the state which he has called distributive justice. But the fact that it is carried on in accordance with the same kind of proportion does not make it identical with or a part of distributive justice, since Aristotle firmly rejects the principle of distribution according to wealth when he is discussing distributive justice. Throughout the *Ethics* and the *Politics*, he uses this term only with respect to distributions of common goods among citizens, and always counterposes his own correct theory of distributive justice, i.e., distribution in proportion to merit, to the mistaken theories of democrats and oligarchs, both of which are also obviously "political" conceptions of justice.

In the second place, Aristotle's conception of corrective or remedial justice is, I think, concerned essentially with transactions between individuals in which one person has suffered harm or injury and the other person, who is held responsible for that harm or injury, is required to make restitution to the injured party or is compelled to undergo punishment to restore the balance between the parties. I see no reason why the concept cannot be extended, though Aristotle himself did not do so, to deal with transactions between legal persons or, possibly, between states, or between legal persons and real persons, so that we could speak, for example, of compensation owed by an individual to a group of partners who own a business. But even this extension in no way blurs the line which Aristotle draws between distribution of common goods among the citizens of a polity and rectification of injustices that occur in transactions among individuals. In every case of corrective justice that Aristotle mentions or discusses, one of the parties involved behaves wrongly or unfairly, and *for that reason* is required to make recompense to the injured party or to suffer punishment.

For all these reasons, I believe that one of the central flaws of current efforts to erect theories of compensatory justice (which state that a society as a whole or a majority of its members owe compensation to some or all of the members of certain minorities because of systematic mistreatment or discrimination extending back for several generations or centuries) is that such theories are attempting to use a principle of corrective justice, not, as Aristotle did, to restore equality where it had been ruptured by the actions of an individual who had acted unjustly, but rather *as a new,* and in some cases possibly a revolutionary, *principle of distributive justice.* Jobs, educational opportunities, perhaps even wealth and income, are to be distributed not primarily in accordance with principles of merit or qualifications or effort, but at least as much—or more—in accordance with membership in a group that has been in the past, and perhaps still is, the victim of discrimination by other groups or by the rest of society. Obviously, at this stage of the argument I am not saying anything about the justice or injustice of the claims for redistributing social goods made by the members of any minority group. All that I am now arguing is that these claims involve new or revised principles of distributive justice, and that nothing but confusion is achieved by ignoring that fact and by talking about these claims as though they were demands for the application of the principles of corrective justice.

II

Apart from this general problem of conceptual confusion, the idea of compensatory justice presents numerous other difficulties of both a theoretical and practical nature. The few that I will mention are in no way exhaustive, but they involve issues which lie at the center of the political process and thus impinge most directly upon the lives and sensitivities of citizens.

The first difficulty is the simple matter of applicability—how far the principle of compensation for past discrimination is to be extended. Would not the members of many ethnic, racial, and religious groups in American society be able to claim that they deserve reparations or compensations on the ground that their present levels of wealth and well-being are markedly lower than they would have been if they had not been systematically discriminated against in the past? Can we ignore the fact, for example, that Jews have been the victims of serious discrimination in many parts of the world, including America, for hundreds of years? Even if it were the case that Jews in America now have average incomes and educational and occupational opportunities equal to or greater than the averages for the population as a whole (and I am not at all sure that this is the case), could they not argue, on principles of compensatory justice, that they are entitled to reparations to make up the difference between their present level of wealth and well-being, whatever that may now be, and the level that they would have attained had they not, for centuries, been the victims of discrimination? And could not similar claims be advanced for the immigrant Irish, Italians, Poles, and Puerto Ricans?

In his defense of the Black Manifesto, Hugo Bedau dismisses this objection—except in the case of the American Indians—as "frivolous," arguing that forced immigration and chattel slavery have "left a heritage of defeat among blacks incomparably more devastating than that suffered by any other initially disadvantaged immigrant minority group." [5] Even if this point is conceded, however—and who would not do so?—one must still insist that the very idea of compensatory justice implies evenhandedness, and that on this account any instance of systematic discrimination, regardless of its duration or intensity, involves the right to claim redress. In all instances of injury, Justice, if not Equity, must keep her blindfold securely in place.

An even more serious problem in the theory of compensatory justice is its attempt to transfer the notion of rectification of unjust actions from the level of transactions between individuals (or between determinate, organized associations, corporations, or other "legal persons") to the level of society as such by requiring that the whole of society or a majority of its citizens pay compensation to classes or groups of persons distinguished by sex, race, or ethnic background. At the individual level there is a person or a determinate group of persons to whom blame or fault is attached for the unjust act that has occurred, and it is from him or them that compensation is appropriately demanded. When we try to apply notions of compensation or rectification at the societal level, however, we are driven to make some rather shaky and indefensible assumptions.

First of all, we seem compelled to assume that every member of the group that has been discriminated against has been injured by the discriminatory practices, and perhaps we must also assume that each member of the group has been equally injured. Yet these assumptions contradict some obvious facts about the differential impact of discrimination on the various members of a group and about the high levels of status, income, and wealth achieved by at least a minority of its members.

Advocates of compensatory justice have recognized the problem, of course, though their solutions do not exhibit an equal degree of imagination and cogency. Several suggestions were evoked by James W. Nickel's careful putting of the question in ethical terms: "Suppose," writes Nickel, "that a characteristic which should be morally irrelevant (e.g., race, creed or sex) has been treated as if it were morally relevant over a period of years, and that injustices have resulted from this. When such a mistake has been recognized and condemned, . . . can this characteristic *then* properly be used as a relevant consideration in the distribution of reparations to those who have suffered injustices?" [6] In short, is not *reverse* discrimination, despite its reparatory aims, nonetheless discriminatory, and thus unjustifiable? Nickel's answer is that while a morally irrelevant characteristic—being black, for example—was the basis for the original discrimination, preferential treatment is justified by a morally relevant consideration—the fact that injuries have been sustained by persons because they were black. Since, however, there is a high correlation between being black and being a victim of discriminatory and harmful treatment in this country, race can be used as a factor in distributing special consideration. "It is sometimes justifiable for reasons of administrative efficiency," he argues, "to use as part of the *adminis-*

trative basis for a programme of benefits a characteristic such as race which would be implausible as a *justifying* basis." [7] What Nickel here attempts is to devise a program of group preference that is established on strict utilitarian grounds but has the moral force and legal effect of corrective justice as traditionally conceived: wrongs are to be righted individually, but justice is to be administered efficiently. Wise as this solution may seem, there is something disturbing in its mixture of principle and utility that runs counter to the American tradition—often a wonder to foreign observers—of enduring great cost and inefficiency to ensure that each citizen is dealt with individually before the law and that no injury is assumed unless established by due legal process. There is also the problem, shared by all defenders of compensatory justice, of assuming that social inequities derive in significant part from social injustices. [8]

But to two other writers who responded to his question, Nickel's care in avoiding the notion of collective injury seemed merely a dodge; indeed, it appeared to blunt, if not fully dissipate, the aim of compensatory justice—the rectification of an *institutionalized* wrong. According to Paul W. Taylor, not to speak of collective injury is to disregard "the most hideous aspect of the injustices of human history: those carried out systematically and directed toward whole groups of men and women *as groups.*" [9] While race was not a morally relevant characteristic at the time of discrimination, he argues, such discrimination made it relevant as a basis for reparation: the group was "*created* by the original unjust practice." [10] Thus, he concludes, "even if society provides for compensation to each member of the group, not *qua* member of the group but *qua* person who has been unjustly treated (for whatever reason), it is leaving justice undone." [11] Similarly, Michael D. Bayles argues that "by using the characteristic of being black as an identifying characteristic to discriminate against people, a person has wronged the group, blacks. He thus has an obligation to make reparations to the group." [12] From this he concludes, however, that since the wronged group is defined intensionally, not extensionally, (1) at best the group, not any specific individual within it, has a right to compensation, and (2) such compensation may be provided by giving benefits to anyone who at the time is a member of the group. [13]

Such arguments for group injury and compensation lead to some curious results: from Taylor's reasoning one could conclude that a subject of discrimination is owed double compensation, once as an individual and again as a member of the wronged group—a rather amazing notion of rectification; while from Bayle's prescription it may follow that a person who has been only minimally or not at all disadvantaged by discrimination receives society's full compensatory attention, while another member of the wronged group whose life has been shattered by discrimination goes wholly unregarded.

The difficulty in speaking of group injuries, however, is only fully manifest when the complementary question is raised: "Who are the persons who bear responsibility for the unjust acts of discrimination, past and present, and who can therefore appropriately be required to assume the burden of making reparation for them?" All the arguments for compensatory justice—even Nickel's modified version—are forced, it seems to me, to impute something very close to collective fault or guilt to the society as a whole (with the exception of the group that has suffered discrimination) or, at the least, to a large class of members of the society, somewhat vaguely referred to as "the majority," and then to argue that the society or the majority should be required to pay the necessary compensation. No attempt is made to show that every individual in the society is now, or has been in the past, responsible for committing acts of discrimination against, for example, black Americans, or that each of them is, or has been, equally responsible for such acts. Individuals now alive who have not engaged in discriminatory behavior and who have indeed spent considerable energy and effort in carrying on struggles for equal rights for black Americans in the areas of voting, employment, housing, and access to public facilities are, if they are not black, held to be just as responsible for discrimination and its consequences as the most bigoted racists, and equally responsible for bearing burdens of reparation. Similarly, white abolitionists and other opponents of slavery in the past, like the men and women who operated the stations of the underground railway, are presumably to be regarded as just as responsible for the horrors of slavery as those who were slave-owners or those who reaped enormous profits from the slave trade.

Conclusions such as these seem to violate our deep-seated feelings about justice and equity and to evoke our strong objections to notions of collective guilt and responsibility. Moreover, these feelings can only be intensified when we recognize that present members of society are being asked to assume responsibility, not only for unjust acts in the present or recent past in which they may have had no share, but also for acts of discrimination which were performed long before they were born, and when, indeed, their fathers and grandfathers may not have been Americans at all but may have been suffering persecution and discrimination, for example, in eastern Europe. We are, in other words, asked to accept not merely the principle of collective guilt but also the even more distasteful notion of a collective guilt that is at least in part inherited from some of the ancestors of some contemporary Americans.

While Taylor in his article simply insists that the wrongs committed were an "integral part of an organized social practice," and that "in this sense the perpetrator of the original injustice was the whole society," [14] Bayles at least shows some sensitivity to the problem of "collective guilt," which he agrees is intrinsic to the very idea of compensatory justice. All sorts of claims, he notes, may be made about how members of the majority "indirectly profit" from the original wrongdoing (Bedau speaks of having "unsought advantages"), but such claims, he continues, are likely to be only generally, not universally, true. "Hence, any principle of justice between groups will probably involve some injustice between individuals." [15] To this I need only add that the social fabric of the United States is far more complex than what is suggested in such facile terms as *majority* and *minority* and that there are sectors of the majority in our country to whom the idea of involuntary advantage may seem little more than a cruel joke.

The attempt is sometimes made, finally, to circumvent the question of guilt by comparing programs of compensatory justice with systems of no-fault automobile insurance, in which a concern with fault is deemphasized and attention is focused on making sure that deserving victims obtain the benefits to which they are entitled and that they obtain them with reasonable speed and without unreasonable cost. This strikes me as an instance in which common parlance obscures judgment, for while we do speak, in the case of no-fault insurance, of "compensation" being given for injury or damage sustained, the system is essentially a loss-spreading scheme in which essential elements of corrective justice, such as the determination that someone is responsible for the wrong or injury and the requirement that he must in some way or another pay for this unjust act, no longer apply. It seems to me preferable, therefore, to remove no-fault insurance systems from the category of corrective justice and to treat them under a different rubric. In any case, it is folly to ignore the important differences between being a participant in a cost-spreading scheme and being a member of a society and a citizen of a government which will use the tax revenues it compels its citizens to pay as the source of the funds it uses to compensate victims of discrimination.

III

If attempts to elaborate principles of compensatory justice and reverse discrimination in order to deal with the problem of rectifying the consequences of past and present discrimination against blacks and other minorities in our society have the morally troublesome and theoretically problematic results that I have indicated, what approaches to this problem can be suggested which may lead us to sounder and less objectionable conclusions? To offer even tentative suggestions on this matter would require another paper and would take me far beyond the topic of compensatory justice. But I should not want to conclude without the reminder that we already have established and widely accepted principles of justice in our society that can be used and extended to enable us to take effective social and political actions to rectify many of the consequences of discrimination and to provide greater opportunities and greater satisfaction of needs for many members of minority groups.

First, we can, of course, use the principles of corrective justice as traditionally understood to guarantee that members of these groups who have suffered discrimination with respect to employment, promotion, wages, housing opportunities, and the like can receive redress or compensation for these injustices from the individuals or corporations, associations, or trade unions that have committed them, and government can assist them in these efforts by setting up administrative agencies and legal assistance programs to make these rights more meaningful.

The main lines of the struggle for greater justice for minorities, however, seem to me to lie in efforts to extend the principles of distributive justice to which American society has been more or less committed for the past forty years. These are the principles of levying taxes on the basis of ability to pay and of distributing benefits—in the areas of education, welfare, health care, housing, and occupational training—on the basis of need. These concepts of progressive taxation and the transfer-payment state are grounded on principles of social justice and fairness in the distribution of costs and benefits that are widely, if not unanimously, accepted in our society. If they were more thoroughly translated into practice and if their range were widened, the results would be major improvements in the situation of members of minority groups, who constitute a large percentage of the poor and disadvantaged in our society. Such measures would move us a considerable distance further in the progress from merely "formal equality of opportunity" (to use John Rawls's phrase) to a large measure of meaningful and fair equality of opportunity. And they would do so without creating the anomalies and inequities that seem inherent in programs calling for reverse discrimination. White Americans, for example, who are at or below the poverty level would be benefited along with black and other nonwhite Americans, and black Americans in the upper ranges of the income and wealth distributions would bear part of the costs along with other Americans.

Instead of appealing to principles that will have the inevitable effect of heightening tensions between racial and ethnic groups in our society, as each seeks special treatment and special compensation for injustices suffered—tensions and conflicts which, as we know from our own and other nations' history, are frequently used by privileged classes in a policy of "divide and rule"—the policy direction that I am indicating will tend to unite those whose needs are greatest and most pressing, and it builds upon values and principles that already enjoy a wide measure of acceptance. In addition, as black leaders like Philip Randolph and Bayard Rustin have never tired of reminding us, policies which call for expanded government programs to improve the conditions and the life-chances of the poorer members of our society, together with efforts to organize poorly paid workers (many of whom remain outside the ranks of organized labor) in order to improve their wages and working conditions, have at least a good possibility of achieving a measure of political success.

Here practical considerations, which the student of politics and the political theorist must take into account (although the philosopher may believe he can ignore them), reinforce the conclusions to which our theoretical discussion has led us. If you tell members of the American working and middle classes that they must pay higher taxes to finance programs of compensation to the members of this or that minority group because they bear the collective responsibility for the discriminatory practices to which that group has been subjected, one may predict, I think, that the effort will be largely unsuccessful and that the results for the poor and disadvantaged members of the minority group will be no gain and considerable disappointment. Counterclaims will inevitably be raised by members of other racial or ethnic minorities, as each group feels compelled to show that it, too, has been victimized and should be given compensatory benefits. "Liberal guilt" may be fashionable among some of the "beautiful people" at the upper levels of the social and economic hierarchy, and it may be found among some intellectuals and some clergymen at the middle levels. It is, I strongly suspect, not widely distributed among the working and middle classes of our society, and cannot therefore serve as the basis for a successful movement for greater social justice.

2

Equality, Race, and Preferential Treatment

PAUL A. FREUND

Benjamin Cardozo once remarked that Justice is not to be taken by storm; she must be wooed by slow advances. That, at any rate, is the way I shall approach our subject, a way virtually compelled where there is a clash of right against right.

Recently, at a conference on civil disobedience held at Ditchley Park, near Oxford, I quoted Paul Valery to the effect that the two greatest dangers facing modern society are disorder and order. The chairman of the conference, a distinguished peer, snapped with Podsnappian impatience, "Oh, that's very French!"

To have said that the two greatest dangers are inequality and equality might have drawn a more sympathetic response. Certainly in nineteenth-century England equality was not a tenet of the liberal cause. "No idea," said Gladstone, "has entered less into the formation of the political system of this country than the love of equality." After E. L. Godkin migrated from England to America he said, in 1872, that the pursuit of "equality of conditions on which the multitude seems now entering, and the elevation of equality of conditions into the rank of the highest political good" would "prove fatal to art, to science, to literature and to law." [1]

In America, of course, all men were seen to be born equal, from 1776 on down—all free men, that is, not counting slaves or women.

Paul A. Freund is Carl M. Loeb University Professor of Law at Harvard University.

(Jefferson's original draft had it "equal and independent," which was changed, Carl Becker has suggested, for stylistic reasons, to avoid a clash of sound with "self-evident.")

Self-evident the truth may be, but its meaning is not so transparently clear. Perhaps it is ungracious and insensitive to lay the heavy hand of analysis to a declaration that may be taken as the expression of a mood rather than an operative mandate. It was not until 1868, with the adoption of the Fourteenth Amendment, containing the phrase "equal protection of the laws," that a general concept of equality was embodied in our fundamental law. To dissect the earlier phrase of Jefferson may be akin to carping at the Golden Rule because its formula would justify masochists in becoming sadists. As Carl Becker movingly put it: "The Declaration taught that beneath all local and temporary diversity, beneath the superficial traits and talents that distinguish men and nations, all men are equal in the possession of a common humanity; and to the end that concord might prevail on the earth instead of strife, it invited men to promote in themselves the humanity which bound them to their fellows, and to shape their conduct and their institutions in harmony with it." [2]

Nevertheless it may be useful to look a little further at the concept of human equality in the Declaration. Its ancestry goes back at least to the Stoic doctrine that no man is so like to himself as each is like unto all. It echoes the concept of the Great Chain of Being, that humanity—every member alike—stands with pride and awe somewhere between the angels and the beasts. Hobbes, seeming to cast about for some less grandiose rationale, observed that all men are equal in their vulnerability while asleep. I should be inclined rather to stress equality of ignorance—ignorance about the questions that matter most profoundly: whence we have come, why we are here, whither we go.

But observe that the idea of equality could lead to a negation of individual claims to entitlement—a negation of individual rights—for if all men are fungible goods, like grains of sugar, then common treatment of all members, and representation of the group by any of its members, would be just and economical. At this point, where individuals are threatened with submersion in a sea of equality, they are rescued by turning the idea of equality itself to a claim of equality of liberties, of "unalienable rights," claims indefeasible by community action, or in more modern terms, defeasible only for very compelling reasons.

These birthright claims or entitlements may be confined to equal liberty before the law, a basic legal and political *persona* that must be respected. Thus in a suit at law there is no John Cavendish, noble-man, or John Taylor, commoner, or John Lilburn, dissenter. There is a person, whose selfhood is to be respected and whose claims to legal redress or defense are to be weighed on their merits. It is very much like the emergency ward of a hospital, where a patient's *persona* is his body, whether that of saint or sinner. Of course these en-titlements, while real and important, and not to be dismissed cyni-cally, are often merely formal: the litigant may lack the wealth or wit to prevail; the patient may lack the resources of pocket or stamina to survive.

The birthright of equality, including equal basic liberties, may then be enlarged, arguably, to embrace entitlements that require affirma-tive provision: the equality of opportunity that depends on not being necessitous, or neglected, or untutored. Obviously there must be a limit to the nature and extent of such claims. If everyone were en-titled to receive according to his need, and if I conceived my need to be education for a career in music, the need could just conceivably be satisfied by an expenditure of the bulk of the gross national prod-uct for this purpose. What is meant, then, by liberty that includes supportive entitlements is liberty compatible with a like liberty for all, or for others.

This bland and familiar formula raises, if it does not conceal, two issues: what liberties are we supposing, and what are the groups designated for comparison as "all" or "others"? The answer to the first question depends on one's ethical perception of the condition of a society at a given time and place—whether minimum entitlements ought to be enlarged, in recognition of innate equality of worth, whether the gap between formal equality of rights and actual capacity to enjoy rights is excessively great because undue weight is given in distribution to achievement rather than need, to competition rather than fraternity. In this perception the momentum of trends will be im-portant. Whether you perceive a pitcher as half full or half empty may depend on whether you are impressed by the plenitude or the paucity of the contents, and in what direction you foresee or wish to see the contents changing.

The other issue—how to identify the "all" or the "others" with whose liberties and entitlements another's are to be compatible—is a problem of defining a group that is relevant for the particular comparative purpose and that is itself representative of its members. The problem is a phase of the persistent interface between uniqueness and commonness.

Consider two cases in the field of labor relations that present issues of relevance and representativeness when a labor union undertakes to supersede individual differences among the workers. The first case is that of the exclusive authority of the union that has been chosen by a majority of the employees in a craft or plant, to bargain with the employer for all employees in the craft or plant, even though individual employees or groups are thereby deprived of what would normally be an important and basic individual liberty of contract. This result depends on the condition—an important one—that the unit is an appropriate, i.e., relevant grouping for bargaining and contractual purposes, reflecting an essential community of interests, and that the majority union has used fair procedures in becoming the representative of the entire unit of workers. A second case is this: under conditions of a closed union shop, the majority of the union members wish to assess every member in order to give financial support to a political candidate who the majority believe will be helpful to labor's interests. May the dissenting minority be assessed for this purpose? We respond rather differently to this case, I suggest, as some judges have done, because of our different assessment of relevance and representation.[3] Election of political candidates strikes us as a function calling for a different grouping than does the negotiation of a collective agreement with management. It is not enough to remind the political dissenter that in many ways it is common to submerge one's preferences in the group consensus. Some things are beyond the stamp of commonness. One thinks of the line in *Hamlet,* one of the most poignant in all of Shakespeare, when Hamlet's mother reminds him that to lose a father has been a common experience, and he replies, "Ay, madam, 'tis common." That piercing sense of uniqueness is not evident to the man from Mars or the census taker.

All this discussion has been rather abstract, and may seem remote from our specific topic. Yet I believe that by stressing the problem of the unique and the common, the centrality of the question of classification, of groupings, in dealing with norms of equality and liberty, we may provide a context for our more immediate issues.

As we approach the specific question of race or color as a legitimate basis for classification, and more particularly preferential classification, we are met with the general admonition that law must reflect "neutral principles." The relation of this general doctrine to the norm of "color blindness" is manifest.

There are two components of the concept of neutral principles: generality and mutuality. The criterion of generality asks, Suppose there are others similarly situated, in like case. The criterion of mutuality asks, Suppose the parties in confrontation were the other way around. Generality is epitomized in the maxim, "What's sauce for the goose is sauce for the gander." But this only restates the problem of equal treatment, namely, are the groupings relevant and representative for a given function? If the sauce is estrogen, or if the diets have been grossly disparate in the past, the maxim may be singularly inapt. The call for mutuality, in turn, is epitomized in the ditty:

> We prosecute the man or woman
> Who steals the goose from off the common;
> But the greater felon we let loose
> Who steals the common from the goose.

But we must be on our guard, too, against an unduly formal symmetry in meeting the criterion of mutuality. Fifty years ago, some judges and lawyers condemned, as a denial of equal protection of the laws, statutes outlawing injunctions against peaceful strikes and picketing, because the laws did not similarly outlaw injunctions that might be sought by labor against practices of employers. A mechanical, angular symmetry was placed above the incommensurable experiences of life. A different vision is given in Wallace Stevens' "Six Significant Landscapes":

> Rationalists, wearing square hats,
> Think, in square rooms,
> Looking at the floor,
> Looking at the ceiling,
> They confine themselves
> To right-angled triangles.
> If they tried rhomboids,

Cones, waving lines, ellipses—
As, for example, the ellipse of the half-moon—
Rationalists would wear sombreros.

To understand preferential treatment on the basis of race as a form of equal justice, it is helpful to wear a sombrero—to look at the problem out of doors, so to speak, not matching it against ideal forms but against the shape of things as they actually exist.

The criteria of relevance, representativeness, generality, and mutuality have to be applied in the context of a particular, concrete form of preferential treatment.

A racial or color (or religious) grouping is not to be rejected out of hand as inherently not *relevant.* The disadvantaged constitute an appropriate group to benefit from preferences, as in taxing the rich to pay the poor. If a racial group is identified as peculiarly disadvantaged, it may be a proper group to receive benefits; the legacy of Hitler attests to this. The real problems are more particularistic.

Is the grouping *representative,* for the purposes of the specific treatment proposed? Or is the grouping overinclusive? If the question relates to financial compensation for German Jews, the inclusion of some who salvaged their property, or who are prosperous descendants of the victims, may not unduly disturb our sense of justice, partly on the ground that the guilt of the polity is being expiated, partly that the beneficiaries are defined in the same way as were the victims, and the benefit bears some, though not a precise, correspondence to the earlier deprivation. A particular mode of compensation may be unacceptable on prudential grounds, because of its detrimental side effects, as if Jews in Germany were allowed to practice medicine without a certificate of fitness.

Whether a racial classification meets the norm of *generality* is to ask whether the class is underinclusive; that is, whether it ought to be extended, and is capable of being extended, to others who are arguably in essentially similar circumstances. In the case of blacks in America, there are at least two factors that may set them apart: they alone were the subjects of slavery, and virtually alone were forbidden to intermarry with whites. Discrimination in portions of the country was official as well as economic and social.

There is, finally, the question of *mutuality:* whether preferential treatment is a one-way ratchet, or instead would be justified if things were the other way around. Of course, to postulate things the other way around, we would have to imagine, at the very least, a black majority and a disadvantaged white minority. What is suspect about a racial classification is its invidiousness, a manifestation of a dominant-subservient relationship. The ethical aspects are not peculiar to race. Suppose that a university establishes a trust fund for scholarship aid, preference to be given to descendants of John Hancock. Now suppose that the university establishes, instead, a scholarship fund, preference to be given to those who are not descendants of John Hancock. One's ethical reactions to these two forms of discrimination differ, and the reasons are not hard to analyze.

The problem of preferential racial admissions to law or medical schools, taken from a pool of qualified applicants, is not finally resolved by these general considerations concerning racial classifications. Closer questions of justice and wisdom remain.

It is argued that the issue is one of compensatory, or—in Aristotle's phrase—corrective or commutative, justice, and that justice is not served because some who will be disadvantaged by the treatment are not themselves guilty of any prior discrimination, and some who will be benefited have not been the victims of such discrimination. The force of this argument is blunted if we accept as just a belated monetary compensation out of public funds to American Indians whose tribe was defrauded or otherwise wrongfully deprived a century ago. Be that as it may, the precision of the fit between past conduct and presently imposed disadvantage, or between past deprivation and present preferment, can be relaxed where no assessment of personal blame is involved. There is a category that is not quite comprehended in Aristotle's dual scheme of corrective and distributive justice. Like his logic, Aristotle's justice may require supplementation. There is a well-understood form of justice in societies that have passed through the stage of formalism in law, known as equitable restitution. If I receive a gift of stolen goods, entirely unaware of the theft, I am under an obligation to restore them to their rightful owner upon receiving notice of their origin. No guilt is imputed to me; but it would be wrong to profit from another's wrong at the expense of the victim.

The application of equitable restitution to preferential admissions depends on how we view the impact of past racial burdens and benefits. If we think of past discrimination as contributing advantages for whites because they were white, in employment, housing, education and health, opportunities that ramified beyond the places and persons and times where specific invidious acts occurred, so that sons of southern planters, for example, carried their advantages to other places and passed them on to their own children, who in turn became the beneficiaries, then we may apply the principle of equitable restitution without the niceties of tracing individual guilt or injury. We may be the more disposed to do so when preferential admissions are not mandated as a requirement of justice, but are adopted by an institution voluntarily as a permissible act of justice.

An institution may be impelled to act on this judgment, that preferential admissions are within the spectrum of justice, because of positive considerations of social welfare, even of public need. Given the existence, or the present danger, of a society of "two nations," measures are to be encouraged that serve to bring about reconciliation. If blacks see the law-enforcement process as racially hostile, a preference in employing blacks on the police force may be an appropriate remedy. If blacks are, in large numbers, distrustful of white lawyers, and if the ratio of black lawyers to the black population is a tenth of the ratio of white lawyers to the white population, a preference in admission to law schools, on a transitional basis, and among applicants who will predictably succeed in the educational process, may be similarly appropriate.

Whether such preferences will be counterproductive, will diminish the self-esteem of blacks, will exacerbate tensions rather than relieve them, are genuine questions of prudential judgment. The immediate question is whether experimentation shall be permitted, to provide a judgment from experience, or whether such programs fall so clearly outside the spectrum of the just that the prudential issue ought not to be reached. A permissive answer appeals to the sense that there are many mansions, some less comfortable than others, within the house of justice.

3

The Costs of Equality

ROBERT A. NISBET

There is in the air in America right now a conception of equality that I take to be committed to a substantial, even massive and revolutionary, redistribution of income, property, power, status—virtually the whole range of what social scientists call primary social and economic goods. The doctrine being so energetically promulgated is not to be confused with equality before the law or equality of opportunity, American values long since institutionalized and familiar. The new doctrine (though it is not in fact new) is commonly referred to by its advocates as *equality of result,* and it bids fair to be regarded by many as the means of America's sovereign redemption. If this doctrine is now discernible in the air, its most compelling statements have come forth in three recent books in as many years: *A Theory of Justice* (1971) by John Rawls, *Inequality* (1972) by Christopher Jencks, and *More Equality* (1973) by Herbert J. Gans. So far-reaching are these books, both in execution and in audience, that I am going to let their authors stand for the New Equalitarians en masse in this inquiry into what amounts to a vision of a new American social order.

There is no lack of candor on the part of the New Equalitarians respecting either their overriding purpose or what in all probability is involved in its attainment. Thus Herbert Gans writes:

Equality . . . cannot be defined solely in terms of opportunity; it must also be judged by *results,* by whether current inequalities of income and wealth, occupation, political power and the like are being reduced. [p. xi]

Robert A. Nisbet is Albert Schweitzer Professor in the Humanities at Columbia University.

Here is Christopher Jencks on the subject:

We need to establish the idea that the federal government is responsible not only for the total amount of the national income, but for its distribution. . . . If we want substantial redistribution we will not only have to politicize the question of income equality but alter people's basic assumptions about the extent to which they are responsible for their neighbors or their neighbors for them. . . . As long as egalitarians assume that public policy cannot contribute to economic equality directly but must proceed by ingenious manipulations of marginal institutions like the schools, progress will be glacial. If we want to move beyond that tradition, we will have to establish political control over the economic institutions that shape our society. This is what other countries usually call socialism. Anything less will end in the same disappointment as the reforms of the 1960s. [p. 264]

It must be said in praise of Jencks that at no point in his book does he pretend that the policy of radical equalization he recommends emerges from anything but his own deeply held moral preferences.

That cannot, alas, be said for John Rawls, whose *A Theory of Justice* is already the most widely reviewed and discussed work in moral philosophy of this century. A personal preference for equality of result is clearly dominant in Rawls's mind too, but he does not at any point say so. Instead, what he asks us to believe is that the ordinary human mind, suitably corrected for prejudices it has acquired from ethnic background, family, language, education, class, and national tradition, will find its own way naturally to a conception of justice rooted in equalitarianism and calling for radical redistribution of social and economic goods. The method employed here—old when the philosophes used it in their assaults on the *ancien régime* in the late eighteenth century—is a stripping away of what is asserted to be secondary and ephemeral in the human mind until one discovers what is alone constitutive, universal, and rational. Needless to say, and as the history of political thought illustrates, almost anything can be proved or defended by use of this device.

High among Rawls's conclusions is that "all social primary goods" must be distributed equally in a social order unless unequal distribution of any of these goods is to the advantage of the least favored. Among "social primary goods," be it noted, are what he calls "bases of social respect." The whole thrust of social policy must therefore be

to nullify "accidents of natural endowment" and "results of social circumstance."

I

It is in full respect for the spreading power, intellectual and political,
of the New Equality that I am obliged to say that on the whole it is a
poor substitute for classical socialism. There is a certain flabbiness of
thought and inattention to implication among the New Equalitarians
that stands in contrast to the best of the nineteenth-century socialist
writers. This may have come about simply for want of the kinds of intellectual challenges the early socialists drew from their contemporaries, challenges which forced them to think their way from premise
to conclusion in a much more disciplined way than have the New
Equalitarians, for whom equality seems oftener to be a dogma or political slogan than an actual intellectual proposition. Indeed, the socialists—and even the self-styled radical or Marxist economists of
late—seem much more aware than do the New Equalitarians that
ideas have more than consequences; they have contexts as well.

To begin with, the socialists never thought equality the majestic
purpose it is now made out to be. Rather, they sought justice in a society purged of its larger institutional conflicts, themselves the results,
it was believed, of private property and profit. The best of the socialists were no more likely to make a fetish of equality than of happiness, contentment, or peace of mind. They knew, too, as the New
Equalitarians do not, that to make an omelet, eggs must be broken.
And, not least important, they knew that if it is an omelet that is
desired, it will not be at one and the same time a full-course banquet.
A certain fuzzy liberalism seems to have engendered the widespread,
dangerous, and ultimately self-defeating view that all things are com-
binable, including freedom and equality.

The New Equalitarians take a similarly naive view of time. No one
can doubt that a great deal of equality can come about in a population, without either social shock or the exercise of repressive power,
when it is a result of historical forces operating over a long period of
time: we need merely look at the history of western European society
during the past two centuries. To the New Equalitarians, however,
time is an enemy. Christopher Jencks uses the word *glacial* to de-

scribe even the pace of change in America in the 1960s! How, we have to ask, can efforts to achieve instant equality invite less extreme uses of centralized bureaucratic power, or be less disillusioning, than efforts to achieve instantly any other large moral perspective? It is a further advantage of classical socialism over the New Equalitarianism that the former candidly accepts the power of time and circumstance. In all socialist writing I have read, there is at least a realization that a very complex evolutionary process, lasting for some time, must precede final revolutionary change. And this evolutionary process touches the moral as well as the institutional sphere.

No such realization is to be found in the writings of Jencks, Rawls, Gans, and other New Equalitarians. Of the many almost childlike assumptions in these writings, none, I think, is more striking than the apparent belief that a moral value like equality can, by intellectual resolve and political enactment, be *pressed onto* a social and economic system, without regard for the contexts of all values in society. We know how complex, how profound were the moral and intellectual changes which accompanied the rise of capitalism. Can we possibly doubt that quite as complex and profound changes would be necessary for the ethic of redistribution to be adopted politically in the kind of society we live in—a society in which economic values are powerful among the advantaged and the disadvantaged alike, one in which the desire for property and for economic power is very great?

II

Then, too, in imposing equality of result one comes up against the immense difficulty of defining economic equity. As the economist Lester Thurow has written:

[I]f there is one lesson in the state of the art of equity economics, it is that there is no way to avoid the problem of specifying economic equity. It is a problem that is not going to fade away. Our political history has been a verbal subscription to the ideal of equality coupled with the practical desire to avoid having to specify what constituted equity (i.e., an acceptable degree of inequality).[1]

The sociologist Nathan Keyfitz has nicely pinpointed the problem by using Christopher Jencks's own formula for economic equity as his point of departure. What Jencks proposes is simplicity itself. If Jones has an income of $10,000 a year, Brown $20,000, and Smith $30,000, the average income among them is $20,000; it follows that half of Smith's departure from the average is to be transferred to Jones, whose own departure from the average is also to be halved. After redistribution, their incomes will be $15,000, $20,000, and $25,000. "This," writes Keyfitz, "is the degree of altruism that Jencks recommends, and it may be indexed by the fraction one-half or 0.50 by which the departures are multiplied. One who was in favor of allowing more incentive might only redistribute one-quarter departure from the average, and his index of altruism would be expressed by 0.25. Thus everyone has a personal index of altruism varying from 0.00 for those satisfied with present distribution to 1.00 for those who want all incomes to be identical." [2]

As Keyfitz notes, at least one major and admitted purpose of Jencks's book is to raise the American public's index of altruism. I would be among the last to say it should not be raised, though I would point out that national economic redistribution is obviously only one—highly putative—way of raising it. No, what is important to stress again is that any such raising must—in the absence of some impending moral or religious change of unprecedented magnitude—occur in a society which, despite its apparent respect for altruism, expresses at all levels an abiding interest in the economic elements of what Rawls calls "social primary goods."

Let us stay a moment more with Jencks's index of redistribution. Let us imagine an achieved national program of redistribution that is working on a thoroughly modest index of 0.25. Such an index has, as noted, the surpassing advantage of simplicity. Computers could fairly swiftly reveal to the Internal Revenue Service just how much a given individual had to make over to the government in the name of those destined to wait for a check instead of write one. But as Keyfitz points out, simplicity once governed the income tax, too: take from the rich and give to the poor, by a schedule simply and unambiguously graduated from 0 to 60 percent. But who, he asks, has not noticed that economic, social, and political interests soon enough modified that simplicity? Municipalities had to be aided, so they were allowed tax-free bonds; farmers have a hard life and must therefore be given variable definitions of income; oil must be searched for, sometimes vainly, hence double depreciation; and university professors must have their sabbaticals funded by foundation income that is ruled not to be income. What was once a simple tax schedule now occupies uncounted volumes, accessible only to the priests of tax law; and we have learned it is possible to be a millionaire and pay no tax at all and to be a laborer and pay tax on every dollar earned.

And so we can conclude that while it is simplicity itself to follow Jencks's or anyone else's index of altruism, taking substantial excess from the rich and— But ah, my foes, and oh, my friends, *there* is a sentence impossible to complete this side of never-never land. For while it might be possible to take from the rich (assuming that they have not yet been able to get themselves legally designated as the poor), all the lessons of history tell us that it is exceedingly difficult for what is taken to reach the poor. Far easier to take from the rich but, instead of giving to the poor, to discover that with all present welfare payments, social security benefits, and public works projects, any further giving to the poor would be (as it very probably would be, indeed) inflationary.

Will Jencks's book soften up the public to accept redistribution of income only to have the public, twenty years hence, discover that while the government can indeed take from the Smiths, there is no record of its having, under any adopted index of altruism, given to the Joneses? Certainly it would distress Jencks, who no doubt both hopes and believes that adequate provisions of law will guard against this happening. Alas, no such provisions are possible. No Congress or other branch of government can bind its successors in any matter. The blunt fact is, once an idea has been set forth with sufficient vigor, it acquires a force of its own; there is no foreseeable relation between it and the intentions which prompted it in the first instance.

III

But such strictly economic consequences of a policy of redistribution seem to me less important than the cultural, social, and psychological consequences. The greatest difficulty is trying to imagine any national policy of equality *that would not create fresh inequalities,* possibly crippling ones, in the vast and complex area of original action, tastes, and ambitions.

Modern evolutionary theory has taught us that the very essence of natural selection is that it produces individuals of varying strengths, talents, and capacities for adaptation. Genetic selection touches individuals, not groups or types or nations, and no system of political, social, or religious power has yet been found which can obliterate this variation. It can only be lightly papered over, and sooner or later, when individual variations are no longer containable, the organization breaks up. Add to individual variation in biology the selective processes of society, and add again the kinds of processes which lead to the formation of groups, and one sees the scope of present opportunities in this country for the eccentric, the bizarre, and the original. One need merely run one's eye down the pages of the telephone book to get some notion of the nearly boundless ingenuity and diversity of social groups in our society, each concerned with advancing some interest. The originality we find in social groups is second only to that found in individuals. And variability, the biologists tell us, is the condition of progress.

I cannot myself see how a national policy of equality of result, based inevitably on the interest, or imagined interest, of the majority, could do other than militate severely against the present variety of social, economic, and cultural interests. As Bertrand de Jouvenel has written, in *The Ethics of Redistribution,* people of uncommon tastes are always at something of a disadvantage in any form of economic order: "But they can and do endeavor to raise their incomes in order to pay for their distinctive wants. And this by the way is a most potent incentive; its efficiency is illustrated by the more than average effort, the higher incomes and the leading positions achieved by racial and religious minorities; what is true of these well-defined minorities is just as true of individuals presenting original traits." [3]

It would be comforting to suppose that the only victims of a national policy of substantial redistribution would be those whose original tastes run to large yachts, Rolls-Royces, and Rembrandts. Alas, the capacity of such people to fend for themselves under all types of economic systems is beyond serious doubt, save possibly during occasional, very brief periods when revolutionary ardor is not yet spent and bureaucracy has not yet taken over.

The original, the bizarre, or the eccentric tastes of the smaller fry will be the chief casualties, those which reveal themselves, not in the yacht, the Rolls-Royce, or the expensive painting, but in the homely experimental device or gadget, the volume of poetry or short stories, or the work of the just-emerging regional artist: those, in short, who occupy the modest but nonetheless vital creative recesses of society. Only an economy which permits saving and sacrifice in however small degree makes possible a society in which the culturally idiosyncratic can exist, with their ever-present possibility of substantially contributing to the human race. Regrettable though it may seem, financial sacrifice is the best measure we have of the intensity of impulse toward the eccentric and the bizarre in such individuals. Government, which sees only bloc interests and majorities, does not, cannot, recognize such extraordinary individuals. By taxing and spending in the interests of an arithmetic mean in the population, government must destroy, as it so relentlessly has in the socialist nations on earth today, the sources of inventiveness and prosperity. In Western nations it is inflation—by now virtually built into the modern democracies with their compulsions to favor politically one or another section of the population through economic rewards—that is fast destroying these sources. A national policy of equalitarianism would quickly complete this destruction.

The individual of special tastes—be these in occupation or recreation, possessions or spirit, sex or art—has enough difficulty gratifying his tastes in any form of society, and I of course include in this the individual member of ethnic, intellectual, social, and cultural minorities. But given those tastes, he is at present still in some position to realize through extra work, cunning, sacrifice, or skill the money to indulge them. I do not see how a national policy of determined equalization can do other than make the survival of original tastes precarious at best and greatly diminished before long. Still less do I see how talented individuals and talented or highly motivated minorities can possibly exist in anything like the number a free culture demands.

IV

But with this argument we still linger close to money and the material things of life, as do the New Equalitarians too much of the time. It is delusory, I think, to contemplate economic redistribution without recognizing the certain spread of the spirit of redistribution to a domain of rewards and incentives which have little to do with money.

Let us be candid: people who seek high incomes rarely do so for the incomes themselves or for the physical gratifications which can be had from them. One desires a high income in order that he may live like an upper-class person. A great deal of European history, especially in late centuries, records the struggle of groups, especially the bourgeoisie, to live like and be thought members of the aristocracy. To an astonishing extent it is the *noneconomic* values, especially those around which the feudal nobility was structured, which have called the tune in much of that modern history. The New Equalitarians seem to me neglectful in not often enough facing up to such social primary goods as personal power, status, and prestige, for it is in this area that inequality is most likely to gall. Thus, for example, the academic's sense of the unfairness of things, his desire for greater equality, is far more likely to be generated by whom he sees elected to prestige-laden societies or appointed to illustrious professorships than by the income attaching to these. For in the academic world, above a certain rather quickly reached point, salaries take on their largest significance as indices of status. It would be shortsighted or naive to suppose that analogous translations do not obtain in all walks of life. Yet it is almost invariably the instinct of the New Equalitarian to suppose that the economic is his only proper concern and that the drive for equality can be contained within the economic sphere.

It cannot. Rousseau saw this accurately, as he saw so much about inequality. Rousseau was well aware that the kinds of gradation cherished by the whole class of intellectuals, scholars, scientists, and artists are among the most galling to be found anywhere. Such was his own passion for absolute equality that he was quite willing in the good society to ban the arts and sciences, or else severely to limit their sway, as had been his preceptor Plato. For as we know from the *Confessions* and a good many of Rousseau's letters (quite apart from the *Discourse on the Arts and Sciences*), Rousseau loathed scholars, artists, and intellectuals generally, seeing in their ranks and their values forces which could not help but be inimical to the General Will and to the equality that must undergird the General Will. Whatever we may think of Rousseau's ideal society, we can admire his consistency and perceptiveness.

Once we start on the road to redistribution, just as we can no longer dodge the need to define equity in economic matters, so we must also confront head on the need for an index of altruism in all the *other* spheres of inequality. Tocqueville wrote that it is possible

to conceive of men arrived at a degree of freedom that should completely content them. . . . But men will never establish any equality with which they can be contented. . . . When inequality of conditions is the common law of society, the most marked inequalities do not strike the eye; when everything is nearly on the same level, the slightest are marked enough to hurt it. Hence the desire of equality always becomes more insatiable in proportion as equality is the more complete.

And, too, envy becomes insatiable. Envy is the cancer that lies within equalitarianism. I suspect it is because of envy that the greatest philosophers and artists have tended to rank equality very low in their hierarchies of values. From Plato's notable discussion of the subject in his *Republic,* through Augustine's *Confessions,* Thomas More's *Utopia,* Montaigne's *Essays,* and a host of other classics down to Faulkner's incomparable rendering of the Snopes family, envy has loomed large among the emotions we are advised to despise and fear in a social order. In one of his letters to Harold Laski, Oliver Wendell Holmes wrote: "I have no respect for equality, which seems to me merely idealizing envy." That may be a shade too strong, and in any event Holmes certainly did not have in mind equality of opportunity or equality before the law. But I suspect that one of the chief reasons equality of result has never been embraced by Western social thought, philosophy, or literature is that it generates envy and resentment like the plague.

Conversely, I think that much of the intellectual decline of the political left in our time springs precisely from its having taken discontent, resentment, and envy in the social order to be ills in themselves demanding direct action, rather than as indices of substantive social conditions. Marx himself seems to have been as uninterested in assuaging envy and resentment as he was in equality as a proper goal of human progress. Only when we come down to the present do we find self-styled socialists, such as Anthony Crosland in his *The Future of Socialism,* proclaiming that the aim of politics is the removal of sources of discontent in society. The mind reels at such an aim! Yet from the expressed desires of Herbert Gans and the clear implications of John Rawls's ideas it is no very long step to a conception of politics which proportions out just about everything in culture from learning and artistic creation to happiness and beauty.

V

But let us come to the crux of it. Most of the interest in redistributing goods in this country is accountable to the clear relation goods have to social influence and authority—to political power. That very real political power goes with wealth can hardly be disputed: it is out of a genuine historical sense of this connection that the New Equalitarians propose extensive changes in our political system and the use of large political powers to bring about their social reconstruction.

They will need all this, and they know it. For there is not the slightest evidence that a majority of Americans, once they understood what was involved, would willingly support the New Equality. I do not mean Americans are passionately fond of the immensely rich or the way of life of the international jet set. But once the majority consciousness fully grasped that most of what was to be given to those earning less than $10,000 a year would have to come from those earning or hoping to earn $20,000 to $40,000 a year, something stronger than the majority itself would be required to effect such economic altruism. Jencks is aware of this, and, acknowledging frankly what his policy of redistribution would do to the family, to inheritance, and to the whole play of what we call, for want of a better word, luck, tells us that an enormous reshaping of American attitudes through political action would be necessary. He is under no illusion that we will reach the New Equality through representative democracy, at least right now. It is no doubt for a similar reason that John Rawls does not include majority representation among the central tenets of his equalitarian doctrine of justice.

The majority, left to their own desires in the matter, would not wish to go much beyond what we now hope to have: reasonable equality of opportunity and equality before the law. Here again, Rousseau is instructive. Rousseau knew that the absolutely equal community he envisioned could never be founded if the people were in fact left to their own desires. Hence his crucial distinction between the Will of All and the General Will, with only the latter invariably right, invariably dedicated to the *true* interest of the people—and fully entitled not to consult the people on important issues. To consult the people, he tells us, "is hardly ever necessary where the government is well-intentioned; for the rulers well know that the general will is always on the side which is favorable to the public interest; that is to say, most equitable; so that it is needful only to act justly, to be certain of following the general will."

It was left to Rousseau's arch-critic in the nineteenth century, Tocqueville, to describe how power uses policies of equality to institute what Tocqueville called "the new despotism." He writes:

The foremost or indeed the sole condition required in order to succeed in centralizing the supreme power in a democratic community is to love equality or to get men to believe that you love it. Thus the science of despotism, which was once so complex, is simplified and reduced, as it were, to a single principle. . . . Every central power which follows its natural tendencies courts and encourages the principle of equality; for equality singularly facilitates, extends, and secures this influence of a central power.

Throughout history the lesson is a vivid one: to centralize power in a society, merely create the masses—through acts which level economic and social strata. To create the masses, merely centralize power. Imperial Rome died, a philosopher of the last century observed, from apoplexy at the head and anemia at the extremities. From the Caesars through the Renaissance despots down to the Napoleons and Stalins of the modern world one finds in nearly all holders of absolute power the compulsion to grind the holders of all rank down to positions of courtierlike equality. But the absolutism of bureaucracy is not different in its effect. Rather, as the immense bureaucracies of the HEW, the Pentagon, the ICC, and the IRS make only too evident, the capacity of bureaucracy to stifle and cripple is greater actually than that of the individual ruler.

If only one could conquer the suspicion that the New Equalitarians prize the state more than ever they prize the individual. It is said that the philosophes, while loving humanity, detested individuals. Certainly this was true of Rousseau. I cannot help thinking that the votaries of the New Equality are less interested in actual living human beings, with their infinite diversity of talents, motivations, and aptitudes, than they are in demographic segments of the population which can be expressed in statistical terms and given representation by immense and ultimately dominating agencies of government. Endlessly fertile in spinning doctrines, ideals, and hope-intoxicating fantasies, the New Equalitarians prove singularly barren in imagining the governmental or judicial execution of even the most elementary of their goals. Think of busing, of Affirmative Action. If equality in education alone can spawn the bureaucracies, coercive uses of power, inquisitorial networks, and severe popular tensions and hatreds it so plainly has, only think what would follow efforts to institutionalize Rawls's Difference Principle.

VI

Whether power is greatly centralized or social rank leveled, the result is the same: the disunited, atomized, pathological mass. From extreme policies of either sort, all intermediate groups and associations inevitably suffer, and none more than the family. The family as an institution has not been fashionable in intellectual circles since approximately the seventeenth century, when the modern Leviathan was first being designed by political philosophers. Indeed, those of despotic inclination have been suspicious of the family since Plato, who banned it from his society of guardians. And Plato and his successors are right: it is quite impossible for a totalitarian political system to flourish where kinship remains a strong value, the family a solid association.

Who can be blind at the present time to the disdain in which the family is held by the New Equalitarians? Christopher Jencks identifies (correctly) the family as one of the prime conduits of social inequality. Rawls considers and then flinches from the prospect of abolishing the family in the interest of equalitarian justice. He writes lamely that perhaps it won't be necessary after all, or that perhaps we should wait and see. He errs in waiting. For no social system as far-reaching in effect as his, and so penetrating of human consciousness, could conceivably be brought into being with the kinship principle left intact. Again, that master equalitarian Rousseau has pointed the way, in the final pages of his *Discourse on Political Economy*. Not only to further equality but to improve education, Rousseau writes, the family should be supplanted to protect the children from the prejudices of their fathers. Indeed, in one of the most charming passages in the whole history of political absolutism, Rousseau goes on to say that the parent should not *mind* his displacement, for he will only be exchanging the authority granted him by nature for the authority flowing from common membership in the General Will.

Seymour Lipset, drawing from Walter Hugins' study of equalitarianism in the early nineteenth-century American working-class movement, describes a document that appeared in 1829 in New York. It asserted that a few hours in school each day was grossly insufficient to overcome inequalities in the American race for success—and flatly recommended that all children, regardless of family, class, and parental wishes, be educated from the age of six onward in boarding schools; thus and only thus would all children partake of the same environmental influence twenty-four hours a day. As Lipset notes, this recommendation to nationalize children must be the most radical doctrine ever to be advanced in American party politics. I have no doubt whatever that it would recommend itself to a good many of the New Equalitarians.

Yet it would be insufficient. For as we have seen, it is equality of *result* that the New Equalitarians care most about, and even the nationalizing of education could do no more at best than to equalize *opportunity,* which is all the equalitarians of 1829 had in mind. Still, proponents of radical equality are always correct in seeing a deep affinity between the family and inequalities of income, property, social status, political influence, and a host of other traits in the social order. What is being described in Mao's China as an assault on the "reactionary principles of Confucius" is only a recognition that so long as the Chinese kinship system stands, even in its currently attenuated form, the official goals of total equality, total communism, and total liberation from tradition cannot possibly be realized.

The question that will have to be faced honestly by all those interested in comprehensive equality of result is whether a free *and also creative* society can exist without the kinship principle. I refer particularly here to kinship-in-time, the interconnectedness of several generations in intimacy and obligation, even if not always in affection or love. Doubtless an orderly—even happy—society can be brought about through the proper uses of power and narcotics with no household or kindred anywhere. But I see not the slightest evidence that a *creative* society can be instigated or maintained when individuals have no place in the kinship group from generation to generation.

If I seem to be laboring this point, it is because the New Equalitarians are themselves profoundly aware of the high correlation between family and individual success—and that if they were not, modern history is already so full of the energetic efforts of utopians and other burning moralists to diminish or abolish the kinship principle that we could be sure our own would get around to the task eventually. They believe, if I read them correctly, that a desirable quality in you and me is somehow less desirable if it proceeds directly from family. For the life of me I am unable to follow them in this, but I dare say it all has to do with the matter of rewarding merit, my final point.

VII

There chews at the minds and consciences of the New Equalitarians the conviction that the distribution of John Rawls's "social primary goods"—which include, let us remember, the ranks, roles, and statuses we enjoy, as well as income and property—should somehow correspond to our individual worthiness. On the surface this is appealing enough, and I don't doubt that up to a certain quickly reached point it is downright creditable. But soon enough it becomes an untenable proposition which, if roundly acted upon, would be devastating to any social order.

Granted that in any just social order—or occupation, discipline, or other walk of life—every reasonable means should be used to prevent the triumph of the illegal and the unethical. The scientist who may be shown to have attained some signal honor through falsifying evidence or other delinquency should not get away with it; the principle is axiomatic. But the obverse case, the ancient and perhaps timeless dream that justice will mean proportionality of reward to moral worth all over the economic, cultural, professional, and intellectual marketplace—that is, alas, to be dreaded in practice. X, let us say, is regarded as a paragon of conscientiousness, industry, fidelity, kindness, charity, and generosity, but nevertheless fails of election to an honored membership or appointment to an honored job because of certain accidents of birth, rearing, or capricious fortune. Y, on the other hand, while standing well within the law and local ethical standards, is not ranked as high on the scale of moral worth, and has moreover enjoyed good luck, good parentage, good schools, and good professional acquaintance. Yet it is Y who has the international name, the renown, and all the prized elections and appointments and awards. Ergo: the system is unjust.

But in his classic *Constitution of Liberty,* Friedrich von Hayek concluded that in a free system it is neither desirable nor practicable that material rewards should be made to correspond to what men recognize as moral worth. Indeed, it is of the essence of a *creative* society that an individual's position should not necessarily depend on the views his fellows take of his moral worthiness. Hayek writes:

The incompatibility of reward according to merit [i.e., moral worth] with freedom to choose one's pursuit is most evident in those areas where the uncertainty of the outcome is particularly great and our individual estimates of the various kinds of effort are very different. In those speculative efforts which we call "research" or "exploration" or in economic activities which we commonly describe as "speculation" we cannot expect to attract those best qualified for them unless we give the successful ones all the credit or gain, even though many others have striven just as meritoriously.[4]

There must, in sum, be a free market to operate wherever individual initiative and creativity are prized; if this proposition were to be replaced by one proceeding from the vision of the Heavenly City, wherein all rewards would be proportionate to one's fund of moral virtues, one's diligence, industry, and conscientiousness, the open society would be doomed. Moreover, to accurately, comprehensively calibrate individual moral worth, nothing less than divine omniscience would serve.

In the end, then, the largest costs of making equality of result our decisive national policy would be, not only the inevitable loss of economic, social, and cultural variability, but the loss of moral pluralism as well. There are, after all, many moral goods: liberty, trust, honor, individuality, truth, justice, beauty, friendship, love. I do not think anything but the most extreme repression and the most extreme banality can come from sacrificing this plurality of goods for one good.

4

The Costs of Inequality

In Response to Robert A. Nisbet

HERBERT J. GANS

Robert Nisbet's inaccurate group portrait of those he calls the "New Equalitarians," in which he includes by name Christopher Jencks, John Rawls, and myself, warns about the dangerous effects of more equality in American life, while somehow not mentioning at all the adverse effects of inequality. In correcting his portrait, examining his warnings, and supplying what he has skipped, I speak chiefly for myself and my own work, although Nisbet was not, of course, always referring to me.

I

Nisbet charges the New Equalitarians with seeking equality above all else, wanting it perfect, and demanding it across the board: economic, political, cultural, and social. According to him, we also wish to impose equality on a population which for the most part does not want it, and we are prepared to recommend such extreme policies as the elimination of the family and the nationalization of children to achieve our ends. Nisbet also describes us as "almost childlike" and intellectually "flabby," but that sort of invective does not deserve rebuttal.

It is some measure of Nisbet's accuracy that my own real position, which I set forth at length in my recent book *More Equality,* is at odds with all of his charges. I do not think equality can or should be

Herbert J. Gans is Professor of Sociology at Columbia University and senior research associate at the Center for Policy Research.

any society's overriding goal or that perfect equality is either desirable or attainable. Like many other critics of egalitarianism, Nisbet chooses to attack the straw man of *perfect* equality; my book advocates only *more* equality, and that only economic and political.

To be more precise, I do not even propose massive redistribution of wealth and income. Rather, I urge that the federal government guarantee every household a minimum income of 60 percent and eventually 70 percent of the median income for its household type. (According to the definitions and data of the United States census, the minimum for the prototypical family of four would currently be about $7,800.) In addition, I favor some additional redistribution to provide modest increases in income to households now earning between 60 and 100 percent of the median (to $13,000 by current figures). I further believe that the minimum income should be provided wherever possible through decent and dignified work, income grants going only to people who cannot work, cannot find work, or cannot earn enough from their work. In short, I believe that income and wealth should be redistributed in this country only to the extent necessary to fund the jobs and income grants required to make up the minimum income. Nisbet would make us all aware (as I am) that some redistribution will be required from households with incomes of no more than $25,000. Nevertheless, most of the necessary funds can be obtained from the wealth of the very rich and the large corporations.

So much for my own standard for economic equality. I should also point out that none of the New Equalitarians I know believes in "instant" equality, or argues that more equality can, in Nisbet's words and italics, be *"pressed onto* a social and economic system, without regard for the contexts of all values in society." Christopher Jencks, for example, writes explicitly in his book *Inequality* that even making the question of income inequality a political issue "will inevitably be a long, slow process, stretching over decades rather than years" (p. 264). But then, none of the New Equalitarians I know believes, either, that more equality ought to be "pressed onto" American society against the wishes of the majority. Redistributive legislation can come about only as a response to economic and political conditions. I do think, however, that changes in the contemporary economy are now creating conditions that will eventually encourage many people, particularly in Middle America, to press for redistribution. As the decline in economic growth, the increasing centralization of our economy, and business-cycle downturns more and more sharply limit the opportunities for individual enterprise by the average American, more and more Americans will conclude that the old laissez faire dream of becoming affluent as the gross national product grows, or through personal entrepreneurial efforts, is no longer realistic. At that point, they will seek to improve their fortunes through political pressure for some redistribution of wealth and income.

Lee Rainwater's new book, *What Money Buys,* provides ample empirical data demonstrating that far more people than Nisbet thinks are ready to support redistributive policies. Indeed, some members of Congress and at least one candidate for the 1976 presidential election are already thinking along these lines.

What Americans will decide to do about redistribution is as yet uncertain; but it is quite certain that none of the New Equalitarians advocates the elimination of the family. Of course Nisbet is right to suggest that the family is an obstacle to achieving *perfect* equality. But it is not an obstacle to the reduction of economic inequality. Rousseau's observations on this topic are not very germane today since the family is no longer primarily an economic unit, many of the incentives to build dynasties having begun to disappear. Even the very rich are spending more of their wealth on themselves these days, and passing less of it on to their children. But, then, surely most American families have for a long time had precious little to pass on to their children: in 1969, when the last data were collected, 50 percent of all families had a net worth of under $3,000, and only 14 percent had over $20,000. As a New Equalitarian who shares Nisbet's concern for the family, I myself favor a "negative inheritance tax" which would allow families of modest means to provide something for their children's future.

As for the nationalization of children, I cannot fathom why Nisbet believes such a measure "would recommend itself to a good many of the New Equalitarians." None has proposed it. Conversely, Nisbet does not even mention the already existing American practice of nationalizing the children of the poor. For at least a century, the state has been taking poor children away from their natural parents on often trumped-up charges of neglect, with total disregard for the rights of due process, and shuttling these children back and forth among foster homes or institutions. Through it all, the state retains total control over the children's future, giving it up only in the rare instances in which adoption takes place.

I I

For the most part, then, in "The Costs of Equality," Nisbet dwells upon what he sees as the effects of more equality (and accuses Jencks, Rawls, and myself of ignoring the whole subject of effects, though much of my own *More Equality* is devoted precisely to that subject). Like other antiequalitarian intellectuals, Nisbet is most fearful of the effects of equality on culture, or, as he puts it, "people of uncommon tastes." But unlike Bertrand de Jouvenel, whom he quotes, Nisbet is not worried about the effects of equality on the artists and audiences of high culture; instead, he is concerned lest we suppress "the original, the bizarre, the eccentric tastes of the smaller fry," among whom he includes "the just-emerging regional artist" and inventors of "the homely experimental device." But he never explains why people with such tastes would disappear or even suffer under conditions of more equality, and I do not myself see why they should. For one thing, their uncommon tastes and interests do not require a large amount of money, and in many the need to create is so strong that they are in fact less interested in economic accumulation than other people are. They may want the rewards of creativity and fame, but fame can be obtained without riches.

Of course, as Nisbet says, in a society with more economic equality, creative people now being financed by the very rich would need new sources of support; even though the very rich would not, as a result of the sort of redistribution I favor, become poor, they would probably be unable or unwilling to invest in risky artistic or technological ventures. Nevertheless, any equalitarian not interested in perfect equality could suggest a host of ways by which funds could be channeled to creative people. Tax incentives or government subsidies are only two.

But then, redistribution may actually encourage more creativity and uncommon tastes, for three reasons. First, if struggling artists are granted a guaranteed minimum income, they will no longer need to take nine-to-five, nonartistic jobs and could then work full time at their creative pursuits. Second, redistribution would spur creativity among those now poor. There are, we may be sure, as many creative people in that population as in any other, and if they no longer had to worry continually about surviving, they too would be able to act on their need to create. Third, an economically more equal society may in general be more creative. Since the values of such a society will discourage moneymaking—at least big moneymaking—as an end in itself, and since even in a more equalitarian society, people will continue to strive for differentiation and status, they will seek new ways of winning differentiation and status, one of which may be original creation. And even if most people were to compete for status only through consumer and leisure behavior, surely some would be likely to develop uncommon tastes along the way.

Nisbet warns us, however, that once a society achieves more economic equality, it will begin to seek equality in other realms because "envy becomes insatiable." In support of this argument, he cites Tocqueville's hypothesis that the human desire for equality increases geometrically as equality increases arithmetically. My own hypothesis is just the reverse: I think people will continue to want higher incomes but will also continue to strive for social differentiation, and will therefore resist measures aimed at achieving status equality. Nor, of course, have the New Equalitarians even proposed that kind of equality; and I cannot imagine any government policies able to implement it, at least in a democracy.

Here Nisbet's argument takes some strange turns. On the one hand, he accuses current equalitarian writers of dwelling unduly upon economic equality, of not worrying sufficiently about the dangers of social and cultural equality; on the other hand, much of his own article is devoted to the dangers of economic equality. His way around this contradiction is to argue that people seek high incomes mainly for cultural and status reasons. Thus: "people who seek higher incomes rarely do so for the incomes themselves. . . . One desires a high income in order that he may live like an upper-class person," and "in the academic world, above a certain rather quickly reached point, salaries take on their largest significance as indices of status." Nisbet seems to forget, however, that it takes a good deal of money indeed to live like an upper-class person, that an upper-class person has always lost status quickly enough when the money ran out, and that it was with money that the nouveaux riches purchased those rights and privileges of feudal nobility Nisbet cites nostalgically. Nor does his description of the academic world apply to the one I know, since colleagues at private universities do not know each other's salaries, and colleagues at publicly funded universities get roughly equal salaries within each rank. And, to linger upon that example, does Nisbet really believe that the current faculty demand for salary increases to keep pace with the cost of living is mere status-seeking, or that college professors are flirting with unionization to obtain prestige?

Nisbet's other worry is that equality would dangerously increase the power of the state, and encourage it to make the population into a "disunited, atomized, pathological mass." But here again, his analysis depends more upon distorting the motives of the New Equalitarians (they "prize the state more than ever they prize the individual") than on empirical evidence or logical argument. I would have thought that the exercise of undue state power quite often leads to a healthily angry and protesting citizenry, rather than to a "pathological mass." Nisbet is probably right that any society which seeks perfect equality, and seeks it as the primary societal goal, can become despotic. But the New Equalitarians do not advocate such a policy.

Still, I freely concede that reducing economic inequality will result in a more powerful government, since only government can bring about the reduction in the first place. And I agree that increasing the power of the federal government is not desirable. But if we must choose between being controlled by large multinational corporations or by a more powerful but democratic government, I would choose the second as the lesser of two evils. Worrying at length about the dangers of immense federal bureaucracies, Nisbet never once mentions the dangers of corporate ones. Nor should it be forgotten that despite the enormous power of at least one federal bureaucracy, the Pentagon, America is hardly a totalitarian society, even if it is not as democratic as it should be.

But need we choose between two evils? More economic equality will reduce the economic and political power of the wealthy, thus creating more political equality—and the likelihood that the American polity will therefore become more democratic. Of course, the reduction in economic inequality will, among other things, complicate decisionmaking because when people have enough economic security they become more interested in using the political liberty which is their constitutional right. Nisbet suggests that more equality can be achieved only at the price of less liberty. But this is the conventional antiequalitarian argument, one that fails to consider whose equality will affect whose liberty, and does not mention that under conditions of economic inequality, de facto liberty is maximal for the affluent but minimal for the poor. In fact, the income and resulting liberty of the very rich are made possible in part by the existence of the poor, who lack most of the liberties which Nisbet—and I—prize. The aim of the New Equalitarians, as I see it, is to increase the liberties of the many, while reducing among the few only the liberty to dominate the rest of us with their economic and political power.

I fully recognize the considerable evidence from eastern Europe and elsewhere suggesting that more economic equality can be accompanied by a denial of civil liberties, especially to intellectual dissidents. But the countries where this is the case, countries which experienced communist revolutions, were formerly very poor and lacked a prior tradition of civil liberties; they cannot serve as examples for the United States. All the same, as these countries become more affluent, their citizens are beginning to demand more civil liberties; their reluctant governments have already made some concessions, and I suspect that in the future they will be forced to make more.

III

Finally, Nisbet displays the same curious blind spot as other critics of the movement for more equality: he pays no attention whatsoever to the effects of inequality. Indeed, he implies in his argument that inequality is preferable. Yet the harmful effects of inequality are not only more dangerous than the effects of more equality; they even touch the lives of the more-than-equal and impair the quality of the entire society. The physical and emotional suffering, the restrictions on liberty and creativity which accompany poverty, these are so obvious as to require no further discussion. But what of the financial and nonfinancial costs that the rest of society must pay for poverty? These include the financial costs of caring for those poor who become victims of chronic mental and physical illness, alcoholism, and drug addiction, and the much higher costs of protecting the rest of society from such victims. There are also the financial costs of abounding street crime and seemingly senseless juvenile violence—and, again, the cost of protecting the rest of society from these and other forms of lower-class anger against an inegalitarian system. Last, but hardly least, come the social and emotional costs of living in a society which forces some people to become members of what nineteenth-century writers called the dangerous classes.

But the hostility between the haves and have-nots affects more than the quality of American life; it affects the polity as well. Economic inequality requires politicians to please the holders of large amounts of economic power more than other constituents. Hence the vicious circle: government is unable to satisfy larger but less affluent constituencies, is therefore mistrusted, and is thus even less able to satisfy them. Economic inequality is hardly the sole cause of undemocratic government; but, as noted above, more economic equality will lead to more democratic practice.

A more egalitarian society will not be more loving and altruistic, and I do not know any New Equalitarians who are seeking or expecting, as Nisbet implies we are, a significant rise in public or individual altruism. More economic equality will not eliminate social conflict, but insofar as it reduces economic differences, it will enable people to negotiate their other differences somewhat more easily. A polity which can deal with differences through compromise will be more effective, more humane—and more like the creative society which Nisbet advocates.

IV

One last word is in order about the role which Nisbet, a social scientist and philosopher, has chosen here. He evidently sees it as his task to warn us that deliberate social change toward more equality will only make life worse, and to predict doom if we do not accept the status quo and pay attention to the lessons taught by classic writers of the past. It is Nisbet's right, of course, to choose his role—although, as he is a sociologist, I have to wish he would occasionally use empirical methods and data to support his arguments. Myself, I see a more useful role for the social scientist and philosopher. If I am right that American society is likely to become economically and politically more equalitarian in the future, whether Nisbet dislikes it or I like it, then it is high time to think about, and as far as possible to *research* the possible effects of more equalitarian policies, and to explore alternative economic, governmental, and social structures—in the current jargon, to build models—which would maximize the beneficial effects of more equality and minimize the harmful ones. Initiating such studies seems to me to provide a more constructive—a more creative—role for social scientists and philosophers than the one Nisbet has chosen for himself.

5

Justice, Equality, and the Economic System

WILLIAM S. VICKREY

An economist writing about justice and equality is a little like a mathematician writing about music and harmony. Just as the mathematician can point to the simple ratios involved in various harmonic relationships and the compromises involved in arriving at an even-tempered diatonic scale, the economist can discuss the measurement of inequality in terms of the maximization of aggregate utility and the difficulties of defining and measuring utility, but he can come no nearer to resolving mankind's age-old quest for justice than the mathematician can explain the satisfaction derived from performing or listening to a symphony. I take this rather abstemious position in spite of all the tomes that have been written on the subject by persons claiming economic expertise.

I

Among economists, positions have been taken at rather extreme ends of the spectrum. At one extreme is the neocalvinist position approached by John Bates Clark, who concluded that ideal economic justice would be reached in a state of perfect competition, where, according to the marginal productivity theory, each would be rewarded according to his contribution to the common product. A

William S. Vickrey is McVickar Professor of Political Economy at Columbia University. From 1974 to 1975, he was Interregional Adviser for the Center for Development, Program Planning, and Policy at the United Nations.

more precise statement of this theory would be that each is rewarded according to the amount by which the addition of his inputs enhances the total product, and it was perhaps fortunate that the possible range of consequences of this proposition under various possible technological circumstances were not as apparent in Clark's day as they are today, for in the light of present knowledge it is possible to imagine and perhaps even to exhibit actual examples of circumstances where the application of this principle could condemn large sections of humanity to extreme misery and starvation.

At the other end of the spectrum is the communist dictum that the goal of society should be to obtain from each according to his ability and to distribute to each according to his needs, a position now maintained in a more sophisticated and rationalized form in John Rawls's widely discussed A *Theory of Justice.* The essence of Rawls's position is that justice consists in so arranging matters that the least well-off member of society is made as well-off as possible. This is perhaps an even more egalitarian goal than the communist one, in that the latter might well consider the "needs" of the least well-off to be satisfied when everything that is reasonably productive has been done for them, without requiring heroic measures to be taken to improve their lot by minuscule amounts, as would appear to be the requirement of Rawls's principle.

Somewhat in between is the utilitarian approach which in its simplest version advocates that society be organized so as to maximize either the mean utility of its members or the aggregate of their separate "utilities." But it is only in a partial and coincidental way that utilitarianism can be considered a criterion of justice. Utilitarianism does argue for a more even distribution of the pie among like participants as long as this can be managed without undue frittering away of the substance of the pie in the cutting-up process. But in cases where the inherent ability of the various participants to generate "utility" differs, utilitarianism and concepts of justice often collide head on. Justice may fairly universally be deemed to require that as much as possible be done to make the innocent victim of another's negligence whole, preferably at the expense of the culprit, while utilitarian theory may require that relatively little be done for the victim if his condition is such that even vast resources devoted to his care produce little improvement in his lot.

The utilitarian principle of maximizing average utility can be thought of as corresponding to the decision an individual would make if he were offered a choice among several alternative societies, each with a constitution that will determine the distribution of income or welfare within it, but with the chooser being unable to determine his role in the respective societies and having to assume that his chance of filling a given role is proportional to the number of persons that will have that role. If all individuals had the same degree of risk aversion, as measured by the curvature of their utility functions, they would all, presumably, choose the same social constitution. Or if it were possible actually to provide a number of alternative communities, each society would be selected by a group of individuals with the same degree of risk aversion (or preference), and in effect each individual would be able to have what he wanted. To a degree, this is merely carrying to the extreme the suggestion of Charles M. Tiebout that each suburban community differentiates itself from others in terms of the amount and variety of public services offered and attracts to itself a homogeneous group of residents for whom this package is relatively most attractive.[1] In such a world, one would be hard put to say that one community was more just than another or to determine which, if any, of the communities was achieving absolute justice.

Rawls's prescription of maximizing the welfare of the worst-off individual in the community amounts to assuming that the selecting individuals are "minimaxers," i.e., that they use what Hans Neisser has characterized as "the strategy of expecting the worst."[2] This may be a rational mode of behavior in a two-person, zero-sum game situation in which each player is faced with an opponent actively interested in maximizing his own winnings at the expense of the other's, but, as Paul Samuelson has written, to behave in other situations as though "nature is an implacable enemy who will attempt to do you in is as good a definition of paranoia as any."

Nor can such behavior be rationalized in terms of a total inability to make even subjective estimates of the relative probabilities of different outcomes: if anything can happen to you regardless of how you behave, then nothing matters and you might as well relapse into morbid indifference; while if you can rule out some outcomes as being impossible consequences of given behavior on your part, you have already taken the great leap of dividing outcomes into those with zero and those with positive probability. In the minimaxing mode, your decision could take a quantum jump as an outcome changes from one deemed absolutely impossible, with probability zero, to one regarded as remotely possible, no matter how unlikely it may be, e.g., with probability $10^{-1,000,000}$, whereas for the maximizer of expected utility the effect on the behavior selected would be commensurate with the change in probability. A utilitarian could, in principle, rationalize minimax behavior by postulating a utility function having values in a Cantorian transfinite realm according to which, if A is preferred to B and B to C, the difference between the utilities of B and C is always infinitely greater than the difference between A and B, no matter how minute the objective difference between B and C may be or how great that between A and B. But this is unattractive as a basis for analysis.

Because of its lexicographical nature, moreover, the Rawlsian prescription of justice is critically dependent on how the community within which justice is to rule is delimited. Is it the nuclear family, the extended family, the clan, the tribe, the polis, the citizens of a nation, the residents of a country, the participants in a culture, the human race, or a yet wider class? *Sub specie aeternitatis,* it is possible to answer "the human race" only at this particular juncture in cosmic time, when the human race is distinct from other primates and not yet in contact with other comparable intelligences. On a less cosmic scale, there is the problem of where to draw the line between the just fertilized ovum, the fully grown adult, and future generations.

There is also the problem of judging which member of the selected class is the worst off, e.g., the paraplegic with the cheerful disposition or the physically healthy but phobia-ridden neurotic. The Rawlsian scheme may require drastically different initiatives in the two cases. It may be relatively easy to upgrade the lot of the paraplegic to a point at which he is no longer the worst-off, so that surplus resources can be spread to others, whereas with the neurotic all efforts to help him may only drive him farther into his misery.

It is not that utilitarianism is free from these many conundrums; rather they tend to make a far smaller difference to a prescribed course of social action than in the case of the Rawlsian principle. Under utilitarianism, one's allocation of resources may vary according to one's appraisal of the relative merits of two individuals or of their susceptibility to the expenditure of resources designed to improve their lot, but it is not an all-or-nothing decision of the kind the Rawlsian approach requires of us.

Even so, it is necessary to concede that utilitarianism is not the same thing as justice. One has only to consider that individuals can vary widely in their efficiency as "utility-producing machines": A, for example, has the ability to generate a great deal of "utility," however measured, out of very little in the way of resources; everything he puts his hand to turns out well and gives him a great deal of satisfaction, whereas B is all thumbs in this respect and whatever he attempts turns out disastrously, not only in terms of physical results, but also in terms of the satisfaction he derives. Understood as egalitarianism, justice might dictate that we make some attempt to compensate B for his innate ineptitude, which is an accident of inheritance and in no way his "fault." Utilitarianism, however, requires that we divert relatively more resources to the more efficient "utility machine" and squander relatively few on the luckless B.

At this point, I personally am inclined to give up entirely on the ideal of justice and treat it as an *ignis fatuus* whose pursuit by humans is bound to lead to perverse results—recognizing the while that a longing for justice appears to be innate in the human soul, but that it is a longing which may be destined to be fulfilled, if at all, only in the hereafter.

II

If there is no salvation to be had in a human search for justice, then we are left free to pursue the goals of utilitarianism. This need not lead us back to apologetics for a pure competitive system—had such a thing ever existed. For even such a system can leave us with a distribution of income so unequal as to be indefensible on any utilitarian grounds, even when a wide range of possible measures of relative individual utility is admitted. There is scope within utilitarian theory for measures designed to mitigate the inequality in the distribution of income produced by a free and unfettered competitive system.

One front on which it can proceed is that of *property rights*. If one were to construct a society from a *tabula rasa,* one could well insist, without departing from a pure competitive approach, that all natural resources be treated as a common heritage and that individual property consist only in what is derived directly or indirectly from individual productive effort. Accumulating large fortunes by seizing, without adequate payment to the community, portions of the common heritage for private gain would no longer be possible. Again, since a city's growth, the concentration of activity in it, and the level of its land rents depend in part on the magnitude of the economies of scale prevailing in the activities carried on within the city, it can be shown, also in principle, that the efficient operation of a city requires that land rents generated by this concentrated urban activity be used to subsidize the activities within the city having economies of scale and hence decreasing costs and marginal cost below average cost. Only in this way can prices for the outputs of these activities be reduced to the efficiency-promoting marginal cost level. Appropriation of any substantial part of these rents by private landlords (over and above that relatively small fraction necessary to motivate efficient management) renders full urban efficiency impossible.

We are, however, confronted with the fact that individual property rights in these resources have been established and legitimized by transfers that have inextricably mixed wealth acquired by individual effort and wealth acquired by seizure. One can take steps to ban further seizures of the public heritage in this manner, though this may mean locking the barn door after nine out of ten of the horses have been stolen.

Another element in the undue concentration of wealth is *inheritance.* Here it is tempting to call for the abolition of inheritance as an institution or for taxing it to the point of virtual confiscation. But it is not as easy as all that. As long as we retain the institution of the family, well-placed parents will have opportunities for ensuring that their children in one way or another are especially advantaged, whether by the cultural amenities of the home, special schooling, or by the handing on of property through gift or inheritance. Some aspects of inheritance, particularly those involving the transfer of property, can be subjected to taxation, but the effectiveness of such taxation is limited in ways that are not always fully appreciated.

In their present form inheritance taxes fall essentially on transfers.

Thus if these taxes become heavy, donors often can still achieve their ends by limiting the number of transfers as by the use of generation-skipping trusts, or, where these are specially taxed or otherwise made unattractive, by making direct gifts to those with longer rather than shorter life expectancies (e.g., grandchildren and great-grand-children). Ultimately there is always the escape of emigration, coupled with the shifting of one's wealth into foreign assets in a country with relatively low levels of succession duties, and for a government to impose controls that would prevent such escape would have serious effects.

This is not to say that a good deal cannot be done to improve matters, such as by basing the tax on the cumulative sum of all amounts received by a given person through gifts and bequests, or by graduating the tax according to the difference in age between transferor and transferee, or even by calculating the tax on the basis of a "bequeathing power" scheme which would establish tables relating the birth date of an individual and the balance of gifts and bequests made and received, such that the amount of bequeathable wealth would diminish, and the corresponding tax increase, as the holder of the "bequeathing power" aged.[3] With such a scheme, the overall burden of the taxes levied up to a given moment depends only on the distribution of inherited and donated wealth at that time, independently of the historical paths by which this result was achieved. Even with this scheme, however, taxing inheritance among the international jet set is something that could probably be handled effectively only by a world-wide agency such as the United Nations.

In practice, death duties, when vigorously applied, are rather more effective than one might expect considering the possible avenues of escape. In most jurisdictions, all that is needed to reduce death duties substantially is a modest amount of concern and planning, and a willingness to make relatively minor alterations in one's financial dispositions. That so much revenue is collected from estate taxes and so little from the various, more lenient kinds of gift tax, however, seems to indicate that most wealthy individuals are strongly motivated to retain control over their fortunes to the very end and are relatively little concerned over the net amount to be received by their heirs. One major exception is a case such as that of the Ford Motor Company, in which the threat that a family business might pass into the control of an indeterminate set of financial interests led to the establishment of the Ford Foundation.

In any case, by far the most important element in income redistribution in most countries is the *individual income tax*. Unfortunately, nearly everywhere the income tax has in practice been so riddled with special-interest concessions (such as depletion allowances) and opportunities for avoiding taxes by combining various provisions in ways that were never contemplated (such as the double allowance for charitable donations of appreciated property), that already in 1938 Henry Simons was talking about "the absurd business of dipping deeply into high incomes with a sieve." [4] In 1972, for example, the average effective rate of the federal income tax in the United States reached only 39.50 percent for the $200,000–500,000 bracket and actually declined to 38.46 percent for the Over $1,000,000 bracket. While some recent reforms have mitigated this situation, the overall results remain very far from what the rate schedule would appear to prescribe.

It is not difficult, in principle, to devise a progressive income tax that would be free from many of the distorting effects on economic activity which result from the existence of the various loopholes—that would make the real degree of progression correspond to what is expressed in the rate schedule—while greatly simplifying the law and freeing the businessman from worry over the tax consequences of his decisions. Briefly, this can be accomplished by a "cumulative assessment" or "cumulative averaging" scheme, the principle of which is to consider all payments with respect to an individual's income tax liability as deposits in an interest-bearing tax account and to compute, at the end of the year, a cumulative tax on the total income realized (including the interest on the tax account) since a base starting date, at rates graduated appropriately for the time period in question; the tax currently payable would be simply the difference between the amount thus computed and the balance in the account. With such a scheme, the computations are much simpler, and the necessary carry-forward of information from year to year far less, than is required under the existing five-year averaging provisions. Adoption of such a scheme, together with a requirement that all receipts be accounted for either as gross income or reductions in the book value of assets, at the option of the taxpayer, would allow the deletion of about half of the present internal revenue code.

If we were to close all of the loopholes in this way, while retaining the present rate schedules, this might leave us, on balance, with a degree of progression that would seem unduly severe to many of those concerned. To a naive observer, the obvious solution would be to reform the tax base and reduce the progression of the rates simultaneously so as to yield a generally acceptable degree of progression, a more equitable distribution of the burden within income classes, and a greatly reduced interference of the tax with the efficiency of the economy. But such a resolution seems hardly to get to the stage of serious discussion, except among a small group of tax reform advocates, and every attempt to legislate reform appears to produce new complexities and new inequities.

Strong forces, indeed, appear to be arrayed against fundamental and thoroughgoing tax reform. There are obviously special interests, such as insurance, oil, mining, lumber, and farming, that draw advantage from particular loopholes that could not be replaced in full by any general reduction in rates. Even when, as in the case of tax-exempt interest, fully adequate compensation of the interested governments is possible, the change is resisted vigorously. Perpetuation of an irrational, discriminatory, and inefficiency-promoting privilege of long standing which appears to be capable of indefinite retention is preferred to the promise of a more explicit subvention, for the arbitrariness and inequity of the compensation payment, though no greater than that of the exemption, would be more obvious and less firmly rooted in past practice and less plausibly sanctioned by spurious legal reasoning.

More generally, there is the perhaps subconscious feeling on the part of the loophole players that it is far easier to hold the line on a set of loopholes than on a level of rate progression. Since the degree to which future governments can be committed in advance to the continuation of any given policy is inherently limited, there is no way of assuring that the rate concessions granted in exchange for the abolition of loopholes would be maintained over a sufficiently long future period to make the bargain attractive. And for those to whom the whole notion of progression of tax rates is anathema because of the potential in such tools for the despoliation of the wealthy minority by the poorer majority, loopholes may appear as the last bulwark against exploitation, so that any loophole, no matter how offensive to equity or prejudicial to economic efficiency, may become a cherished element in this defense. Nor can one altogether discount the possibility that the large vested interests of tax lawyers and accountants, in their knowledge of the intricacies of the existing code, lead, subconsciously at least, to a preference for dealing with tax problems by introducing new complications rather than by basic reform that goes to the heart of the matter.

III

There are of course other forms of taxation that can serve as means of achieving a better distribution of income and wealth. Among these are the net-worth tax, the expenditures tax (with rates graduated according to total expenditure on personal consumption), and the negative income tax. These taxes, in combination, can bring about fairly substantial changes in the distribution of after-tax income, small as their effect has been so far. What is needed, however, is a criterion for determining how far this redistribution process should be carried.

The extreme answer of John Rawls is, in effect, that a social dividend should be distributed according to need to establish the welfare floor, and that this floor should be as high as possible, meaning in turn that the tax system should be operated so as to extract as much revenue as possible from those who have incomes above the floor, in order to have as much as possible to distribute to those below, thus raising the floor itself. Retentions above the floor are allowed to the extent needed to give the incentives that will enhance the tax revenues.

The utilitarian answer to the question of rate progression finds its most persuasive explicit expression in the theory of "minimum aggregate sacrifice." In its naive form this theory was once held to require the confiscation of all income above a set limit and was rejected as being obviously impractical, but when account is taken of effects on incentives its implications can be almost the reverse of this. Indeed, if the utility yielded by an additional dollar of consumption by upper income groups is considered negligible or very small relative to the utility conferred by an additional dollar of consumption for the poorest income groups, then the main concern in the taxation of the upper income groups is the amount of revenue to be obtained with which to enhance the consumption of the lower income groups by reducing their taxes or possibly, in the more recent formulations, through a social dividend or a negative income tax. The proper tax rate to apply to a given bracket of earned income can then be thought of as determined by balancing the effect of increasing the tax rate in obtaining more tax from a given amount of income against the effect of a higher tax rate in discouraging the production of the income that will be subject to tax.

Now for the top-bracket rate, all incomes that pay tax at this rate will also be subject to this rate at the margin and will feel a disincentive effect, while for the rates applicable to a given lower bracket—while all incomes falling in or above this bracket will generate tax revenues from this rate—only those incomes falling within the bracket or close to it will find their incentives impaired if this rate is raised. If this rate is raised and other rates kept the same, incomes in the higher brackets will pay more tax, but their incentives at the margin will remain unimpaired. Balancing the higher revenue effect of the rate on the lower bracket relative to its incentive effect against the relatively smaller revenue effect of the rate in the higher bracket, it is possible to conclude that the marginal rate in the top bracket should, for the greatest possible equalizing effect on the distribution of income, be lower, rather than higher, than the rates in the intermediate brackets. This somewhat surprising and indeed paradoxical result applies, to be sure, only to the taxation of "onerous" income derived, directly or indirectly, from personal effort where there is an effective choice between more untaxed leisure and more taxed income: it does not apply to "rentier" income derived from passive investment where the question of more or less individual effort is largely irrelevant. The result also depends on the not unreasonable assumption that the distribution of talents as measured by earning power has a rather blunt upper tail. Also if the difference in the utility of income between the highest incomes and those somewhat lower down were considered to be a sufficiently important consideration, this could operate to justify progression. But there is no reason to suppose that the change in the relative importance of the incentive effect cannot outweigh this decline in the marginal utility of money, and the rate schedule that maximizes total utility can well include declining marginal rates over this range of incomes. Somewhat paradoxically, Rawls's principle would

be even more likely to produce this effect of declining marginal rates
with increasing income, since for Rawls there is no consideration
given to possible differences in marginal utility of income at different
income levels: the only thing relevant for the application of his principle is the aggregate revenue to be derived. The fixed payment or
credit element in the tax, of course, would still assure that for low incomes the average rate of tax will increase as income increases.

Political realism, however, forces one to admit that few, if any, industrialized communities would give even majority assent to such redistribution. In economies with a broad base of the impoverished, it is
conceivable that the poor might constitute themselves a majority and
unite politically to lawfully despoil the rich. But in communities with a
dominant middle class, a relatively small poor segment, and an elite
that has made itself master of the game of loopholes, the mass of the
middle class tend to balk long before welfare payments have reached
a utility-maximizing level.

On the other hand, personal empathy can at times become strong
enough to motivate direct person-to-person, or even person-to-
group, philanthropy. There is, moreover, an "externality" to such
voluntary redistribution: third parties also are favorably affected by the
gift to the extent that they are concerned for the lot of the beneficiaries. From independent philanthropy it is but a short step through
voluntary gifts to organized philanthropy "in consideration of the gifts
of others," and from such philanthropy to voting in favor of taxing
oneself along with others in a group in order to confer benefits on
another group deemed sufficiently worthy of help to motivate each
member of the donor group to vote in favor of group effort, though
not to undertake any part of it independently. Such "Pareto-optimal
redistribution," while a far cry from Rawlsian justice or even utilitarian maximizing, does serve to mitigate inequality resulting from unfettered competition.

Such redistribution, however, is likely to be somewhat subjectively
oriented, since some degree of empathy between contributor and recipient is required, and this is likely to be the weaker the more remote the recipient, either in distance or in culture. The donative urge
can be thought of in terms of a relative neediness factor and an empathy factor. As the difference in income increases, the donative urge
resulting from the interaction of these factors will start from zero, rise

to a maximum, then taper off again. Thus the natural objects of a contributor's most intense concern will not generally be the poorest elements of the human race, but a group somewhat closer in culture or location: the wealthy tend to support universities, the poor to support the Salvation Army. Likewise, political realism leads one to expect a group of voters to support redistribution in favor of local residents rather than foreigners. One must also admit that many a program nominally designed to help a class of deserving poor turns out to have very substantial fallout in favor of moderately well-to-do groups, even when the benefits are not largely intercepted by intermediaries.

In sum, as an economist I cannot help but regard justice as an elusive and fickle guide to social policy. Economic efficiency gives far clearer indications of where to go, though it cannot be denied that in some cases the going may become unduly rough for some, in which case the palliative of redistribution is to be applied with a degree of vigor that will have to be left to a mixture of philanthropic and political considerations over which individuals may legitimately differ. Even here, the economist can advise, from the sidelines, against those redistributive measures that tend to impair efficiency to a greater degree than is strictly necessary to achieve the desired degree of redistribution.

6

Equality and Fraternity

A Note on Subjective Realities

BENJAMIN DeMOTT

> The sentiment of equality is not an old sentiment, a perpetual sentiment, a universal sentiment of first magnitude. At determined periods it appears in the history of humanity as a peculiar phenomenon.
>
> Péguy
> *Les Vérités Fondamentales*

The assumption in what follows is that The System, any system, is a grain of feeling; it is not separate from or superior to or independent of individual beings; understanding its nature requires close reading of movements of response within persons.

I

An extended example: My student unhappy about his grade is slow to bring up the matter when he calls at my study. I am his adviser, and he speaks for a time about options under consideration for the coming term. I am patient, I counsel with him about these choices, observing him the while, knowing to a certitude that when he finishes laying out these curricular problems . . . I am without meanness or irritation during this waiting. I understand manners oblige him to approach his major business offhandedly, diffidently; he must demonstrate that he is moved by concern for a principle, not by concern

Benjamin DeMott is Professor of English at Amherst College.

for his grade; he must ensure that I am not spoken to as a person supplicated (in order to protect himself from being spoken to as supplicant).

At length my student broaches his subject. He does not look at me at this moment, and I for my part do not seek his eyes; we are advancing not toward combat or danger—this is an academy—but rather toward that which stands in for both in academic settings, namely, awkwardness. My student picks up the pace of his speech, anxious now to be done with it. It appears that from various sources he had gained the impression that the freshman seminar in which he was enrolled with me was to be regarded as a different affair from an ordinary course, something in the nature of an introduction to higher academic study, a shared learning experience in which an instructor would be better described as adviser-counselor-friend than as teacher. No grades, he reminds me, had been distributed during the term. The "entire atmosphere" gave him confidence and he had enjoyed the "relaxed feeling," the "absence of pressure"; he had assumed that if grades were necessary they would be worked out in consultation between teacher and student, wherein personal self-estimate would be a factor. But then . . . his grade—the very border of failure—pulled him up abruptly. My student wonders whether there were some special deficiencies, some uniquely bad performance or mistake committed at the end, that had caused this surprising and depressing defeat of expectation.

I am not chilly when the lad subsides. I nod, signifying my understanding of his protest and my regret that he's had to lodge it. Moving to the ground of principle with him, I agree that if the sense of informality—the belief that grading, assessment of performance by a uniform standard—was irrelevant to our enterprise, if this had been pervasive, so that the group as a whole had arrived at his conclusion, then plainly I was at fault and the fault was not trivial; it is wrong to create an air of security, of easy sociability, and thereafter "pull the rug out" from under people. Wrong, dishonest, vindictive. In answer to my direct question, however, he hesitates only briefly before assuring me of what I already know, namely, that the others in the seminar did not share his surprise, or his bitterness. I thank him for his honesty and tell him that while I cannot do anything about his grade, the complaint he has made has value for me. Next time—with a

rueful smile, the first in this conversation, I grant that what I'm about
to say can't rejoice·him—next time I shall remind each seminarian of
the limits of informality, of the teacher's obligation in all courses here-
abouts to report a letter-grade judgment to the registrar, and of my
intention to arrive at this estimate without consulting others. I then
address his question about special deficiencies, taking this matter too
up to high ground. The weakness I detected, I inform him, was an
unaggressive critical stance, an absence of eagerness to challenge.
. . . He is young, a beginner (my manner is a shade more benign
and expansive as I remark these facts). He will have time to learn.
But he should be aware that people trained to be leaders are expected to show a spirit of independence; their teachers hope to encounter a zest for challenge in the first words on a page, and unfortunately this zest did not appear (that was my observation, at least) in
his written work.

Perhaps, I add, returning to the earlier theme (it now seems slightly
less painful to us both), perhaps once again it was the atmosphere, it
was this feeling that he was not being called upon to exert himself or
to press forward, that led to the somewhat yielding or—would he let
me say?—rather flat quality of his writing. We agree that this is a possibility. Before he leaves I lean forward to him, seriously, and tell him
that if after a little time he continues to feel resentment, I would inter-
vene to secure a different adviser for him, for this is too important—
at once he cuts me off. My suggestion is rejected. He wishes to continue with me. This is definite. We shake hands firmly, pleased that
each of us has conducted himself without loss of dignity; he goes his
way.

As a human encounter under the present system, this is (I submit)
unbarbaric. I am aware of a better self, a happier and more generous
character, a more admirable role than that which these circumstances
thrust upon me. Scrutinizing my behavior in retrospect, I miss signs of
insecurity—redemptive hints of self-criticism. There was no harshness
in my voice or manner, I was, in fact, "sorry" for the lad's misfortune, and felt a seemly embarrassment in having been placed even
momentarily among the nearly supplicated; yet at no instant was I
doubtful of my position or of its justice. As a student of history, I am
aware that from time to time men have imagined cities purified of rivalry or of competition, wherein functions are distinguished from
each other, yet not arranged in hierarchy, wherein workers donate

their labor to the community and the community donates its products to the citizens, wherein men cease to regard each other as enemies ready to devour one another in pursuit of a mouthful of bread or a job with an exploiter. And I am clear in my mind, when the matter comes to my mind, which is seldom, that competition does set us ineluctably against each other, commanding us to work against each other instead of for the good of the whole.

Yet, as I say, these awarenesses were remote as I dealt with my nonsupplicating student. I masked my judgmentality, but at every moment knew the lad to be in the wrong, and knew, furthermore, that his punishment was not unjust. My student had assumed without warrant an interruption in the system, a *lusus naturae*, a gap, an interval of grace. He had assumed that an exemption from the mode of competition could occur without explicit announcement. He had failed to grasp (as he is obliged to, as all are obliged to) that precision about the terms of relatedness of members of a group is a personal responsibility; guessing about the degree of remission, if any, is unpardonable. I myself had been, throughout, correct. Whatever my affability, my bonhommie, I had said nothing to this class to imply a commitment to an alternative, more pliant world. And it is a fact that under the system as it exists, those who misread tones, who fly on from a raffish idiom or other intimation of unbuttonedness to the conclusion that the war is over and the reign of fraternity has begun—these truants must pay. The system is a Serious Thing. When by some accident teacherly conduct brings suddenly freshly home to the young the impermeability of the system, its essential seamlessness, the appropriate response is gratitude. —We must divide ourselves against each other, we must labor each for our own sole good, we must not imagine surcease . . . thus to teach is to serve the good; guilt goes with the soft; I am innocent.

II

Now then, says a teasing voice, if such truly is the grain of feeling under the system, then—putting aside for the moment the value of equality and focusing on the higher value, fraternity—then surely there can be no sense of human solidarity in our world. How do you speak to this?

I should disagree. But I should also grant that the modes of connectedness with others which do here and now exist—the bases of my belief that I am bound into them as they are bound into me—differ considerably from the solidarity of other times and places. My solidarity resembles in no way that of people living together in poverty short of destitution—I think, say, of the figures who populated Oscar Lewis's San Juan and New York studies. Neither is the solidarity I know related to the unity of the brave as in Homer. Nor is my solidarity comparable to that experienced by those united through the workplace, through feeling for quality of material, solicitude for tools, joy in *ouvrage bien faite.* I turn the pages of George Sturt's *Change in the Village,* pause at a few words on a master wheelwright, and at once am lost in strangeness:

Truly it was a liberal education to work under Cook's guidance. I never could get axe or plane or chisel sharp enough to satisfy him; but I never doubted, then or since, that his tiresome fastidiousness over tools and handiwork sprang from a knowledge as valid as any artist's. He knew, not by theory, but more delicately, in his eyes and fingers. Yet there were others almost his match—men who could make the wheels, and saw out on the saw pit the other timbers for a dung-cart and build the cart and paint it—preparing the paint first; or, if need be, help the blacksmith tyring the wheels. And two things were notable about these men. . . . The first . . . is, that in them was stored all the local lore of what good wheelwright's work should be like. The century-old tradition was still vigorous in them. They knew each customer and his needs; understood his carters and his horses and the nature of his land; and finally took a pride in providing exactly what was wanted in every case. So, unawares, they lived as integral parts in the rural community of the English. Overworked and underpaid, they none the less enjoyed life, I am sure. They were friends, as only a craftsman can be, with timber and iron. The grain of the wood told secrets to them.

I as liberal educator have some fastidiousness about tools but small knowledge of my customer's needs, and no natural object speaks secrets to me. Sturt's sentences on craftsmen as tribalists further define my remoteness:

The men, unlettered, often taciturn, sure of themselves, muscular, not easily tired, were in a sort of way an epitome of the indomitable adaptation of our breed to land and climate. As a wild animal species to its habitat, so these workmen had fitted themselves to the local condition of life and death. Individually they had no special claim to notice; but as members of old-world communities they exemplified well how the South English tribes, traversing their fertile valleys, their shaggy hills, had matched themselves against problems without number, and had handed on, from father to son, the accumulated lore of experience. If one could know enough, one might see, in ancient village crafts like that of the sawyers, the reflection as it were of the peculiarities of the countryside—the difficulties and dangers, the daily conditions—to which those crafts were the answer. . . . They themselves, you found, were specialists of no mean order when it came to the problem of getting a heavy tree—half a ton or so of timber—on a saw pit and splitting it longitudinally into specified thicknesses, no more and no less. What though the individual looked stupid? That lore of the English tribes as it were embodied in them was not stupid any more than an animal's shape is stupid. It was an organic thing, very different from the organized effects of commerce.

Men perfectly continuous both with each other and with nature; man and nature perfectly coextensive with culture: these solidarities I cannot conceive.

Yet, to repeat, in my world I do know connection, I do feel bonds. A second tale—this time brief—puts the feeling before me: a homely episode still evocative, after twenty-five years, of the true terms of our linkages. Imagine a young married man with two children; at twenty-five, after some amount of what he perceives as struggle, he is a beginning student at the university and, part-time, a taxicab driver. Beating one noontime for a signaling fare at a Pennsylvania Avenue intersection, he is outrun—another white Veteran's Cab darts in front of him, cuts him off, steals his client. Hull by hull, the two cabs sit for seconds then at a red light, and the driver who had beaten our young man to the fare looks over, grins, lifts a shoulder: "Dog eat dog," he says. Our young man feels no repugnance, rather a shiver of recognition. *The other driver is saying this will happen to him too, and that he knows this as I know this. We are locked in an imperative, all of us together, Dog eat Dog: we must beat each other to the light. The laws of competition are irrevocable, we are up against—up against gunsels, sharpies, pacesetters, all the voraciousness on earth? No, we are acting through the deep rule of life itself: be for yourself.*

If to be thus linked with others can cause one pain, that pain invariably intermingles with other stuffs, as for example, with fantasies of endurance. I am not nothing, after all, when I see myself as provisioner against odds. Hang by your teeth and write when you find work. I confront the harshness of things, the hard loneliness of things: neither I nor my brother who beats me out and will himself be beaten is a patsy; the furies shriek, and in answer we shred each other but we are conscious. And together we recognize the inexorabilities. And we have our tenacity to keep us warm.

And also we have our wariness. That is a second mode of solidarity worth a word. I am together with those who are doughty in their struggles against each other, and, in addition, I am together with those who must be wary. This union is composed of people who, because they believe nothing, cannot be fooled. —Brother if you *knew* what goes on down there. . . . My communion with my children is richest, I sometimes think, when together we taunt our deceivers. Families that negate together stay together. I watch the news with my younger daughter; spontaneously we register our disgust at the dancing man who says *I smell clean cha cha;* together we see he's a deceit, we see that America the Beautiful is not, we see the spacious skies are polluted, we see that Jefferson was a slave-holder, we see that cynical disbelief is, like bitterness, like shared helpless anger at the unremittingness of competition, among the blessed ties that bind. We are far from Sturt's Surrey village; we are in the global village of the unsuckered; but we are at one.

III

As these remarks will have suggested, the relationship between the system and justice and equality—the general subject of the preceding essays in this section—seems to me less consequential than the relationship between the system and fraternity and charity. I have postponed thus long a consideration of grains of feeling relevant to equality because I wished to make a gesture of resistance to the continuing national obsession with objective circumstance. I believe equality and justice bemuse us because they seem to have to do with "the arrangement," with affirmative action Out There in the manipulable world. Presumably a calculus of equality and of justice is conceivable, presumably we can develop criteria for measuring progress toward equality, economic and other; presumably a scale better than Astrea's can be manufactured on which to weigh Judge Sirica's sentences and pardons. Whereas ambiguity (subjectivity) attaches to fraternity and to charity; it is inconceivable to "turn these out" or otherwise deny their essential inwardness.

Still it is undeniable that equality too is a configuration of feeling and attitude or else it is nothing; certainly it is not merely a congeries of distributable rights, privileges, goods. At the center of the configuration lie the responses we point to with the word *respect*. And, as was the case with that structure of feeling which I have argued is related to the value of fraternity, the structure of feeling hinged to equality is not without positive content. Its nature is most strikingly embodied in the tone of our letters. In the present as in the past, the common cause of American writers has been to establish the appropriateness of disregard for the authorized and established, and to continually infuse stimulants to irreverence into the culture.

The quietest, weightiest and most elegant irreverence has been, predictably, that which has said unsaddened farewell to established religious orthodoxy. Consider for one example the poet Wallace Stevens. No echo here of the noisy preening gestures of village atheists. This man's hero is notable, not for the disposition to taunt or negate, but for a grave readiness to wear loneliness as a dignity—as a distinction too long declined owing to the ease of self-deception. We are obligated, said this poet, to exercise the full powers of imagination, to grasp that when the gods are dead and the religions are seen to be empty, people have to create meaning on their own:

in an age of disbelief, when the gods have come to an end, when we think of them as the aesthetic projections of a time that has passed, men turn to *a fundamental glory of their own* and from that create a style of bearing themselves in reality. (*Opus Posthumous;* emphasis supplied)

The power to do this, Stevens asserted, lies within us, and the discovery that we don't have any choice except to use it cannot diminish anyone: to rise to a consciousness that imagination is all is to achieve serious elevation, equality with the nonexistent gods. The great poem "Sunday Morning" ends with stanzas summoning a nobility achievable not by social ascent to official honor or rank but by a new consciousness, a facing of fact:

> *Supple and turbulent, a ring of men*
> *Shall chant in orgy on a summer morn*
> *Their boisterous devotion to the sun,*

Not as a god, but as a god might be,
Naked among them, like a savage source.
Their chant shall be a chant of paradise,
Out of their blood, returning to the sky . . .

The mode of presumption ultimately becomes heroic.

Even when it is less than heroic—a matter of political uppityness rather than metaphysical audacity—American presumption, the American refusal of deference, is wonderfully energizing. Consider for a second example the most effective contemporary baiter of authority, Norman Mailer. The Mailer version of the writer (a socker, an athlete, an infantryman, the friend of pugs, dangerous chap at parties, a *macho* caesar) derives from a romantic literary past, from Byron, from Hemingway. The qualifications as an outsider or bottom dog are shaky: a cosseted Harvard boy, Mailer was subsidized into the life of letters—*à Paris*—by loving parents after World War II. And his avowed attitudes toward the weak and unlucky of this society are hostile. Yet, as an embodiment and an endorsement of the energies of truculence, Mailer has exceptional value. His posture as a journalist, his mode of address to the great and near great in politics, is furiously participational. Nixons, Kennedys, Goldwaters, Kissingers—all our chieftains—are perceived as small potatoes, figures commanding no awe, people to be accepted putatively or socially as my equals perhaps, but under no circumstances as my superiors. And this manner of address, filtering through the media and into the attitudes of university teachers and students alike, has nourished a trilling nerve of impertinence in the culture. The power of presumption thus transmitted can issue in loutishness, or produce only dreams of glory that are bad poison when diffused in men and women unprepared to bear the responsibilities of governors. But when the power *penetrates*—I am thinking of Mailer's portrait of Henry Cabot Lodge as a person who understands himself to be in some way "necessarily superb"—it is difficult to scorn. To be released from meachingness, invited to a transcendent self-assertiveness, exhorted to vilify pretension—this is exhilarating. And the resultant grain of feeling clearly "belongs" to the system: *Be for yourself alone.*

But, as must be added, the triumphs of presumption are dangerous triumphs: their force is beyond containment. The refusal of deference ultimately diminishes respect for self, diminishes respect for the conditioning terms of our existence, and at length brings respect for respect itself to an end. The process is in no particular mysterious. In another system my capital equals my private knowledge, my reticence, my conversancy with the content of my feelings, my wounds, my humiliations, my attachments, my fantasies. I spend this capital cautiously, rarely give it away, for it is not only what I possess, it is what I am. But in the market economy, assigning worth to one's innerness, unless it can be sold—as a star sells her "frank talk of infidelities" to the talk-show interlocutor—is difficult. My secrets are valueless—I hurl them recklessly before strangers, whether in words or in public embraces, and behavior that began as splendid Emersonian defiance of the claims of others ends in self-contempt. An episode in Saul Bellow's *Herzog* shows forth the process:

At the hack stand [Moses Herzog] found a Puerto Rican driver who was touching up his sleek black hair with a pocket comb.

Moses knotted his tie in the back seat. The cabbie turned around to look. He studied him.

"Where to, Sport?"

"Downtown."

"You know, I think I got a coincident to tell you." They ran eastward toward Broadway. The driver was observing him in the mirror as he drove. Herzog also bent forward and deciphered the name beside the meter: Teodoro Valdepenas. "Early in the morning," said Valdepenas, "I seen a guy on Lexington Avenue dressed like you, with the exact same model coat. The hat."

"Did you see his face?"

"No . . . the face I didn't see." The taxi rattled into Broadway, and sped toward Wall Street.

"Where, on Lexington?"

"Like the sixties."

"What was the fellow doing?"

"Kissing a broad in a red dress. That's why I didn't see the face. And what I mean *kissing!* Was it you?"

"It must have been me."

"How do you like that!" Valdepenas slapped the wheel. "Boy! Out of millions. I took a guy from La Guardia, over the Triboro and the East River Drive and left him off at Seventy-second and Lexington. I seen you kissing a broad and then I get you two hours afterwards."

"Like catching the fish that swallowed the queen's ring," said Herzog.

Valdepenas turned slightly to look at Herzog over his shoulder. "That was a real nice-looking broad. Stacked! Terrific! Your wife?"

"No. I'm not married. She's not married."

"Well, boy, you're all right. When I get old I'm going to be doing just like you. Why stop! And believe you me, I stay away from young chicks already. You waste your time with a broad under twenty-five. I quit on that type. A woman over thirty-five is just beginning to be serious. That's the kind that puts down the best stuff. . . . Where are you going?"

"The City Courthouse."

"You a lawyer? A cop?"

"How could I be a detective in this coat?"

"Hombre, detectives even go in drag now. What do I care! Listen to me. I got real burned up at a young chick last mont'. She just lies on the bed chewing gum and reading a magazine. Like she's saying, 'Do me something!' I said, 'Listen, Teddy's here. What's this gum? Magazines?' She said, 'All right, let's get it over.' How's that for an attitude! I said, 'In my hack, that's where I hurry. You ought to get a punch in the teeth for talkin' like that.' And I'll tell you something. She was a no-good lay. A broad eighteen don't know even how to shit!"

Herzog laughed, largely from astonishment.

But what cause for astonishment? The academician's shameless public embrace in the street has brought on the shameless disclosures of the cabdriver. Not the solidarity of Dog eat Dog; but the solidarity of folk who daily auction off their privacy for attention, i.e., assurance that he who drives and he who rides, he who studies and he who performs, are equal.

IV

For the rest—the loss of respect for the conditions of life—not much need be said. The pride of the political left is that political arrangements can be perfected by means of which to enact the reign of charity; the voice declares itself (through its tone) unimpressed with tragedy, undepressed with the darkness of our comings and goings. "If we want substantial redistribution," says Christopher Jencks briskly in his *Inequality,* "we will not only have to politicize the question of income equality but alter people's basic assumptions about the extent to which they are responsible for their neighbors and their neighbors for them . . ." (p. 264). Grace has failed to "alter people's basic assumptions," but let us go to work on this material in new ways; nothing can stand up against us.

The pride of the right is more complex perhaps but not less troubling. I feel it at its finest in Robert Nisbet's telling critique of the Jencksian simplisms. In "The Costs of Equality," Nisbet sensibly and persuasively links the contemporary equalitarian attack upon inequality in the name of moral worth or merit with "the ancient, perhaps timeless dream that justice means proportionality of reward to moral worth in the economic, cultural, professional and intellectual market place." Shrewdly the writer completes his case:

> X, let us say, is regarded as a paragon of conscientiousness, industry, fidelity, kindness, charity, and generosity, but nevertheless fails of election to an honored membership or appointment to an honored job because of certain accidents of birth, rearing, or capricious fortune. Y, on the other hand, while standing well within the law and local ethical standards, is not ranked so high on the scale of moral worth, and has moreover enjoyed good luck, good parentage, good schools and good professional acquaintance. Yet it is Y who has the international name, the renown, and all the prized elections and appointments and awards. Ergo: the system is unjust.
>
> But . . . Hayek concluded that in a free system it is neither desirable nor practicable that material rewards should be made to correspond to what men recognize as moral worth. Indeed, it is of the essence of a *creative* society that an individual's position should not necessarily depend on the views his fellows take of his moral worthiness.

But a bit of a tonal problem nags the reader. The critic appears pleased that circumstances and evidence enable him to show forth his enemy's beamishness. But his satisfaction needs leavening with regret; the inevitability of injustice, the impossibility of escaping maldistribution, our quandaries, our contradictions, our helplessness to contrive the fully humane arrangements that we nevertheless ceaselessly imagine for ourselves, the perpetual gap between our reach and our grasp—these matters too deep for tears cannot be the substance of "winning arguments." . . . When respect lived, men did not pitilessly marshall these terrors to establish the foolishness of the political opponent; they prostrated themselves, they sought a voice for prayer.

Several routes of exit from these gravelling subjects propose themselves. A question might be posed: can Justice exist where respect has perished? A thesis could be urged, namely, that no imaginable system could do more to nourish the value of equality than democratic capitalism. (This is, indeed, my view.) Or a glance at areas of feeling left out of discussion because of a lack of space (for example, stiffness and detachment in human relationships, absence of capacity for intimacy and sympathy); or a glance at those elements of competitive capitalism that are, on their face, antiequalitarian, hostile to sharing, possessive, committed to earned privilege differentials. Or a battle could be joined, pitting "Blessed are the meek" against the musing that a proud man is perhaps the greatest joy of the gods.

My own preference, as already indicated, is for turning away altogether from these matters, or at least far enough from them to permit full representation of—the word in this context bears its own ironies—full representation of the competition. I wish to end, in short, with a reminder of what can be said against our key value and for its opposition. The text once again must be Péguy's *Les Vérités Fondamentales.* First a complaint against equality:

Equality often reaches only men loving the limelight, men loving publicity, and men of government. . . . Sentiments of equality are artificial sentiments, sentiments obtained by formal construction; bookish, scholastic sentiments.

Finally these memorable words:

From age to age, fraternity, whether it puts on the guise of charity or the guise of solidarity; whether it is practiced towards a guest in the name of Zeus Hospitable; whether it welcomes the poor as an image of Jesus Christ or whether it establishes a minimum wage for workmen; whether it invests the citizen of the world, introducing him by baptism into the universal communion; or whether by the improvement of economic conditions it introduces him into the international city, this fraternity is a living, deep-rooted, imperishable human sentiment. It is an old sentiment which, maintained from form to form throughout transformations, is bequeathed and transmitted from generation to generation, from culture to culture. By far in advance of the civilizations of antiquity, it has been maintained in the Christian civilization and remains and will doubtless flourish in modern civilization. It is one of the best among good sentiments. It is a sentiment at once deeply conservative and deeply revolutionary. It is a simple sentiment. It is one of the principal among the sentiments which have made humanity, which have maintained it, which will doubtless free it. It is a great sentiment, one of great moment, of great history, of great future. It is a great and noble sentiment, old as the world and it has made the world.

Private Rights
and the Public Good

The problem is to find a form of association which will defend and protect with the whole force of the community the person and property of each associate, and in which each, while uniting himself with all, may still obey himself alone and remain as free as before.

Jean-Jacques Rousseau
The Social Contract, 1762

7

Private Rights
and the Public Good

CHARLES FRANKEL

The problem of adjusting the individual case to the general rule is as old as Parmenides and Heraclitus. The oddity of life in society, which requires that individuals live by rules and also that they evade, suspend, or modify them as a regular matter, provides a basic theme of jurisprudence and philosophy, and one of the forms in which this problem presents itself is in the interplay of two commonplace and not entirely clear phrases—"private rights" and "the public good." The arguments that go on about these ideas and about the discordant human claims for which they stand are close to an obsessive motif of contemporary debate in our society.

Consider these familiar problems around which angry controversy rages daily: the rights of individuals to engage in conduct regarded by large parts of the community as indecent, immoral, or simply very irritating; the rights of individuals engaged in activities whose direct consequences seem harmful only to themselves to be let alone no matter how much harm they are doing themselves; the rights of people in local areas to resist urbanization, down-zoning, or the busing of children to promote school desegregation; the rights of people to be parents as often as they wish, regardless of population pressures; the rights of individuals and groups who believe themselves victims of intolerable injustices to disobey the law, the rights of public service

Charles Frankel is Old Dominion Professor of Philosophy and Public Affairs at Columbia University and President of the National Humanities Center. From 1965 to 1968, he was Under Secretary of State for Cultural Affairs.

workers to go on strike; the more general problem, of which both civil disobedience and public service strikes are marginal examples, regarding the limits within which corporations, labor unions, professional organizations, and other powerful groups vested with private rights should be required to behave.

The list, plainly, could be many times as long. It suggests that the issue of private rights and the public good is a central and urgent one. I cannot, of course, deal with the specifics of all the problems that illustrate it, but I shall try to suggest a general framework for considering them.

I

The first step is to examine the word *right,* one of the most conspicuously trouble-making words in the language. It is systematically ambiguous in the meanings people attach to it, yet at the same time it is a fighting word, used when someone is demanding strong moral and political medicine.

If someone says that he has a "right" to do something or to receive something, he normally means to say that he need give no further reasons for doing it or receiving it, that other people are wrong to get in his way, that they ought in fact to facilitate his attainment of what he desires, and that there are really no grounds for compromise, concession, or bargaining. A right is not subject to being annulled or even trimmed except when it conflicts with other rights.

Claims that we have rights would thus appear to be incitements to battle, limitations on social planning, invitations to turn social problems into theaters for moral disputation. It is not to be wondered, therefore, that philosophers like Hobbes, who have put peace high on the calendar of social virtues, have looked with a jaundiced eye on the language of rights and have been inclined to think that claims of rights should be strictly limited and never honored at all except when embedded in a determinate system of positive law. What are the social and intellectual functions of this concept of rights, and what is the rationale for employing it?

Let us begin by taking a closer look at the various meanings of the word *right.* Two principal senses can be distinguished.*

* In what follows, I lean heavily, as will be evident, on the writings of Wesley Hohfeld and Hans Kelsen.

1. We often use the expression "to have a right" to mean that there is no relevant moral norm or legal rule under which one has the duty not to do or have what one specifically proposes to do or have. The policeman asks me what I think I am doing walking on the grass in the park, and I say that in the absence of any Keep Off the Grass sign, I have a right to walk on the grass. To have a right in this sense means only not to be under the pertinent restrictive duty.

Probably it would be better to use the expression *privilege,* as Hohfeld suggested, or the word *liberty,* as Glanville Williams has recommended, to designate this wholly negative conception of a right.[1] For when we use the word *right* in this way we use a strong word for what is in fact a weak claim. The premise, "I have no duty under the law to refrain from addressing this club-meeting," does not automatically imply the conclusion, "I have a right under the law to address this club-meeting." The law permits me; it does not entitle me. To make the slide from one to the other is to presuppose the general principle that what an individual is permitted to do he is entitled to do. But this by no means follows from the notion of "right" as such. It comes from a separate assumption unconsciously compressed into that notion.

2. For the expression "to have a right" is also used in a much stronger sense, and it is the unconscious glide from one to the other that makes the use of the expression in its weak sense dangerous.

I have signed a contract with you to pay me $100 on the delivery to you of a manuscript. I deliver the manuscript; I have a right to the $100; you have the duty to pay me. And my right is simply your duty. In fact, I do not need the word *right* at all to describe the relation in which I stand to you. It can be completely described simply by listing your duty to me, along with the duties of others, particularly officials designated by the law, to assist me in seeing to it that you do that duty. Max Radin has suggested that the first kind of right be called a "privilege-right" and this second, stronger kind a "demand-right."[2] The reason, obviously, is that in this stronger sense of the term *right* we are making claims on other people's behavior. We are, indeed, talking more about others than ourselves. It is *they* who must do or omit certain actions; we are merely the beneficiaries and the claimants. This is different from privilege-rights, like taking a walk, eating four meals a day, or daydreaming, which require no mention of others' duties.

There is, however, a special twist in this second meaning of the term *right*. All rights, in the second, strong sense, are other people's duties. Many rights are exhaustively describable in these terms; the word *right* adds nothing at all to what we are brought to understand by a description of the duties of the relevant individuals. My right to be secure from physical assault, for example, is nothing more nor less, legally speaking, than other people's duty not to assault me. However, there are some rights which, though describable in terms of other people's duties, have also a special technical meaning.[3] They are "rights" because the individual who possesses them has the authority to put into effect or to suspend the rules that define them.

A right, in this special sense, is not simply a protected claim, definable in terms of the legal duties others have. It is a portion of the law which, as it were, belongs to the individual designated as the possessor of the right in question. For it is up to that individual to activate the norms and sanctions that demarcate that right. Thus, if I make a contract with Jones and then fail to pay him the fee I promised although he has done everything he has committed himself to do, it is entirely up to him to decide whether to call on the law to enforce the contract. Although the failure to keep promises is a matter, surely, of interest to the entire society, and though it could be treated, like theft, as an offense against the society as a whole, a legal system which recognizes private rights leaves the option of enforcement or nonenforcement of this and other legal duties to particular private individuals.

It is possible to think of a legal system constructed quite differently, in which no one possessed any rights in this sense except a class of persons specifically designated as lawgivers empowered to institute, apply, or suspend legal rules. Under such circumstances, all breaches of the rules would be offenses against the state; the procedures of civil law as we now know them would be annulled. One of the consequences of having a system of extensive legal private rights is that participation in lawmaking is broadly distributed in the population. A right so understood is a ticket of access to the managerial level of the legal system: to the extent that one has such rights, one is not merely a consumer but a producer of the law.

I I

These considerations, I believe, throw our topic, the relation of private rights to the public good, into a fresh light. If the analysis I have offered is accepted, the conclusion follows that private rights are *public* in at least three well-established meanings of the term.

They are public, first, in that they are protected by collectively supported legal institutions: by taxes, policemen, officials, the machinery of the state. It is not true that my having a private right is my business alone. There is a social investment in it; and it seems reasonable to ask, therefore, what society gets out of it.

Second, private rights are public in the sense that they are relational, and represent forms of control by one individual over other individuals' behavior. One man's right is another man's duty or restraint. Since most of our most important liberties are rights, it is misleading to describe a system in which liberty is prized as one in which restraints are fewer. To speak of a "free" society is, crucially, to speak of one in which *certain selected liberties* are protected, or, in other words, one in which certain kinds of restraint are distributed in certain kinds of ways.

Third, private rights are public in that, in the terms of this analysis, the line between the rights accorded by so-called private law and the rights accorded by so-called public law—namely, political rights—fades. A political right like the right to vote guarantees access to the law-making function; a private right, like that of recovering damages from a manufacturer for his negligence, confers a similar, though more individualized, access to the law-making function. The function of private rights, like the function of political rights, is to reduce public officials' control over the law-making process. It makes private citizens agents of the public process of lawmaking.

In a society in which there are private rights, we are thus confronted with what, at least on the surface, seems to be a puzzle. We are dealing with rules that are public in their operational meaning and consequences. Yet they seem also to be rules which collide with notions of public good in at least two ways.

First, to speak of having a right is to announce that one does not need to look at consequences. Shylock has a right to his pound of flesh and he will have it; what Portia asks, it will be recalled, is not that he be just but that he be merciful. Second, rights fall into two classes, fundamental and functional. A professor considering the promotion of a junior colleague has the functional right to ask him about his scholarly plans, but he does not carry this right out into the streets where he can make similar demands on every passerby. His right is in his office or function, not in him as a person. But his right not to testify against himself or to practice a religion or to marry any competent person of the opposite sex whom he chooses goes with him everywhere. His fundamental rights are his by virtue of his general membership in the legal system and not in consequence of some particular performance in which he has engaged, like signing a contract, or of any specific office or profession into which he has entered.

Thus, private rights, though public, seem to be doubly immune from public scrutiny. First, they may be exercised without producing any reasons. As a permit to do something, one merely has to show that one has a *right,* and that is enough. Second, the most important of these rights are individual, not functional. One claims the right by invoking a general rule, but this rule seems not to be subject to appraisal in relation to a clear and identifiable function. We can understand why a doctor should not be compelled to testify against his patient: the practice of medicine would be seriously embarrassed. But why, it may be asked, should it not be held as presumptive evidence against a man that he does not take the stand in his own defense in court? The values that are relevant to the answer to this question are not as easily circumscribable as in the case of clearly functional rights.

Our discussion thus brings to the surface a state of affairs which I have known from personal experience very greatly to perplex visitors from countries that take a radically different view of the relation between law and the individual. At considerable expense to all of us in money, pain, frustration, ugliness, discord, inefficient government, and the daily frustrating sense that injustice is afoot and there ought to be a law, our society maintains the principle that individuals, short of the most extreme necessities, ought to have protected spheres of activity and ought to have them no matter what they do with them. Private rights, maintained by public force and empowering individuals to shape the public force, apparently define a sacrosanct area within which the public force may not intrude, and considerations of the public good are relevant, if at all, only in the last extremity.

Moreover, almost everyone I know holds this view in some form and usually holds it passionately. There is disagreement, to be sure, about which rights should be on the list. Some think that property deserves more consideration than it has been receiving; others think that the little lady from Peoria who is a victim of psychic aggression when she walks past the marquees of the theaters on Forty-Second Street has not been getting the legal protections to which her rights entitle her. But almost all of us agree that there are some rights which no government ought to infringe except under the pressure of clear and present danger and other rights which ought not to be impaired even then. What, then, is the basis for drawing this line between the private rights of individuals and the public good?

We do draw the line confusingly. In the nineteenth century it was said of the Germans that they would accept the most constraining forms of government control but would rise in arms if the price of a glass of beer were raised two pfennigs. Similarly, we Americans said hardly a word a few years back when the government prohibited the sale of cyclamates to us, but when, farther back, the American government banned the sale of alcoholic beverages, millions of patriots cried, "Tyranny!" We think that a tax on church property is an infringement on the private right of religious expression, but we do not think a tax on theater tickets is an infringement on the private right of artistic expression. Is it possible to produce any basic philosophic premise which can explain why such puzzling distinctions are made and perhaps help us to correct the faulty ones?

III

To establish such a premise was the notable effort, as we know, of John Stuart Mill. Mill's classic essay *On Liberty* was the attempt, in his words, "to assert one very simple principle, as entitled to govern absolutely the dealings of society with the individual in the way of compulsion and control." In Mill's well-known words: "That principle is, that the sole end for which mankind are warranted, individually or collectively, in interfering with the liberty of action of any of their number, is self-protection. That the only purpose for which power can be rightfully exercised over any member of a civilized community, against his will, is to prevent harm to others. His own good, either physical or moral, is not a sufficient warrant."

Appealing as this may seem to common sense, Mill's principle assumes a conception of the relation of the individual and society that does not survive scrutiny. For it is difficult to find an action so wholly individual and self-enclosed that no other individual is possibly affected. How much time one gives to sleep affects one's health and energy, which affects one's work, which affects other people; what one reads in the privacy of one's home affects one's tastes, which affect one's companions. And if we can imagine any act wholly evanescent, inconsequential, unremembered, and unmemorable, would it not be, in a poor and wasting world, a flagrant misuse of time and an abuse of social resources? The line between private and public cannot be the line between what affects other people and what affects oneself alone. It is not a discovered line but a social construct. Not that it is arbitrary, or that no line should be drawn between the public and private domains. But the line is not to be found in the way we find the line of a stream through a forest. It is more like the line between two countries; it is drawn, not found, and it is drawn for specific reasons. It is conditioned by historical and cultural biases and represents a kind of compromise among a variety of considerations and claims. And there is almost always room for argument that it could have been drawn a little farther in one direction than in the other.

Yet there courses through the pages of Mill's work, beyond and beneath the arguments for his "very simple principle," a pattern of thought much in line with this position. In building his case for liberty, Mill says: "I regard utility as the ultimate appeal on all ethical questions; but it must be utility in the largest sense, grounded on the permanent interests of man as a progressive being." It is an idea of man as a progressive being that stands in the wings as Mill's discussion of liberty unfolds. I take it that he does not mean Man whose historical destiny is to illustrate God's logic or fill out His Idea. He means a being who, in his collective character as a participant in civilized activities, develops and articulates ideals, criticizes and rectifies them, and attempts to move toward them.

In any case, if this is not Mill's intent—though I think it is—it is mine. In what follows I want to sketch out the considerations which, I believe, play a part in helping us draw the line between private rights and the public good. What I say is surely not, for the most part, to be found in Mill, except implicitly. But I should like to believe that it is in the spirit of his ideas and tradition. It is a filling in of that notion of man as a progressive being which is implicit in liberal civilization.

Rights are the logical consequences of the institution of rules that leave with private individuals the option to demand performances from others and that leave these individuals free to exercise these options without giving further reasons other than that such rules exist. What, then, establishes the validity of such rules? An examination, I believe, of their consequences with respect to certain long-term aspirations and values of the society provides an answer. With respect to our historical commitment to private rights, I suggest that there are, cabbalistically, seven of these. Perhaps there are more, but the seven I shall name seem to me to cover a good portion of the issues that are likely to be raised. Specific private rights, of course, are not to be neatly categorized under one and only one heading. They often serve several values, and sometimes these values conflict. This is part of the reason why specific rights become problematic and need to be redefined. In listing these seven values, I would also emphasize, I do so in no particular order. I am not sure that there is a fixed order of preference among them. The need to balance and reorder them in different contexts is precisely what makes the preservation of liberty and private rights a matter for intelligence and imagination and not merely for brute courage.

1. One value is what Montesquieu, in *The Spirit of Laws,* called "political liberty." Probably it is less misleading to a twentieth-century audience to call it either "security under law" or, in the honorific sense, "the rule of law." Montesquieu describes this value as follows:

A government may be so constituted, as no man shall be compelled to do things to which the law does not oblige him, nor forced to abstain from things which the law permits. . . .

The political liberty of the subject is a tranquillity of mind arising from the opinion each person has of his safety. In order to have this liberty, it is requisite the government be so constituted as one man need not be afraid of another. . . .

Political liberty consists in security, or, at least, in the opinion that we enjoy security.

I take it that Montesquieu is describing the security of expectations which social authority provides only to the extent that there are clear and general rules, strictly limited in the interpretations that can be placed upon them, which are promulgated in advance of people's actions. This is a necessary condition if people are to be able to make long-term intelligent life plans for which they can reasonably be held responsible. To be ruled by the arbitrary, to be the subject of decisions taken without a general principle to justify them, is to live in fear and to have the conditions for reasonable and voluntary choice annulled. Private rights, among their purposes, serve as a corrective to the gravitational tendency in human society toward the arbitrary. In Montesquieu's words, "Political liberty is to be found only in moderate governments; and even in these it is not always found. It is there only when there is no abuse of power. But constant experience shows us that every man invested with power is apt to abuse it. . . . To prevent this abuse, it is necessary from the very nature of things that power should be a check to power."

2. But there is hardly any value, as Aristotle noticed, of which it is not possible to have too much. A perfected law, with all the *i*'s dotted and all the *t*'s crossed, which covered all possible future contingencies, assimilating them automatically to past cases, is impossible; and if it were possible, it would be stifling. No idiosyncrasy, nothing with a vitality of its own, would be in a position to assert its claim for recognition. Or, as Aristotle put it in the *Nichomachean Ethics,* "all law is universal but there are some things about which it is not possible to speak correctly in universal terms. . . . And this is the very nature of the equitable, a rectification of law where law falls short by reason of its universality." Private rights serve this Aristotelian cause of equity, the corrective of formal justice. They register the social recognition of the fertility of nature, of its extraordinary powers of individualization, and of the limitations of human beings in keeping up with it. A justifying assumption on which they rest is that a legal and institutional system that spreads autonomies is better designed to create a setting for human life and work that has an individual fit.

On one side, liberty in society depends on the establishment of uniform laws, applicable impartially to all; on the other side, it depends on the pluralities, experiments, and diversities that allow different people to find their own different ways. The constant tension between these two poles and the difficulty but necessity of accommodating their joint claims are illustrated by the centralizing and standardizing tendency of Supreme Court decisions on school desegregation and by administrative decisions of the federal government on the hiring of minorities. Without a system of rights centrally established and enforced, individuals in many fields would undoubtedly be subject to arbitrary discrimination. With such a system of centralized enforcement, homogenizing tendencies take over.

3. A third value incorporated in the concept and practice of private rights is that of the pursuit of truth. It is important that this value and the requirements for the respect of private rights which it imposes not be confused with other values and their requirements, which are also important but different. Justice Holmes, in a celebrated opinion, wrote that "the best test of truth is the power of the thought to get itself accepted in the competition of the market." [4] If he meant these words literally, one need only look at the contents of the average newsstand to refute them. What counts in the market is money and popular tastes, neither of which is a touchstone of truth.

The negative point that Holmes makes, however, is valid: the truth of an idea is not tested by a government's opinion of it. The truth of an idea is tested, in the end, by competent critics and observers, specially trained and selected by those already accredited as competent critics and observers—in short, the nonpolitical, professional community of scholars and scientists. And the rules by which this community performs its functions cannot be identical with the rules of the society at large. It is a fallacy to believe, for example, that universities ought to offer a spectrum of all existing ideologies.

A system of private rights makes the autonomy of such a community possible. Indeed, it makes possible the autonomy of other professional communities entrusted with key social values—for example, lawyers and doctors. The "truth" of which I speak has its siblings—justice, health, human welfare, and the like.

4. Once again, a certain balance is needed. Professionalism easily becomes pedantry, a rationalization for the exploitation of laymen, or a collection of insular prejudices and abstractions. To correct this, and, more broadly, to control the exercise of power generally in the society, another requirement emerges as a long-term value—the competition of political interests. The marketplace to which Holmes's doctrine properly applies is the *political* marketplace. Freedoms like speech and association are justified in that marketplace on the

ground that they permit an open competition for power and a public control of the exercise of power.

It is when we see freedom of speech as an arm of vigorous political competition and not as an instrument for the pursuit of truth that we can understand the latitude it is given in our courts and the difficulties that American law puts in the path of any public official seeking redress for what he considers libelous personal accusations against him. Behavior quite out of place in a university may well be necessary and desirable in the political arena—though it should be added that in the political arena there are limits, too. The open competition of interests depends on civility and, in the broad moral sense of the term, parliamentarism. These are liberty's spine.

5. Closely allied to the interests of political competition are the social values of entrepreneurship and innovation. The degree to which the economy should be based on decisionmaking by private right depends on one's estimate of the relative efficacy of decentralized as against centralized decisionmaking. This efficacy, further, has to be measured with respect both to economic purposes and to the general diffusion of power in society. The matter is complex and not entirely a matter simply of competing *Weltanschauungen.* Factual matters are involved, and very probably the answer differs from country to country and from period to period.

Economic entrepreneurship is not the only kind of inventiveness which society encourages through a system of private rights. Artistic creation is also a beneficiary of such rights. Tyrannies have often nurtured extraordinary works of architecture, painting, music, and drama. They have rarely done so, however, if artists have not possessed either a certain quasi-professional standing or aristocratic patrons—and thus certain de facto private rights. It is such artists, endowed with this special status, personages not merely persons, who are the principal source of the modern ideal of moral and esthetic individualism. They belong, I think, to the same family as the great business entrepreneurs, the creators of their own names, the founders of great houses. This kind of ambition, this image of human possibility, is also implicit in the doctrine and practice of private rights.

6. Clearly related to this value and yet rather different from it is another—self-expression. The man or woman who has a statement to make, who wishes to wear his or her religious creed, moral out-

look, or sexual identity out in the open where everybody can see it, is not necessarily an ambitious artist. Nor is he or she necessarily trying to make a dollar or win a vote. Here, I think, is a value close to the value of simple individualization, of having a small piece of the world tailored to one's form, but one more assertive in its quality. It is, probably, the democratization of the value of artistic individuality. It is easy to sniff from on high at this value, but its importance in a society of mass production and consumption and of urban anonymity cannot be easily exaggerated.

7. Still, it is a value that can easily be overdone. Self-expression can flow into an expression of scorn or hostility; the assertion of one's own authenticity often invades the privacy of others. The balance to self-expression, and perhaps the most precious value of all, as Brandeis thought, is privacy, the simple chance to be let alone. The function it serves, the long-term social interest it represents, is a complex of basic goods: family, free friendship, love and personal communion, solitude and self-confrontation, religious or philosophical integrity, the intimacies and intensities that are only possible for rational people when they can choose their company and keep the prying world away.

Indeed, this seventh value reveals most clearly the moral aspiration, the commitment to an ideal of personality, which runs through a system of individual rights and gives a kind of family resemblance to the several values that underlie it. It is the ideal of the conscious self, mobile, transcending any particular social role, objectified for itself so that it is its own critic, defining itself through its choices. A liberalism like Mill's is pluralistic; but it does not hold together, I believe, unless it contains a preferred model of human excellence. This conception of the self is, I think, that preferred model. The self that transcends any particular role, that is not wholly submergible in class or religion or cultural group, is the prerequisite for the hope that a public interest can emerge from the conflict of private interests, that there will be a spirit of accommodation and communality in the air. The self which has choice forced upon it, the self which, by pressure of circumstance, is forced to be free, has, in addition, the opportunity to be critical. This self is the prerequisite for the maintenance of the institutions of self-appraisal on which a liberal polity bases its claim to legitimacy.

This ideal, to be sure, can be turned into something grotesque. There is abroad in our society today a view that the only norm an individual should accept for his guidance is the norm of being independent of all norms inherent in his social roles. He becomes "inauthentic," it is suggested, when he allows his personality to be cut into pieces, a bit of him a worker, a bit a parent, a bit a citizen, a bit a man or woman. This view confuses an absence of social identity with a discovery of self. Even the purest self-expression, if it is to be expressive, must meet some public standards of communicability. To keep a core of principle or purpose intact in a variety of roles, and to judge what one does in these roles in relation to such principles and purposes, is independence; to seek a quintessential self, floating about with no links to the world, is to mistake solitary confinement for freedom.

Even when the ideal of the conscious, independent self is not stretched this far, it can still be misapplied. The ideal of an independent personality, critical, mobile, never wholly absorbed in or defined by its environing web of circumtances, is one that probably nobody can achieve constantly, and perhaps only a minority can approximate as a general rule. If such an ideal of personality is construed as a standard imposing requirements on everyone, it can generate stereotyped caricatures of individuality on one side and widespread feelings of failure and self-contempt on the other. Educational, psychiatric, and political doctrines preaching the goal of moral and intellectual independence, universal, complete and constant, are prescriptions for self-deception. They lead to that parody of individuality in which people en masse adopt a certain convention in dress, speech, and manner to show their independence of convention.

It is useful to distinguish in morals and in law between norms that are mandatory and norms that are advisory and orientive. The law may permit gambling, but it may show its preference for other social occupations by refusing to enforce wagering contracts. It may not require people to marry, but it can offer encouragements to those who do. Similarly, the ideal of personality that underlies the structure of private rights which liberal societies have developed is best understood not as a command to all but as a guide to the formation of social policy—an indication of what should be specially prized and guarded. It is not to be supposed that most people, at any time and in any place, will seek the values of disengagement, skepticism, and personal choice. But a liberal culture must do what is necessary to see that it has a saving proportion of people who do seek such values.

We come, then, to a kind of answer to the question we have posed—the relation of private rights to the public good. The answer I would suggest is that the considerations I have discussed define some considerable portion of the public good as it should be understood in liberal society. We make the best sense of what we are doing in our kind of society when we see that the line we draw between the private and the public is drawn, in the end, for the sake of perpetuating and embodying a complex ideal of civilized life. The problem of private rights and public good turns out to be that of weighing different social values, involving different aspects of liberty, against one another and of constantly readjusting the relations among them. And of course there are other values—for example, equality and fraternity—against which the considerations I have mentioned must also be weighed.

IV

A final question must briefly be considered: What is the basis for that complex of values and that ideal of personality to which, in my view, the conception of private rights is attached?

They are tested, I believe, by their fit with other values we cherish and by their capacity to guide practical decisions to results compatible with these values. I gladly admit, speaking for myself, that I do not derive these values from *a priori* principles. Nevertheless, I do not think myself a creature of caprice in hitting upon them. I have not "hit upon them"; I have found them when I have looked for the roots of my being and for the integrating principles of what, in our society and our society's past, we regard with the greatest pride.

Why be committed to these values? Homer noted that the Achaeans were different from other people: they never bent their knee to any man. They expected reasons to be given for their loyalty, not mere commands. That aversion, that impulse, extended and deepened, is what we preserve in the modern notion of private rights. A great deal of history, a great deal of blood, has gone into the nourishing of it. We know the cheapness, vulgarity, banality, and cruelty of the social systems which in our time have denied the values that private rights serve. These private rights are threads in the fabric of our moral system so placed that if we pull them out we cannot be sure what will be left. All that is clear is that we will look with shame on what remains.

8

Public Rights
and Private Interests

In Response to Charles Frankel

HANNAH ARENDT

The area of agreement between Charles Frankel and myself is large, and so the questions I shall raise will be of a general nature concerning matters on which there exists an almost universal consensus which I would like to challenge.

The first concerns the title [of this section of papers, a title which Frankel borrows for his essay.] "Private Rights and the Public Good" assumes that our rights are private and our obligations are public—as though there were no rights in the public realm. [It is true that this has been the standard rhetoric in the West for many centuries, even into our own, but it is an assumption that I must challenge. For it is necessary, I think, to distinguish the *private* rights we have as individuals from the *public* rights we have as citizens.] Every individual in the privacy of his household is subject to life's necessities and has the right to be protected in the pursuit of his private interests; but by virtue of his citizenship he receives a kind of second life in addition to his private life. These two, the private and the public, must be consid-

Hannah Arendt was University Professor of Political Science at the New School for Social Research. Before her sudden death in December 1975, she had approved the publication of this essay but had expressed hope of being able to give fuller statement to what she called her "rather apodictic" remarks: for in its original form, the text was little more than an outline of the response she gave orally at one of the conferences. In accordance with this wish, we have included in brackets at appropriate junctures of the text edited sections from the transcript of her remarks.

ered separately, for the aims and chief concerns in each case are different.

Throughout his life man moves constantly in two different orders of existence: he moves within what is his *own,* and he also moves in a sphere that is *common* to him and his fellowmen. The "public good," the concerns of the citizen, is indeed the common good because it is localized in the *world* which we have in common *without owning it.* Quite frequently, it will be antagonistic to whatever we may deem good to ourselves in our private existence. The reckless pursuit of private interests in the public-political sphere is as ruinous for the public good as the arrogant attempts of governments to regulate the private lives of their citizens are ruinous for private happiness.

In the eighteenth century, this second life in the common was characterized as being capable of affording "public happiness," that is, a happiness which one could attain to only in public, independent of his private happiness. [The possibility for enjoying "public happiness" has decreased in modern life because] during the last two centuries the public sphere has shrunk. The voting booth can hardly be called a *public* place; indeed, the only way in which a citizen today can still function as a citizen is as member of a jury.

Hence, my first question is: assuming that the private individual and the citizen are not the same, do we still use our public rights? The Constitution provides for one such right in the First Amendment, which concerns the "right of the people peaceably to assemble"—a right deemed cognate to the right of free speech and equally fundamental. This right still survives in "voluntary associations," of which the civil disobedience groups of the sixties were an outstanding example, but it has also degenerated into lobbying, that is, into the organization of *private* interest groups for the purpose of public, political influence.

It should be clear that my distinction between private and public depends on the locality where a person moves. No one will doubt, for instance, that a physician has different rights, obligations, liberties, and constraints in the hospital, on the one hand, and at an evening's social gathering, on the other. Or take the case of a juror. Once made member of a jury, a person all of a sudden is supposed to be impartial and disinterested, and we assume that every individual is capable of this impartiality regardless of background, education, and private interests. [To be constituted a jury, a number of persons must be *equalized*, for people are not born equal and they are not equal in their private lives.] Equality always means the equalization of differences. [In society, for example, we often speak of law as the great equalizer: we are equal before the law. In religion, we speak of God before whom all are equal. Or we say that we are equal in the face of death, or that the human condition in general equalizes us. When we talk about equality, we must always ask what equalizes us.]

Jurors are equalized by the task and the place. [Though drawn from quite different backgrounds and social strata, through their task they are made to act not as party members or as friends but as peers. They must deal with something which is of no private interest to them at all: they are interested in something in regard to which they are disinterested. What jurors share is an interest in the case—something outside themselves; what makes them common is something that is not subjective, and this I think is quite important.]

My second question is connected with the first. The working premise of this second series of essays was that the enduring challenge of a society is to care for the public good without violating the rights of individual citizens and that, happily, the two interests often coincide. This coincidence is indeed the basic presupposition of any "harmony of interests." The assumption is that "enlightened self-interest" automatically reconciles opposed private interests.

If we understand "enlightened *self*-interest" as the "interest in the common good," I would argue that such a thing does not exist.

The main characteristic of the common with respect to the individuals who share it is that it is much more permanent than the life of any one individual. [There is an intrinsic conflict between the interests of individual mortals and the interest of the common world which they inhabit, and the source of this conflict lies in the overwhelming *urgency of individual interests*. To recognize and embrace the common good requires not enlightened self-interest but *impartiality;* such impartiality, however, is resisted at every turn by the urgency of one's self-interests, which are always more urgent than the common good. The reason for this is very simple: such urgency protects that which is most private, the interests of the life process itself. For us as individuals, the privacy of our own life, life in itself, is the highest good, can only be the highest good.

[Until very recent times, whatever belonged to life's necessities was hidden in the obscurity of privacy. But we seem to have decided of late—how wisely I am not sure—that everything should be made public. The life process, however, particularly the group process of children, requires a certain obscurity. Whatever advantages it may have, public space exposes one mercilessly, in a way none of us could stand all the time. We need a private space in order, among other things, to hide; we need it for all our private affairs, with our families and our friends. And we have acquired, since the eighteenth century, an enormous space of intimacy which we consider sacrosanct. And rightly so. Yet it is precisely this space which we are asked to sacrifice when we act as citizens.]

For public interest always demands a sacrifice of individual interests which are determined by life's necessities and by the limited time which is given to mortals. The necessary sacrifice of individual interests to the common weal—in the most extreme case, the sacrifice of life—is compensated for by public happiness, that is, by the kind of "happiness" which men can experience only in the public realm.

[My second question, then, is this:] what about the private rights of individuals who are also citizens? [How can one's private interests and rights be reconciled with what one is entitled to demand as a citizen?] Among the most important private rights is "the right to be let alone" (Brandeis). This right is by no means a matter of course. It is Christian in origin: "Render unto Caesar the things that are Caesar's" (Matt. 22:21); "No thing is more alien to us than the public thing" [*res publica*] (Tertullian); or "Mind your own business." [For the Christians were those who minded their own business. Their reason, to be sure, is no longer our reason. Their reason was that saving one's soul required all the time one could spare so that politics to them was a luxury. And indeed, freedom, political life, the life of the citizen—this "public happiness" I've been speaking of—*is* a luxury; it is an *additional* happiness that one is made capable of only after the requirements of the life process have been fulfilled.

[So if we talk about equality, the question always is: how much have we to change the private lives of the poor? In other words, how much money do we have to give them to make them capable of enjoying public happiness? Education is very nice, but the real thing is money. Only when they can enjoy the public will they be willing and

able to make sacrifices for the public good. To ask sacrifices of individuals who are not yet citizens is to ask them for an idealism which
they do not have and cannot have in view of the urgency of the life
process. Before we ask the poor for idealism, we must first make
them citizens: and this involves so changing the circumstances of their
private lives that they become capable of enjoying the "public."

[But many people today, and not merely those who care about the
salvation of their souls, demand to be let alone. This is, in fact, a new
freedom that is being demanded—the right to be free from any mandatory participation in public life, be it something so basic as the duty
to vote or to serve as a juror. If we wish to spend time painting pictures and let the whole community rot, we feel we must have this
freedom. But again, this freedom is by no means a matter of course,
and perhaps not even a matter by which we should judge the relative
freedom of governments. Consider, for example, the case of Aleksandr Solzhenitsyn. Solzhenitsyn didn't mind his own business and
for that reason came into conflict with his government. In other
words, the Soviet Union is no longer Stalinist. Under Stalin, if one
minded his own business, he was sent to a labor camp (they were actually extermination camps), just as if he had opposed the government. Indeed, Stalinist terror acquired its full momentum only after
all political opposition had been liquidated. In Soviet Russia today,
however, the private individual who minds his own business can live
without any conflict with the government. The government remains
tyrannical—that is, it does not permit political life; but it is no longer
totalitarian—that is, it has not simply liquidated the whole sphere of
privacy. Solzhenitsyn's difficulties arose when he demanded *political*
rights, not private rights.]

The notion that private rights are sacrosanct is Roman in origin.
[The Greeks distinguished between the *idion* and the *koinon*, between what is one's own and what is held in common. It is interesting
that the former term has become in every language, including the
Greek, the root for the word *idiocy*. The idiot is one who lives only in
his own household and is concerned only with his own life and its
necessities. The truly free state, then—one that respects not only certain liberties but is genuinely free—is a state in which no one is, in
this sense, an idiot: that is, a state in which everyone takes part in one
way or another in what is common.

[But the Romans were the first to claim, through the tall walls demarcating their properties, the sanctity of the private sphere; the notion that private rights are sacrosanct] sprang from the Roman feeling for the sanctity of home and hearth, [their insistence that the private cult of hearth and home was as sacred as a public cult.] Indeed, only he who owned his own home was deemed able to participate in the public life; that is, private ownership was the condition sine qua non for participation in politics. This implies two things: (1) life's necessities are private, not fit to be seen in public, and (2) life is sacrosanct. The chief value of this life was precisely that it was protected from the glaring lights of the public realm. [Whereas the value of the public is illumination (it is like a light that exposes everything to all sides so that, for example, all sides of a question can be seen), the value of the home and of privacy is precisely obscurity. Both these values are basic to a good society, but] obscurity is one of the necessary conditions of life itself.

As the public realm has shrunk in the modern age, the private realm has been very much extended, and the word that indicates this extension is *intimacy*. Today this privacy is very much threatened again, but the threats arise rather from society than from government.

In brief: while governments threaten our public rights, our right to public happiness, our private interests and rights are threatened by society. Moreover, given the necessities of modern production, some of these private interests are organized and effectively influence the public realm.

The primary condition of privacy is ownership, which is not the same as property. Neither the capitalist system nor the socialist system respects ownership any longer—inflation and devaluation of currency are capitalist modes of expropriation—although both, in different ways, respect acquisition. Hence, one of our problems is to find a way to restore ownership to private individuals under the conditions of modern production.

What is necessary for freedom is not wealth. What is necessary is security and a place of one's own shielded from the claims of the public. What is necessary for the public realm is that it be shielded from the private interests which have intruded upon it in the most brutal and aggressive form.

9

On Privacy and Community

EMILE CAPOUYA

Privacy is not loneliness, my friend says at once. And come to think of it, Crusoe alone on his island is not enjoying privacy but suffering from unrelieved loneliness. Privacy is an aspect of community, available only to persons living in society.

That appears to me to be so by definition—by adequate definition—but not to be a vulgar tautology of the kind that connects, in common parlance, opposites like peace and war, wherein each is presumed to be a condition of the other. If you want peace, prepare for war, the Roman adage says; but we know the limits of that bit of wisdom. Privacy arises in community, but not in the manner of a truce or armistice in the midst of battle.

So I think it is out of keeping with our sense of the value of privacy to regard it simply as the relief we find in retreating from public exposure—or as a kind of counterirritant, a pain milder in degree, perhaps, than the strain of publicity, but more a restorative by contrast than a positive good. That homeopathic view would suggest that public and private life are at bottom similar nuisances, though they may be rendered harmless if we establish a nice balance between them. It reminds me of the apologue of the man who had one foot on a block of ice and the other in a pan of water just off the boil: on the average, the story goes, he was quite comfortable.

No, privacy is not one of those dangerous drugs that are useful in small doses. It is a positive good, and one of the positive goods of community. If I am wrong in thinking so, then the questions that surround the issue of privacy are not philosophical—that is, not properly

Emile Capouya is Literary Editor of *The Nation* and Associate Professor of English at Baruch College, City University of New York.

the concern of everyone—but questions that fall within the province of technicians and experts, possibly of the mythical social engineer. Such persons, real and imaginary, should be able to tell us how many decibels of privacy are tolerable, or how many milligrams can be absorbed without dangerous side effects, and the rest of us can then stop talking as exact science replaces mere opinion. But I mean to go on talking, because the matter is not one for experts but falls within our competence as human beings. Plato set down the governing principle more than two thousand years ago: if you want to know something about the art of making statues, ask the sculptor; about seamanship, ask the pilot; about truth, ask the first passerby.

I

If privacy, then, is a feature of community, shall we say that it flourishes wherever two or three are gathered together? I think not, because community itself is not simply a function of numbers. It is rather an achievement or a gift; it is society in the honorific sense, a group whose members are effectively in communion. Nor does it appear that we can establish beforehand a numerical limit beyond which communion is impracticable, say two or three thousand, or million, or hundred million. It depends on the persons involved, their relations, the means of communication and the possibilities for communion that are available to them.

It is true that one kind of sympathy lessens as the square of the distance from our pocket, but there are other kinds of sympathy that obey a different law. Accordingly, the expression "face-to-face relations" must be understood as a metaphor. Its real significance has nothing to do with locality. When Buber speaks of the "I-Thou" relation, which is clearly the ultimate sense of what is implied by "face to face," he means that the parties, man and man or man and God, are in effective communion, not that they are in the same room. And though "I" and "Thou" are two persons only, at least by the conventions of our grammar, "We" is expansible without limit. For a long time, the Athenian city-state was taken for the archetype of the successful polity, and it was somewhat hastily assumed that the size of the Athenian assembly represented the upper limit of political practicability. But larger assemblages have since shouted yes and no on various matters, showing themselves to be as wise or as misguided as the Athenians. The question does not lie there. What is important is not how many can be gathered in the Agora, but what is the disposition of the citizens and what facilities they have for social intercourse. If these are inadequate, their actions when they are assembled—for example, at polling places distributed over a large modern nation—will not be edifying.

The history of modern states suggests that the citizens' disposition and their facilities for intercourse have been generally inadequate. At least that is so in terms of an ideal standard. I appeal to such a standard without apology, for if we wish to discuss privacy and how it may be protected against threatened invasion, and even to determine the extent to which it may be proper to protect it, we can scarcely shrink, while engaged in so quixotically ideal an undertaking, from considering an ideal standard. The ideal in this case is social happiness. And that must be a somewhat larger category than is represented by Hannah Arendt's term, "public happiness," since it means the happiness that can be achieved in society, and I take it that for human beings such a definition is necessarily exhaustive.

It is useful to take up the matter of social happiness in the spirit of a remark of the late Justice Black. He said, approximately, "When the Constitution says, 'Congress shall make no law . . . ,' it doesn't mean that Congress may make some law in the prohibited area, or any law, or perhaps a little, inconspicuous law." In that spirit, let us mean what we say when we say social happiness, not "cabined, cribbed, confined," but "broad and general as the casing air." Our practice now is to congratulate ourselves upon such crumbs of social decency, justice, generosity—happiness, in a word—as fall to us within a social order designed to achieve things of a rather different kind. That means, in effect, only so much happiness as we already have. So our practice now expresses two attitudes which appear to me to be inconsistent with an inquiry into the value of privacy for the United States in the latter part of the twentieth century. Each of those attitudes has its time-honored shibboleth: "Pull the ladder up, Jack, *I'm* on board" and "Don't rock the boat."

But in terms of a standard of social happiness generously conceived, we know our own way of living to be a very middling success—not so much with regard to our expectations, for we have contracted our hopes to fit the rigors of the climate, but with regard to our needs. The plant, man, does not flourish in our country as it should. We all know it. Some time ago, one of our presidents was inspired to create a commission to define our national goals—as it were, to search in broad daylight with Diogenes' lantern. I do not recall that anything of moment was discovered by the commissioners. Lanterns of the kind are useful when they are the outward signs of an inward light, and ours has been so baffled for so long that it is a very murky one. "What?" the president seems to have said to himself, "we have no national goals, expressing a National Purpose like those that served us so well in the past—Manifest Destiny, the White Man's Burden, 54° 40′ or Fight? Let us look around for some." Charles Frankel used to insist on one particular *distinguo,* and perhaps he still does when occasion offers. He used to say, "Rational—why, sure. But shouldn't we be reasonable, too?" If we have no soul, it may be rational to look for one, but it hardly seems reasonable. I trust that our attempt to define and protect privacy is not merely rational in this meagre sense.

For we cannot be certain whether privacy, as a function of community, is an endangered species or an extinct one. Our interest in it has certainly diminished within the last generation, and what survives has possibly had another object substituted for the original. A good many members of the youngest generation seem to assume that privacy is the polite term for mistaken feelings of guilt, or at least embarrassment, about personal matters—with a consequent desire to conceal those things from the world. Privacy in this sense is held to be an antonym for frankness, and perhaps for honesty. I take it that that is not the quality we are concerned with here, though it does have a rough correspondence with a value more easily hedged about by means of legislation, what Brandeis meant by "the right to be let alone." The literary record, which is among other things a record of manners, suggests that in past ages privacy as a function of community had a more vigorous existence than it has in our own day. But such a judgment must be offered most tentatively, for all we can be sure of is that, in our present state, privacy is more the substance of things hoped for than something clearly identifiable in our experience. We may have lost the ability to distinguish, as well as the disposition to prize, the value it represents.

Let me explain what I mean by so gloomy-seeming a suggestion.

John Jay Chapman wrote around the turn of the century that "Misgovernment in the United States is an incident in the history of commerce." It seems to me that our later experience has fully ratified that pronouncement and that we are justified in offering a corollary: "Reform in the United States is an incident in the history of commerce." Not that commerce has its powerful synod that regulates whatever is regulable in our public and private existence—of course it has, but the possible evil that represents in itself can properly be regarded as superficial. It is rather that as a people we have grown even more disposed to conduct all our affairs in keeping with the prudential wisdom of a commercial culture than we were in Chapman's day. We should not be discussing privacy in the ways we discuss problems of conservation; to do so suggests to me that our notion of the subject is likely to be defective. Rightly understood, privacy can hardly be consistent with a prudential ethos—a tendency elsewhere so valuable, even indispensable, but demonstrably indifferent to spiritual goods of every class. I arrive at this indictment of us all through introspection. It is as much a confession as a charge.

To submit some gross evidence bearing on the question: some time ago it occurred to me that it would be convenient to have a credit card, and I filled out the required form of application. The company that offered the card questioned me, by means of that form, with some particularity, though solely in my character of would-be borrower. That was fair enough: I wanted credit, the company wanted to know if I was good for the money. So I wrote down my income, whether I owned or rented, and also some information that would enable the company to check my statements in case my memory was at fault. Among other things, I supplied the name and address of my employers.

Some weeks later, I received a letter (also a form) from the college where I teach, part of the City University, saying that credit inquiries had been made about me, that it was the policy of the college not to give out the information requested in such cases without the permission of the person most interested, and would I be good enough to indicate yes or no. A glow began to steal through my frame at the idea of the college standing so staunchly between me and persons who—well, persons with whom I might yet have business relations but with whom I was not in communion.

Then I read further. If my answer was yes, the university would take that as authorization to respond freely to all future inquiries about my finances. If I said no, or yes but just in this case, or chose any possibility other than the yes unalloyed that would settle the matter once and for all, I would be required to present myself at the college and make the appropriate declaration—on this and every subsequent occasion. I thought the matter over, and said no.

Now, here are two parties to a tragicomedy—or rather two actors in an incident of commerce, for the other phrase is unforgivably subjective. First, there is the City University, which has thousands of employees and must receive thousands of inquiries of the kind. There can be no question of the university's not responding at all, for that would not be in the interest of the employees (not in their commercial interest), and it would set obstacles in the way of transacting business. But if every such inquiry were to generate a letter to an employee, the university would find the work and expense burdensome—again we must invoke a commercial calculus. Perfectly natural, when you think it over.

And yet, I remember that when I was a student in England, and ran away from my college for good and bad reasons of my own, and sent the authorities belated notice of what I was up to, I received a letter from the provost, in which that worthy scholar said, among other things, "It occurs to me that you may be in a position that no member of the college should be in. Do you need money?" The college in question, in the person of its chief officer, regarded even a scapegrace like me as a member of its community. I must conclude that the City University of New York, where I am, I trust, in good standing, is by contrast a community only in a Pickwickian sense.

As for the other actor in the incident: when I learned that I must say uncle at once or look forward to unnumbered tedious journeys to deliver my instructions in person, I was stung by the threat to my convenience, and thought to capitulate. Then I sensed that there was a deliberate attempt on the part of the personnel department—good term, that—to coerce me into pronouncing a blanket yes that would minister to *its* convenience. That stung me further. Finally, I decided the question on a purely commercial calculation, that it was not prudent, as the world wags, to have the university give out my financial statement to all inquirers. And that calculation drowned my pique. I was not ready to argue with commerce. Not only did I know that if I were to stand in the way of progress I could look to get run over, but I had been trained to see perfect justice in that state of affairs. All that was left me to call my own was some residual annoyance that what we are wont to refer to as "my business" was in this instance being intruded upon.

Clearly, this drama took place only in the anteroom of privacy. It hardly impinges on the realm that is, or should be, the object of concern in this inquiry. But it is not certain that the anteroom leads to any further chambers; at least I am not sure that my own soul has any such apartments. And if I proceed, it must be with a sense that the privacy that by its nature is a feature of community and cannot be a feature of the history of commerce may now be for us as elusive as grace and perhaps altogether beyond our attaining.

II

Let me return to John Jay Chapman. Having raised the issue of misgovernment, he went on to ask how we might best deal with it. And having had some personal experience with reform politics, he had observed that movements of that kind have a natural history that could be summarized like this: (1) corruption; (2) a platform of unconditional reform; (3) compromise—since politics is "the art of the possible"; (4) corruption. Shirtsleeves to shirtsleeves in four movements. Accordingly, he looked upon reform movements as morally obligatory but woefully inefficient. And he concluded that there was only one way for us to separate ourselves from the body of this death. He proposed that we reform ourselves as individuals, and do so publicly. He said, more or less, we know that A is a corrupt politician: why do we not refuse to shake his hand, turn our backs when we meet him at our club, neither invite him to dine nor accept his dinner invitations? No man, however corrupt, he said, could long hold out against such clear hints of our displeasure.

In its time and place, Chapman's suggestion was practical politics. If we think it quaint, that is because our own situation is wholly different. We live in a world whose political temper is determined in large part by a singularly novel condition, one of the great social innovations arising from the industrial revolution—that most people who read and write are disfranchised. The process was not quite complete in Chapman's day. At that time the people who could read and write in the full sense—those persons, for instance, who were sufficiently educated to be equipped to read Chapman—had a good deal of political initiative. Indeed, they were the ruling class. Most of us, by contrast, could not send a corrupt dogcatcher to Coventry; the levers of social power are not directly within our reach, and those who do have access to them are very nearly beyond our influence. Unless favored by a disastrous war, an alarming inflation, a stock market in which the important traders are sustaining daily losses, we can scarcely expect our sentiments to prevail even against manifest misconduct on the part of the princes who govern us. That is, we are not in communication, let alone in communion, with the persons who discharge our public business in the spirit of the commercial calculus, and, so doing, create the atmosphere in which our private selves live and move and have their being.

If our disfranchised state, if our political powerlessness were far from being as absolute as I have suggested, that condition would still, I think, be intolerable in the light of an ideal standard, which is to say a reasonable standard, of social happiness. When I say our state, I am referring, of course, to the political tutelage that we members of the middle class suffer, but still more to the condition of helotry suffered by the nation as a whole. We are relatively privileged; we know the ropes, we have some shadow of recourse, we can meet, discuss, petition. To the greater part of the nation, these consolations are socially unavailable. When I was in the army some twenty years ago, I was surprised to find that a number of the young men in my company did not know the name of the then-current president. They were not defective. They were rational beings, and reasonable beings, too. I thought it a paradox, until it dawned on me that knowing the name of the president was of no earthly use to them. They were not of the class that meets to discuss public business, that lobbies, draws up petitions, frames model statutes for the guidance of the legislators. Remarkably little of our store of social tenderness had been expended upon them, and they had drawn the conclusion suggested by their experience: You can't fight City Hall. If it is regrettable that members of a relatively privileged order live under a ban, excluded from the councils of the nation, and have learned to regard that condition as normal, it is still more regrettable that no account is taken—except by way of demagoguery, when some ambitious clown sets out to harvest votes with a clam-rake—of the experience, the ideals, the practical virtue of the nation as a whole.

How does that affect the issue of privacy? Indirectly. Or rather in so direct a fashion that, measured against the tortuous course of public institutions, the practice of corporate bodies, the tenor of legislation, and the issues framed for public debate by those who do have political initiative, it must seem indirect to the point of abstraction.

Let us take the example of one traditional public concern in our country, that the citizens be protected against unreasonable searches and seizures. It is of course an essential condition of civil decency that there be effective guarantees against such practices, and to the extent that law and administrative policy can secure us against those affronts, we must, in this conditional world, be in favor of the appropriate statutes and attitudes. Further, surveillance over letters sent to this person or that, telephone taps, and the varieties of electronic surveillance have been assimilated to the issue of searches and seizures—and these measures, too, while less brutal in their outward aspect, are equally obnoxious and retrograde by civilized standards of public policy. But I think it is misleading to classify these breaches of civilized manners as invasions of privacy. Rightly considered, the search of a person's dwelling, whether on suspicion of wrongdoing as the law may define it at any given time, as a fishing expedition, or as a consequence of official whim, is, from the point of view of the victim, indistinguishable from attempted burglary—and, if papers or other objects are seized in the course of the search, indistinguishable from burglary. But the law does not regard breaking and entering, felonious entry, second-story work, the accomplishments of step-over men, and other such technical or picturesque categories, as invasions of privacy. This follows, I imagine, from the distinction made between offenses against the person and offenses against property. Nevertheless, if we consult our feelings in the matter, any offense against property that we, as individuals, are likely to regard as serious is primarily an offense against the person: it breaks the bond of community and leaves the victim feeling betrayed and violated.

That is, indeed, the prime emotion experienced by anyone whose home has been burglarized. Concern about the damage assessable in money that we may sustain from the loss of our goods and chattels may be the coloration we give to our feelings on some occasions, for such is our social custom. But that does not express our real loss, the sense that the web of trust—gossamer in texture and invisible against the light, but which nevertheless sustains our life in society, let alone such community as we may have achieved—is rent, and has dropped us into a state of orphanage in the world. Any crime worthy of that heavy name has the same effect. I have known persons who were so unfortunate as to have been beaten on the streets by muggers, and they have said that the lasting effect is spiritual depression, and a loss of confidence—not in themselves so much as in the fabric of the human world. For, prudent and cynical as we may think ourselves, we all live by faith in the good intentions of our fellow human beings. Even in so trivial a matter as asking directions of a stranger, we make the assumption of his perfect good faith. We should be profoundly shocked to learn, not that we were given faulty directions—that is commonplace—but that we had been deliberately misdirected. And, practically speaking, such conduct is unheard of.

The real offense, then, of searches and seizures and other burglarious conduct, official and amateur, and of assault upon the streets, is that it disrupts the passive form of community—passive only because it is constant and unregarded as our heartbeat—of the communion that is a spiritual requisite without which we droop and pine. Physicians speak of an insult to the physical system, something that disrupts the organic integrity of the body. The offenses we have been discussing are insults to the soul, and from a more philosophical viewpoint than the customary one it seems arbitrary to classify such injuries differently as they arise within doors or out. In any case, such a classification does not shed much light upon the nature of the things themselves as measured by their real effect upon us. And if categories of the kind are a convenience for the framers and administrators of the law, it does not follow that the law and its administration are any the less wrongheaded and misconceived since they are addressed to other effects than those which we feel most feelingly.

III

The so-called invasions of privacy are misnamed, for privacy cannot be invaded or infringed. But it can be extinguished, by the destruction of society that the offenses we have been discussing clearly do effect, of the society which is the ground of community. Privacy and community are not antonyms, but the systole and diastole of a single living process. Privacy is communion, passive and active, with our fellow beings when we are not physically in their presence. When, in privacy, we talk to ourselves, we use the common speech, feel the preferences and aversions common to the race, and are then perhaps least personal because least self-regarding, for when we are alone we stand before no audience but man. To prevent through practical measures the spiritual insults that annul privacy by cutting us off from communion is a proper object, certainly, for our efforts. But to make this society, this polity, fit soil for community will require another kind of husbandry, one that seeks to shape the laws and their administration to the prior law that governs our growth and flourishing. That law, so different in spirit from the prudential impulse, is that love and trust, and the habit of acting upon our generous instincts, are our necessary aliment.

To secure these as regular conditions of our existence is surely our object as social beings. Yet it seems clear that such an object is not envisioned by our present social machinery, which sets out to administer us to within an inch of our lives as the materials for the further history of commerce rather than to advance our bosom interest—the humanizing of mankind. The youth of the nation are demonstrating in a hundred ways—some of them inspired and some of them deplorable—that the politics to which we have been treated for decades cannot command the respect of reasonable beings. If we understand ourselves, we can scarcely maintain the habit of submission to institutions that have come to be patently out of keeping with our ideals as Americans. This is, after all, the nation that for a hundred years was justly regarded as the light of the world.

The political goal of America, generously, rashly envisioned at its founding, is the enfranchisement of the Americans. Its social goal is the community that gives meaning to the idea of privacy, to that self-respect that may arise when we are again respectable. The unfulfilled promises of democracy are our peculiar heritage and burden. To assume them is to pursue the American idea to its humane conclusion, to that direct democracy in which the association of the citizens is the autonomy of each person.

10

Do Rocks Have Rights?

Thoughts on Environmental Ethics

RODERICK NASH

> We abuse land because we regard it as a commodity belonging to
> us. When we see land as a community to which we belong, we may
> begin to use it with love and respect.
>
> Aldo Leopold
> *A Sand County Almanac*

Highway 64 leads north from Williams across the Coconino Plateau.
Drive it on a clear day and the sense of space is overwhelming. To
the southeast the San Francisco Peaks loom a mile above the bench-
land. Ahead, north, you feel more than see another mile of topo-
graphical variance, but this time *down:* the Grand Canyon. It is a setting in which billboards loom for miles. One reads: "Arizona's Fastest
Money Maker: Land!"

Such signs not only advertise but presume to *define* land. Measured by the beliefs of other cultures at other times, the definition is
not only unusual but incomprehensible. If our Indian precursors on
the Coconino had been in the practice of erecting billboards, the definition might have read: "The Land Is Our Mother." But the sign on
Highway 64 says it all about traditional American attitudes. Economic
or utilitarian criteria have dominated American thinking about land
and land use. Occasionally, aesthetic factors are brought to bear on
the question, but here too anthropocentrism prevails. It is *man's*

Roderick Nash is Professor of History and Environmental Studies at the University of California at Santa Barbara.

sense of beauty that matters. The land is tailored to man's taste. The human interest dominates. Few have ventured to challenge this fabric of anthropocentricity, replacing it with a conception of "community" that extends to all forms of life and to the environmental setting. From this perspective, economic and even aesthetic considerations are not the basic determinants of man's relation to land. The ultimate criteria become moral—questions of right and wrong. Environmental ethics spring from such an orientation to life and land.

Two assumptions underlie the following discussion. The first concerns the relevance of ethics (and of the humanities in general) to what is called the environmental crisis. Many well-meaning persons downplay mental as opposed to technological solutions to environmental problems. In their opinion, concern over ethics is like worrying about the position of deck chairs on the *Titanic.* Why fiddle over rights and wrongs while Rome burns? On the other hand, it could be argued that the most serious kind of pollution we experience today is pollution of the *mind.* What we do, after all, is a product of what we think and, more precisely, what we value. It follows that the humanities are vital to understanding and solving environmental problems. Ethics in particular are a vital part of what Robert Heilbroner calls a society's "internal capacity for response" to external threats such as the deterioration of environmental quality. Ethics answer the questions haunting all environmentally responsible people—*why,* after all is said and done, should anyone be concerned with proper land use? Why should anyone persevere when personal sacrifice is involved? Unless we can find acceptable answers, I submit, conservation is built on a foundation of sand.

The second matter concerns the idea of "public good." I suspect that my understanding of this concept is not widely shared, for I would define the *public* as no less than the totality of life and the earth itself. To limit its scope to "other people" or "society," I submit, is to concede at the outset the very points at issue in this essay.

I

Many recent converts to the idea of an ethical posture toward nonhuman life and the nonliving environment are surprised to discover that the roots of American thinking about environmental ethics extend beyond the burgeoning ecology movement that caught public attention in the late 1960s and early 1970s. Up to that time, they assume, conservation entailed economic, not ethical, relationships. But a closer look at American intellectual history brings into view an American who died on April 21, 1948, while fighting a brush fire along the Wisconsin River—Aldo Leopold. Leopold's formulation of a land ethic must rank as one of this country's most significant contributions to the history of thought. For those courageous enough to face their full implications, Leopold's conceptions for man's relations to the environment are revolutionary.

Although his formulation of a land ethic begins in seminal essays written in 1933, 1938, and 1939, Leopold did not weave the various strands of his thought together until a decade later. The resulting paper, "The Land Ethic," appears as the final essay in *A Sand County Almanac and Sketches Here and There,* and from this statement most of Leopold's basic concepts can be extracted for analysis.* Leopold begins his argument with the concept of sequential ethics. Figure 10.1 may provide a useful illustration of what Leopold meant although it extrapolates from his thought (and thus represents my own) in several particulars.

The first point to make about this graphic representation of the evolution of ethics is that it applies to ideal, not necessarily actual, conduct. As a depiction of the history of thought it is also abstract and hardly universal in its application. At particular times and in particular situations many people are still mired in the lower ethical echelons. Human beings, after all, still kill each other, even members of their own families. Although we fail as a society consistently to live up to our ethical principles, we do *conceive* of right and wrong with respect to other people. It is the standard according to which we construct and enforce our laws, which are nothing more than the institutional expression of the ethical conviction of the group. A husband might kill his wife but because of the existence of a social ethic and its attendant laws, he is punished if apprehended.

* New York: Oxford University Press, 1949. The quotations from Leopold are from this volume, with page references given in parentheses. Additional statements of his land ethic can be found in *Round River: From the Journals of Aldo Leopold,* Luna B. Leopold, ed. (New York: Oxford University Press, 1953). A substantially complete bibliography for him appears in the *Wildlife Research News Letter* 35 (3 May 1948): 4–19, a publication of the Department of Wildlife Management of the University of Wisconsin.

Figure 10.1

THE EVOLUTION OF ETHICS

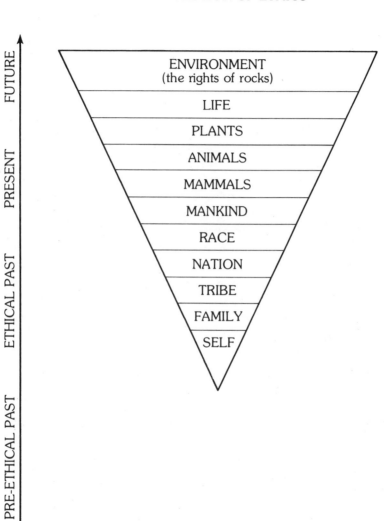

The central idea expressed in Figure 10.1 is the evolution of ethics. The time scale at the left of the diagram expresses Leopold's assumption that at some point in the past ethics did not exist. The reason is simple: life existed before man developed the mental capacity to think in terms of right and wrong. In what is labeled the "pre-ethical past," living things interacted on a strictly utilitarian, tooth-and-claw basis. The "ethical past" began when one form of life evolved mentally to the point where it was possible to conceive of an action as being right or wrong on grounds other than those of simple utilitarianism. For eons thereafter, it seems logical to assume, ethics applied only to the self (the diagram's lowest tier) and were, in fact, hardly an improvement over the preethical Hobbesian stage, in which each being struggled against the other for existence. Under pressure to survive, a person at this rudimentary level of ethical development might cannibalize his mate or offspring without remorse or punishment.

At the stage denoted by the figure's second tier, ethics had evolved to include families: a mate and offspring were encompassed in the envelope of ethical protection even though outside the charmed circle of the family hearth or cave all was a dark tangle of unethical (really, a-ethical) relationships. The extension may well have been prompted by the impulse to sustain one's kind. Ethics became aids in the struggle for existence. The social contract had begun. Leopold recognized this when he wrote that ethics prompted instinctively competitive individuals to cooperate "in order that there may be a place to compete for" (p. 204). This realization, according to Leopold, sprang from the individual's recognition that he was "a member of a community of interdependent parts" (p. 203). As Leopold the ecologist saw it, all ethics stemmed from this recognition of community. In this sense, then, Figure 10.1 traces the expanding definition of community (alternatively, "society") as well as of ethics. The implication that ethics are based ultimately on enlightened self-interest remains one of the most plausible explanations of why we formulate ideas of right and wrong.

The extended family marks a transition to the tribe. In this stage, which lasted, we can assume, for many thousands of years, were the seeds of society. The members of a tribe would respect and protect each other, but otherwise right and wrong functioned much as they do in urban street gangs today. To appreciate the force of ethics, consider a chance meeting among members of the same tribe on a forest path far from the check of shame or punishment: the meeting would involve no rape, robbery, enslavement, or murder. But a similar meeting between members of *different* tribes would certainly involve violence and probably death. Ethics help explain the difference.

But ethics continued to evolve. Tribes occupying the same region gradually discovered the benefits of mutual respect and joined with each other in defining a broader-based ethic. The roots of nations lie in such associations: we can cross state borders from New York to California, confident that we will not be raped, robbed, or killed—at least not because we hail from out of state. Today, during wars, we see the power of nation-based ethics. William Calley, the soldier convicted of killing twenty-two Vietnamese civilians at My Lai, has been excoriated for his total lack of an ethical sense. This misses the point entirely. Calley was *perfectly* moral within his ethical frame of reference, which was the nation. He was not, for instance, in the habit of machine-gunning residents of his hometown in Georgia. But Vietnamese were beyond the ethical pale for him just as Jews were for Adolf Hitler and rattlesnakes for most human beings still. In Calley's eyes his actions at My Lai did not involve a question of right and wrong at all. This explains the mystified reaction of Calley and his apologists to the criticism and trial that followed his return to the United States. For him, "gooks" (military slang for Vietnamese) simply did not figure in moral codes.

A sense of racial identity marks a transition from nation-based ethics to species-based ethics in the same way that the extended family led from familial ties to tribal allegiances. Most black people, regardless of nation, share a sense of community. So do white and red and yellow people. Ethics expands with this expansion of brotherhood. Christian concepts like the Golden Rule and its parallels in other faiths show the potential of an ethic embracing all mankind.

Aldo Leopold took a particular interest in the ethical blind spot that permits slavery. His opening paragraphs in "The Land Ethic" describe how Odysseus killed a dozen slave girls on his return from the Trojan wars. It was not that Odysseus believed murder was right: slaves simply did not fall into the ethical category that protected Odysseus' wife and fellow citizens. The slaves were property; relations with them were strictly utilitarian, "a matter of expediency, not of right and wrong" (p. 201). For slaves, the achievement of ethical identity awaited the attainment of a consciousness represented on the diagram as the level of "mankind." In the West, such an extension of ethics did not come until the nineteenth century. In the United States, many historians feel, it required a civil war. The environment, one could argue, is still "enslaved."

I I

What concerned Aldo Leopold thirty years ago was the possibility of evolving beyond an ethic that halted with homo sapiens. The land ethic, he explained, "simply enlarges the boundaries of the community to include soils, waters, plants, and animals, or collectively: the land" (p. 204). The upper tiers of Figure 10.1 represent this enlargement. For most Americans today, the first step is not too difficult to make. We are accustomed to include cute mammals in our ethical consciousness. We make them into pets and often treat them as carefully as children. We would defend them from abuse with our lives. Useful animals, like horses and chickens, also have a secure place in ethical constructs. For example, most people in the United States today would be shocked at the sight of someone needlessly killing a horse or a dog on the street. They might even call the police or at least the Society for the Prevention of Cruelty to Animals. Somehow they feel that killing a dog is as morally reprehensible if not yet so legally punishable an act as killing a person.

Despite these promising beginnings, ethical blindness begins soon after horses and kittens and Bambi's mother pass from our view. To extend the earlier example, few today would be offended at the sight of someone stoning a rattlesnake or trapping a gopher or spraying an insecticide on a column of ants. For most people, these forms of life are beyond the pale of ethics. It is not that such people are unethical; they would probably protect their dog's life with their own. It is just that they have an ethical cutoff point. Dogs are inside the magic circle; snakes, worms, and potato bugs are outside. Of course there are exceptions. Everyone knows someone who keeps pet snakes and twines them around his neck. But what does the snake lover think about the mice he feeds to his snakes? What does he think of amoebae? Once the pet stage is passed, man's capacity for ethical relationships with other life forms declines rapidly. Only a few feel that plants of any kind deserve inclusion in the ethical fold; still fewer extend ethics to protista, monera, and similar primitive organisms. To do this is to accept the sanctity of life itself. It is to approach what Albert Schweitzer called "reverence for life."

It is important to note that Aldo Leopold carefully avoids anthropomorphism. The tendency of man to attribute human characteristics to other forms of life before granting them ethical respect defeats the whole purpose of the land ethic, which is to affirm the dignity and sanctity of life *apart* from man. Anthropomorphism, of course, is actually anthropocentrism. The way to extend ethics is not to convert animals into people but to recognize the worth of animals as animals. Anthropomorphism is like that inadequate ethic that accords respect to black people because they look or act white: real ethical extension lies not in making black men white but in affirming blackness. Similarly, a genuine extension of ethics requires more than an enthusiasm for Lassie or Snoopy or Smokey the Bear.

The highest level of ethical evolution involves man's relation to parts of the environment not commonly regarded as alive, such as air, water, and rocks. Leopold clearly had this extension in mind in "The Land Ethic" since he includes within the community inanimate matter like soil and water. "Land," in fact, was Leopold's shorthand term for the entire environment—its living parts as well as those to which we commonly do not ascribe life. In no relationship, then, was man excused from ethical responsibility. "A land ethic," Leopold explains, "changes the role of *homo sapiens* from conqueror of the land-community to plain member and citizen of it. It implies respect for his fellow-members, and also respect for the community as such" (p. 204). The "fellow-members" are clearly other forms of life, but Leopold is careful to recognize "the community as such," indicating his extension of ethics to habitat, system, process, and place.

Passing from the living to the nonliving environment is the most difficult transition in ethical evolution. Even Leopold hedged a bit in ascribing ethical identity to the nonliving environment—in valuing it in its own right and not simply as life's necessary setting. But it is possible to conceive of the rights of rocks. From such a perspective, strip-mining would be as heinous a crime as the rape of a neighbor's daughter. The extermination of a species would rank with genocide.

Several approaches can be made toward an all-inclusive environmental ethic. Subscribers to religious faiths of the Far East have made the ethical leap for centuries by minimizing the significance of life compared to that of the divine spirit which permeates all things living and nonliving. From this perspective a rock can be as eloquent as a tree, a bear, or a baby in revealing universal truths and harmonies. Another possibility is to argue that rocks, rightly seen, *are* alive and thus deserve the full measure of ethical respect accorded to all life. One can make this point by arguing that rocks contain all the raw materials of things we normally regard as living. A little reorganization, a lot of time, and who is to say that the inanimate does not live? Leopold recognized that "land . . . is not merely soil; it is a fountain of energy flowing through a circuit of soils, plants, and animals" (p. 216). Loren Eiseley put it well when he observed that man and all life are but compounds "of dust and the light of a star." It is also entirely conceivable that our present definition of "life" is greatly abridged—a mere fragment of a spectrum that extends to things like rocks. We know that sounds exist which man cannot hear and colors which he cannot see. Perhaps there are ranges of life that also transcend our present state of intelligence. Again, Eiseley expresses the concept eloquently: stones, he has written, are "beasts . . . of a kind man ordinarily lived too fast to understand. They seemed inanimate because the tempo of life in them was slow." Such ideas resist traditional kinds of proof, but they are the essential underpinnings of an environmental ethic.

There are problems with this extension, of course. By the time we ascend the inverted ethical pyramid to the level of rocks, we are well beyond the point where the beings having ethical identity can speak for themselves. Slaves demand and, more importantly for the present purposes, define freedom. Rocks don't. This means men must. As a consequence, inappropriate action is likely. What, after all, do rocks want? Are their rights violated by quarrying them for a building or crushing them into pavement or shaping them into statues? Anthropocentrism is inevitable in any answer we might make as well as in any corrective system we might apply. For the time being, as Supreme Court Justice William O. Douglas opined, the *people* who care must plead the case for trees, and rocks, and swamps.[1]

Before turning to some implications of a land ethic, it is interesting to dwell momentarily on the intriguing possibility that the quest for an environmental ethic represents recovery rather than invention. Figure 10.2 is another abstraction which does some violence to the actual history of thought and does not hold for all societies at all times. It may, however, provide a basis for general discussion.

The essential concept illustrated in Figure 10.2 is that so-called primitive man possessed an ethic that extended well beyond his fellowmen. There is substantial evidence from anthropology and the history of religion that pretechnological man believed other life forms and even mountains and rivers worthy of respect and subject to ethical restraints. But, as the diagram indicates, this broad ethical horizon gradually shrank. To the right of the lower pyramid are listed some of the forces that figured in this ethical contraction. We might imagine (only somewhat facetiously) that ethics reached their narrowest definition in the person of the nineteenth-century "robber baron"—the archetypical American individualist who fancied himself an invulnerable island accountable only to himself. Recovery of a broader ethical perspective draws some of its energy from the forces noted to the right of the upper, inverted pyramid in Figure 10.2. Admittedly oversimplified, the graph has some use in suggesting that in moving toward a land ethic we are not creating something new so much as trying to regain what we once had and lost. It is encouraging to know that some men at some time had the capacity for a notion of ethics that extended to the limits of the environment itself.

III

The land ethic has a revolutionary impact on land use. In the first place, it makes environmental planning a matter of what people think and, more precisely, what they value. Every kind of environmental pollution can ultimately be traced to pollution of the human mind by ethical myopia. It follows that our educational and religious—not our economic, political, or technological—institutions hold the keys to success in establishing responsible relations between man and his environment.

But this realization immediately gives pause. With only a few recent exceptions, schools and churches in the West have not been

Figure 10.2

THE DECLINE AND RECOVERY OF ETHICAL PERSPECTIVE

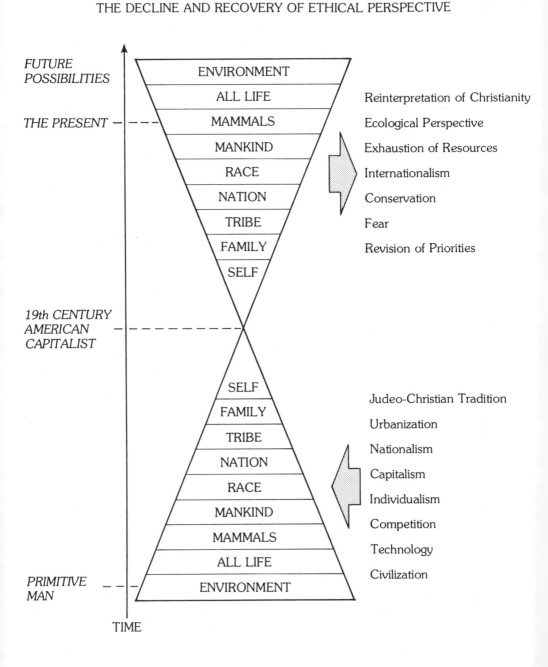

committed to teaching environmental ethics. In our churches, the neighbors we are taught to respect are nearly always human. The community in which we are admonished to be ethical extends no further than to homo sapiens. Aldo Leopold, for one, deplored this situation, arguing that conservation would forever be "trivial" unless it touched the "foundations of conduct" through such institutions as religion (p. 210). As the moral leaders of society, holy men have a responsibility to the environment they have thus far largely ignored. Environmental ethics are a pointed reminder of this failing, and a challenge to explore, explain, and establish on an ethical basis man's relation to land.

Similarly, education must embrace the challenge of a land ethic if the taproots of thought and conduct are to be significantly altered. Schools, from elementary levels to the university, have a clear mission in this respect. I wish to call attention, however, to the more subtle but enormously important educational process called socialization. Surely most of us have on occasion been summoned to destroy an insect or other allegedly loathsome creature that suddenly appeared in the sink or bathtub. The procedure is stereotyped: a scream and a frantic cry, "Get it!" are followed by squashing and flushing. I recently observed the reaction of a three-year-old to the "get it" performance. There was a flash of sadness in the child's eyes as the bug turned to pulp. After all, here had been something moving by itself, without batteries or wind-up keys, something alive and instinctively intriguing. The three-year-old had not yet been taught by society to regard insects or snakes or mice as loathsome. The squashing, of course, ended all that. A process of socialization which would teach the child to distinguish between man and nature was well under way. An alternative scenario, however, is conceivable. A bug might be picked up lightly and placed outside on the grass—so Albert Schweitzer treated the worm he found crawling hopelessly on a pavement after a rain. Children could thus be reminded that bugs as well as people share life and should share the environment. It would be a mere gesture, of course, but in such small ways are revolutions of the mind begun.

The realization, implied in the land ethic, that all forms of life have equal rights to a finite environment is one way to approach the problem of growth that is basic to the question of land use. The term *growth* refers, of course, to human growth, and a moment's reflection reveals that it cannot take place without infringing the rights of other forms of life to flourish or even to exist. Subdivisions are often built at the expense of frogs and field mice. An environmental ethic that included frogs and mice could be used to check such anthropocentric expansion just as the social ethic now acknowledged checks the desire of one person to grow at the expense of another by robbing him. Ethics act as restraints on growth: we say we will not grow if growth entails destruction of our neighbor. What is needed is an ethical posture that protects life and the planet itself from the egocentric impulses of a single species.

Implicit here is the challenge of an ethical as opposed to an economic view of man's relation to his environment. The typical approach of the economist to environmental problems is to internalize externalities—to make polluters pay enough to clean up their mess. A land ethic would not permit the mess in the first place. It would deny the right to rape at any price. Most people can accept this principle in terms of human relations—every woman does *not* have her price. Some things, in other words, are simply wrong and intolerable. But as Leopold knew so well, this attitude has made few inroads on our relations with the environment. "Land ethics," he declared, "are still governed wholly by economic self-interest, just as social ethics were a century ago" (p. 209). The valuing of land, he explained, had to extend beyond economics to such things as "love, respect, and admiration" (p. 223). The essential attitude could be phrased quite succintly, he thought: "Quit thinking about decent land use as solely an economic problem. Examine each question in terms of what is ethically and aesthetically right, as well as what is economically expedient" (p. 224). And then, coming as near as he ever did to actually defining the content of ethics, Leopold declared that "a thing is right when it tends to preserve the integrity, stability, and beauty of the biotic community. It is wrong when it tends otherwise" (p. 224). On these grounds alone, the unnatural, cancerous growth of man's population and civilization can surely be condemned.

The urgency of finding ethical rather than economic ways of ordering man's relationship to his habitat is increased by the probability that in the not too distant future man will have the technological power to completely order the earth according to his own designs.

For centuries man's appetites have been checked by his inabilities. It is possible to argue that the Indian did not clear-cut the redwoods simply because he couldn't. But with an unlimited technology making nearly complete exploitation of the environment a possibility, the need for restraints based on ethics is indisputable. The situation which man is rapidly approaching compares to that of a 150-pound kindergarten student: having the power to smash his 40-pound class-mates makes his possession of a social ethic all the more essential. Kindergarten teachers, to extend the metaphor, strive to build an ethic (it is usually called "courtesy" or "respect for others") in their pupils. Their efforts are never fully successful; fights break out daily. But the ability of any one student to hurt another seriously is limited by his small physical power. A discourteous five-year-old with the strength of a professional prizefighter would be quite different and quite frightening. With mankind possessing this kind of power and with an environment unprotected by ethics, the potential for serious problems is readily apparent.

Finally, the land ethic provides an entirely new justification for national parks, wildlife refuges, wilderness preserves, and open spaces. Traditionally these institutions were established and valued as sources of human pleasure. Aldo Leopold's philosophy suggests that they night be seen as valuable for the life and land forms they contain regardless of man's interests. Parks, for instance, can be considered islands in the environment where man has exercised restraint, where he acknowledges the rights of other life and of rocks themselves. To create a park, after all, is to establish a limit. If we cannot prosper without another dam or a mine or a lumbering operation or a city, then we limit our prosperity. We put other considerations before growth. Parks could be understood as symbols of a revolutionary attitude toward the whole environment. It was not a coincidence that Leopold, the philosopher of the land ethic, was also a prime mover in the United States Forest Service for wilderness preservation. Wilderness engenders humility and respect, and these are precisely the qualities needed to nourish environmental ethics.

If the billboard on Highway 64 stands at one end of the spectrum of man's attitudes to land, at the other are the words that greeted visitors to the United States pavilion at Expo '74 in Spokane, Washington. Carved in broad letters over the main entrance was a statement from the 1854 speech of Chief Seattle: "The Earth Does Not Belong to Man; Man Belongs to the Earth." A land ethic is implicit in the Chief's words, but, significantly, he spoke from outside the mainstream of American economic, legal, and religious thinking. In 1854, he was frighteningly alone—a survivor of a dying way of life and thought. But today there is an increasing number of Americans prepared to look critically at our dominant traditions, if not yet ready to move beyond them. We have seen the term *counterculture* come to prominence in this connection. And Aldo Leopold has become something of a guru in the new gospel of ecology that eschews old utilitarian rationales for a new definition of conservation that emphasizes man's harmony with the land. From such a perspective, responsible land use is a matter of ethics, not economics or aesthetics or even of law. The land ethic demands that we be concerned about the condition of the environment not because it is profitable or beautiful and not even because it promotes our own survival as a species, but because, in the last analysis, it is simply right. The point is that "private rights" no more insure a liberty to rape land than they guarantee a freedom to rape people. If man ever evolves morally to the point at which environmental ethics acquire the status of a general consensus, to rape the land will be as wrong as to rape a neighbor's wife. Indeed, the entire community of life and the land itself will be perceived as man's neighbors, which in fact they are.

Technology and
the Ideal of Human Progress

It is the extremities of heat and cold, the inconstancy and badness of seasons, the violence and uncertainty of winds, the vast power and treachery of water, the rage and untractableness of fire, and the stubbornness and sterility of the earth that wrack our invention, how we shall either avoid the mischiefs they may produce, or correct the malignity of them and turn their several forces to our own advantage a thousand different ways, while we are employed in supplying the infinite variety of our wants, which will ever be multiplied as our knowledge is enlarged, and our desires increase.

Bernard Mandeville
The Fable of the Bees, 1723

11

Living with Scarcity

ROGER L. SHINN

Scarcity—meaning hunger, pain, and deprivation—has always been a human problem. It has been the normal situation of most of the human race. For a short time in recent history mankind enjoyed the entrancing dream that technology was about to overcome scarcity. The swift fading of that dream requires new attention to some old social-ethical issues.

I

The new sensitivity has caught the attention of the world with startling suddenness. Thus Secretary of State Henry A. Kissinger told James Reston of the *New York Times* on 13 October 1974:

In 1969, when I came to Washington, I remember a study on the energy problem which proceeded from the assumption that there would always be an energy surplus. It wasn't conceivable that there would be a shortage of energy.

Until 1972, we thought we had inexhaustible food surpluses, and the fact that we have to shape our policy deliberately to relate ourselves to the rest of the world did not really arise until 1973, when we did call for a world food conference.

How, we may wonder, could such monumental problems as the energy and food crises slip up on a world unaware? One might ask whether political leaders are vocationally unable to look more than one or two elections ahead. That is part of the story. There were scientists and humanists, a decade and more ago, concerned about

Roger L. Shinn is Reinhold Niebuhr Professor of Social Ethics at the Union Theological Seminary in New York.

population, food, and exhaustion of resources. Yet the amazing thing is that so much of the world took for granted continuing triumphs of production, perhaps combined with social revolutions, which would bring to all societies the wealth already attained by a few.

The reason was the stunning achievements of agricultural and industrial technology. The industrial revolution, after its initial agonies, achieved successively new stages of prodigious productivity. The successes of technology burst one barrier after another until people ceased to believe in any barriers. Even war, often a destroyer of economies, became a stimulator of production. The British economist E. F. Shumacher, pointing in 1971 to the "continuous acceleration" of the economy of the Western world for twenty-five years, warned: "There have never been twenty-five years like this before, and there may never be twenty-five years like this again." [1]

Yet people and nations, not only in the Western world but throughout most of the planet, assumed continuous expansion of production and consumption. John Kenneth Galbraith, describing common opinion without endorsing it, wrote: "The belief that increased production is a worthy social goal is very nearly absolute. . . . That social progress is identical with a rising standard of living has the aspect of a faith." "No other social goal is more strongly avowed than economic growth." [2]

The faith, even then, had its problems. It became common knowledge that the gap between rich and poor on a worldwide basis was increasing. The predicted spillovers of wealth seldom kept up with schedules.

Then came the new sensitivity to human disruptions of the intricate natural systems that sustain life on earth. Mankind, said powerful voices, is pushing at the limits of the capacity of the planet. The human race—with its fecundity, its hunger, its craving to consume irreplaceable resources, its pollution of air and water—is straining the ability of the globe to support the race.

Today a worldwide debate centers on the issues of continued growth in production and consumption of food and industrial goods. A series of forceful essays, books, and manifestoes has argued that continued exponential economic growth is suicidal and that mankind must adopt other expectations and social policies.[3] A comparable series of replies has sought to refute the case.[4] An issue unknown five years ago except in esoteric circles has become so impressive as to dominate numerous international discussions.

The controversy hinges in part on issues so complex and technical that the wisest experts disagree. How perilous are nuclear breeder reactors? Are nonradioactive fusion reactors feasible within twenty-five years? Can the Green Revolution go into a second phase as effective as the first? Are chemical fertilizers and pesticides doing irrevocable damage to the global environment? Are supersonic aircraft and aerosol cans depleting the ozone surrounding the earth and hastening cancer and genetic disaster? Is it true that the period of 1931–1960, on which most projections of food production have been based, had the most abnormally favorable climate of the past five hundred years? Is the increasing carbon dioxide in the air producing a "greenhouse effect" that will melt polar icecaps and inundate the coastal regions of the industrialized world?

Uncertainty about such questions imposes some modesty on anyone who tries to think and live responsibly. But modesty need not mean silence. In the face of the uncertainties two generalizations seem valid.

First, if all the unknowns should turn out to be unfavorable, the human situation will be worse than the most vehement alarmists are saying. If all should turn out to be favorable, then consternation is premature. But to assume that *all* will find a happy resolution is contrary to most human experience and is morally irresponsible.

Second, critics of "the limits to growth," using the phrase to refer to a theme and not solely to the book by that title, usually concede much of the case. For example, one whole set of critics quickly grants that limitation on population is desirable and necessary. That is itself an interesting choice. Latin Americans and Africans frequently respond that technologists betray their value-premises when they choose to center attention on population rather than on the excesses of industrial societies with their extravagant consumption of resources and pollution of the environment. An ideological bias has warped the argument.

On the basis of these two generalizations, I am assuming in the rest of this essay that the problem of scarcity is and will continue to be real, although I do not know the precise pressure points or the time schedule of the coming confrontations with scarcity.

II

The human race in its long experience has devised many methods of handling scarcity, nearly all of them still in effect somewhere.

One of the commonest is a system of caste, class, or slavery. Part of the population escapes the worst of scarcity by imposing it on another part of the population. In much of the world through much of history, high cultures were available only to privileged fews—royal families and their retainers, hereditary castes, economic and intellectual elites. The classical philosophers, of both Platonic and Aristotelian traditions, assumed that the good life was available only to a few and that the few did not include slaves or those engaged in manufacture or trade. Biblical and Stoic writers challenged that motif but rarely conquered it so far as the organization of society was concerned.

A second method of meeting scarcity has been war, which may remedy scarcity in some societies by imposing it on others. Powerful nations have often conquered other nations or intervened in their internal affairs for economic advantage. The time has come when it is more embarrassing than in the past for knowledge of such activities to leak to the world. The role of the CIA in the overthrow of Muhammed Mossadegh in 1953 brought more congratulations than its comparable role in the overthrow of Salvador Allende in 1973. In the current petroleum crisis, both military conquest and internal intervention against the petroleum-producing countries, though not unthought of, are generally considered to be counterproductive. Even so, there is occasional talk in military and political circles of future wars aimed at control of sources of scarce materials.

A third method is sharing. Most tribal societies develop modes of sharing so that scarce resources may be distributed among the tribe. When scarcity is a serious threat, as it often is, tribal societies may combine painful methods of population control—sometimes including exposure of infants and the aged—with rigorous rituals of sharing life's necessities. In large-scale technological societies, sharing takes such highly rationalized forms as governmentally enforced transfer payments (Social Security, relief, Medicare and Medicaid, among others in our society); rationing, especially in times of emergency; and socialization of goods and services (few in American society, many in Chinese society).

A fourth method is reliance on the market. The existence of a market assumes scarcity. That is, only items that are scarce are bought and sold. Nobody buys things that are not scarce, in the economist's use of the term. The merit of the market is the freedom it affords. Rather than accept a governmental allocation of food, housing, clothing, books, theater tickets, and the like, the consumers decide how to distribute their income, perhaps choosing to scrimp on some items in order to splurge on others. The problem is that the market obviously favors the rich. Some people have to scrimp on everything while others can splurge on everything. Hence all societies regulate the market at some points and usually increase regulation when scarcity is most severe. Presumably the modern concept of the market does away with such medieval metaphysical abstractions as the "just price"; but the persistence of such terms as exploitation and profiteering, as well as cries to government whenever shortages (e.g., of fuel oil and gasoline) become painful, show that the doctrine is not quite dead.

A fifth method involves the increase of centralized planning so that both abundance and scarcity may be widely distributed. If some countries have more petroleum than they can consume and others have more food than they can eat, and if the combination of free markets and cartels means that the poorest societies lack both petroleum and food, an international system might so allocate petroleum and food that everybody would have a minimal share. The alternative is likely to be misery and unrest that are morally intolerable and threaten the safety of the world. Hence many scenarios for a future of scarcity sketch a highly organized world in which people and nations give up some cherished freedoms for the sake of an efficient distribution that may contribute to human survival. Such a world might be morally preferable to the present world in its increased human equality; it might be morally costly in its surrender of spontaneity, idiosyncrasy, and personal freedoms.

A sixth method is the opposite of the fifth. It calls for self-sufficiency of local economies, with people accepting considerable scarcity rather than enduring the dependency that goes with greater consumption. Scenarios drawn on this model envision a reversion to locally autonomous economies enduring great inconvenience for the sake of relative invulnerability to external hazards. Those local economies may be national in scope (witness the increased interest in the idea of a Fortress America) or may be small communes of like-minded people trying to minimize their material needs and reject the highly technological culture that, in their judgment, has made the world so perilous.

Since history never acts out purely theoretical models, it is safe to assume that no one of these conceptions will dominate future history. The reality may be a muddled combination of them all—and of others that I have not thought about. But probably all six methods will come into play in some times and places, and they signal some of the ethical choices likely to confront people and societies.

III

The problems of social ethics generally involve conflicts of interest. Hence they are inherently hard to solve. For that reason societies will usually, so far as possible, seek technological answers to ethical problems. That method often succeeds.

For example, a shortage of food presents a conflict of interest familiar to mankind. The distribution of the food becomes an ethical problem. Shall some feast and others starve? Shall food be distributed equally? Shall there be special rations for warriors or for pregnant women or for the sick? Shall all endure some hunger in order that all may eat? If some must starve, who will these be? Societies have met this problem in many ways. Modern transportation and communication have begun to create a world society, at least to the extent that mankind can begin to see food scarcity as a world problem, presenting moral opportunities and burdens unknown in past centuries.

If technologies of food production succeed in overcoming the shortage, then a technological answer can be given to the ethical problem. That makes life easier. Usually the technological answer ameliorates rather than eliminates the problem. Instead of asking who shall eat and who shall starve, we may ask who shall eat steaks and who shall eat hamburger (or rice or soybeans). The latter question is still an ethical question, perhaps beyond the ability of mankind to handle, but easier to live with than the former.

Industrial society has seen some spectacular successes in finding technological answers to ethical problems. But times come when the method won't work. Increasingly, people and societies face problems for which there are no technical answers. Garrett Hardin in his now-famous essay, "The Tragedy of the Commons," has written:

An implicit and almost universal assumption of discussions published in professional and semipopular scientific journals is that the problem under discussion has a technical solution. A technical solution may be defined as one that requires a change only in the techniques of the natural sciences, demanding little or nothing in the way of change in human values or ideas of morality.

But, continues Hardin, the world faces "a class of human problems which can be called 'no technical solution problems.' "[5]

In endorsing this idea, I quickly add that ethics needs technology. The achievements of technology have much to do with setting the problems of ethics, with shaping alternative possibilities, with influencing values, with contributing to solutions. Ethical prescriptions, indifferent to technological change, are often destructive. It is the interaction of the technical and the moral that needs exploring today.

Because technology has triumphed beyond any expectations of past ages, it has sometimes been assigned a messianic role. Journalist Kermit Lansner voices a popular mood when he writes of the energy crisis: "The chief instrument of salvation will have to be science, for only science can open the new sources of energy that will finally resolve the crisis." [6] The trouble with this reasoning is that there is no reason to believe that anything will "finally resolve the crisis." Whatever help comes from scientific technology will be welcome. But people who expect salvation from technology are likely, when disappointed, to demonize technology. That revulsion against technology accounts for the almost hypnotic effect of Jacques Ellul's *The Technological Society* on significant numbers of people, especially younger people, in technological societies. My own judgment is that the demonizing of technology is as foolish as its idolizing. I want all I can get from technology: help in meeting scarcity and opening new possibilities, more effective contraception, correction of former technologies that have been destructive of nature and humanity, innovations that will ease the pressures and give more working time for solving problems of food, energy, raw materials, and pollution. But I do not expect technology to remove the necessity of ethical decisions that are becoming more urgent and intricate.

But how can decisions on social ethics be made in a pluralistic world? Conflicting interests become rationalized in conflicting value systems. These become imbedded in ideological, metaphysical, and religious visions, which are bearers of meaning for ingroups but have no authority for outgroups. There are signs of convergence among many disparate cultures on some basic human values or sensitivities. There are other signs of divergence that heightens conflict.

In recent thinking about the ecological crisis, survival has become a conspicuous value. It comes as close as any nameable value to being universal. In a pluralistic world, all peoples have a stake in it and recognize that it is a prerequisite to many other values. It transcends some conflicts. For example, it affects contemporary thinking about war: there is no satisfaction in clobbering the enemy if the clobberers do not survive. So survival often becomes the universal value to which policy projections must refer.

One trouble with ethics of survival is that most groups remain more interested in their own survival than in a universalized concept of survival or in the survival of other peoples. Without slandering the American people, I might suggest that considerable evidence indicates that right now this society is more concerned with its convenience than with the survival of thousands of families in India or in Africa south of the Sahara. I do not ascribe this interest to sheer perversity. The magnitude of the problem, its geographic distance, and the sense of helplessness in meeting it—all these make it easy to put out of mind most of the time.

A second problem with ethics of survival is that it is not very inspiring or liberating. It may even lead to a sense of desperation that inhibits imaginative and heroic action.

So I suggest that survival, although a real value and one that may be rationally worked into policy decisions, is far from adequate. It may prompt some restraints against destruction, some acts to improve social orders, some willingness to accept discomforts rather than rebel blindly. But life can survive in misery and injustice.

A value that affirms more than survival is justice. It is a nearly universal value, in the sense that hardly anybody in any culture wants to deny it. But hardly anybody knows how to define it, and the definitions do not agree—not even in the abstract: does justice mean equality, as in equal justice before the law? or does it mean rewards for efforts and ability coupled with punishment for sloth and crime? or distribution of resources with regard to need, as in Medicare and Medicaid?

IV

The problems become more difficult as they become concrete.

To look at a society of high consumption facing unexpected scarcities is to realize how little we know what justice is. The obvious example is American society facing the petroleum shortage of 1973 – 74—a very mild shortage compared with what may come, but severe enough to show up the issues.

One answer has been to exhort citizens to conserve energy. That has some effectiveness, not much. Ethically it has the value of voluntary as against coercive action, the disvalue that hardships are imposed on the conscientious rather than on all alike.

Another answer has been to raise prices. It may be just to give individuals the choice of economizing on gasoline and fuel oil or of economizing elsewhere in order to keep warm and mobile. But caprice rather than justice determines that some colleges go broke, some businesses fail, some people lose jobs, and some people shiver while more affluent persons and organizations share none of the hardship.

Another answer has been a few timid steps toward allocating resources. The federal government ordered some apportionment of crude oil between gasoline and fuel oil and some regional distribution of supplies. Various governmental agencies issued some orders limiting gasoline sales and heating temperatures. But government stopped short of rationing. That was probably an ecological mistake, but it was politically understandable. Who knows what a fair system of rationing is? What are the rights of those who drive fifty miles a day to work compared with those who drive fifty miles a week for fun? To get to work is more important than to get to a ski resort, but ski resorts are some people's work. What are the relative rights to gasoline of the physician, the insurance salesman, the bookie, the farmer, the cab driver? What about the comparative rights to heat of arthritic old people, of infants, and of healthy youths?

How does justice relate to regional needs? In New England, where winters are cold and distances short, it seems just and reasonable to reduce consumption of gasoline and maintain consumption of fuel oil. In southern California the opposite logic prevails. Montana and North Dakota might ask, in justice, for more fuel *and* gasoline than New England or southern California. Perhaps it is just to forbid electrical heating, with its immense energy consumption, in new homes. But what is just to families that in a more innocent time bought such homes and cannot now afford to heat them?

So we may marvel at how little we know what justice means. We may also marvel at how little justice enters into policy decisions. We may think of the whole patchwork of federal taxes and subsidies, of housing and agricultural policies, of trade-offs between unemployment and inflation whether in classical situations or in the new stagflation. Instead, I will mention a single vivid example. Mayor John Lindsay once used it on television, but I shall substitute my personal version. When I first came to New York in 1938, the five-cent subway fare was almost a sacred symbol. The fare for driving through the Hudson tunnels or over the George Washington Bridge was fifty cents. Now the subway fare has been multiplied by ten. The Hudson crossing has risen by only one-half. A demonic power, designing policy to bear down on the poor and to maximize ecological damage, could hardly have plotted a craftier scheme. But none of this was planned. It just happened in response to practical necessities and political pressures, largely indifferent to justice.

I am not about to conclude that justice is meaningless. It is a concept, abstracted from some deep human intuitions and experiences of the meaning of selfhood and community. Even the abstraction has some clarifying value and some power to influence behavior. At a minimum, we are all capable of some sense of outrage over flagrant injustice—and that implies some notion of what justice might be. But justice remains an exceedingly vague concept.

If the problems of justice in relation to scarcity are perplexing within a high-consumption society, they become baffling on a world scale. We can start with a conventional figure. The United States with something like 6 percent of the world's population consumes something like 40 percent of its production. (Estimates range from 30 percent to 50 percent.)

There was a time when liberal groups in this country urged that

1 percent of the gross national product be put into economic aid to other countries. The assumption was that an economy growing at perhaps 5 percent a year could give up 1 percent of the GNP without any real loss. The target of 1 percent proved unrealistically high, and in days when the economy is not growing (except in inflationary terms), it is out of sight.

Let us get more specific. In a time of massive food shortages, how much food should the United States export for the sake of justice? We have been exporting about $21 billion worth of food a year for the sake of profit and foreign exchange. But in times of reduced harvests, the government has put some brakes on exports—not because we don't want the foreign exchange and not because farmers don't want the markets, but because exports raise American prices at a time when inflation is already too high. Inflation need not mean real hunger for anybody in the United States, except insofar as we mismanage things horrendously in this society. It does mean inconvenience, maybe some hardship. How does our own inconvenience relate to starvation elsewhere?

A related example is petroleum. The Organization of Petroleum Exporting Countries has adopted policies leading to a quadrupling of the cost of petroleum. From the viewpoint of the gods, that might be taken as a step toward justice, because it means a huge flow of money from high consumption economies into areas filled with poverty. But this may be a case where the mills of the gods grind swiftly but exceedingly roughly. The real victims of the economic tilt caused by petroleum are not the industrialized nations but countries like India and several of the African and Latin American countries.

The rise in petroleum prices has meant that Japan imports less petroleum for manufacture of fertilizers to sell to India. Urea, the synthetic nitrogen fertilizer, advanced in price from $50 per ton in 1972 to $250 in early 1974. According to estimates for 1974, India was short a million tons of fertilizer which meant a loss of about ten million tons of crops. The petroleum shortage may have contributed to twenty million deaths that year, in a world where 400–500 million children face the threat of severe malnutrition or starvation.

Those of us who grumble about the high cost of gasoline, fuel oil, and meat rarely bother to reckon that our consumption of these items contributes to starvation in Asia and Africa. Is it valid reasoning or simply rationalization that tells us that our acts of restraint would really do Asia no good apart from an ambitious political plan, which our political leaders show no signs of devising? Is it cogent thinking or indifference to justice that tells us our efforts to feed hungry populations will do no good unless those peoples do something about birth rates? Questions like these remind us that intellectual rigor is feeble when confronting the question of justice. We know little of what justice really is; perhaps we care less.

Uncertainties about the nature of justice become compounded when the question arises about responsibilities of the present generation to the future. Have we today an obligation to bequeath a habitable planet to our descendants? Should we restrain consumption now in order that they may have more later?

Conceivably, since human history is not eternal in any case, contemporary society might set out to maximize present values, accepting no responsibility for a future that at best is problematic and temporary. But such an ethical course is dubious in a world with as low confidence in its values as ours. And a future-be-damned attitude is unlikely to enhance ethical sensitivity in the present.

But if affluent people are barely concerned about human need in the ghettos of the same city and are almost totally unwilling to sacrifice for hungry folk across an ocean, how likely are they to put themselves out for people who do not yet exist? Men and women will change their behavior for the sake of children, born or unborn. They will do so for grandchildren. So far there is little evidence that they will do so for the children of the world three generations hence.

Yet just such an ethical challenge has been raised. Jørgen Randers, one of the coauthors of *The Limits to Growth,* addressing a study group of the World Council of Churches in 1971, said this:

It is, however, possible to change the objective function [the objective of human living]—in the same way as Christianity changed the objective of man from selfish gratification to consideration for the welfare of all other people living at the same point in time.

Today we could change again, and for example adopt as our cardinal philosophy the rule that no man or institution in our society may take any action which decreases the economic and social options of those who will live on the planet 100 years from now. Probably only religion has the moral force to bring such a change.[7]

Some of the church people listening to Randers commented that only an atheist or a skeptic could expect so much of religion. But they could not dismiss his question: what obligation has the present to the future?

The argument for an ethical sensitivity to the future may cut in different, even opposing ways. The hard-nosed form argues against emergency measures to prevent starvation now, because the population explosion means that every thousand persons saved from present starvation are likely to mean two thousand to starve in the future. The gentler form argues that people should moderate their consumption and recycle resources so that, to the extent possible, the planet is left no poorer a decade hence than now.

The foregoing three inquiries on justice—within high-consumption societies, within world society, and in relation to the future—are offered here, not to endorse a program, but to support one somewhat melancholy conclusion: mankind knows rather little and, worse, cares rather little about justice. Yet the claims of justice—like the claims of survival, which may or may not be related to justice—will not be utterly silenced among people who want to be human.

V

Chances are, only a few of the most extravagant of the world's consumers will ever see this essay. Only a small fraction of earth's people ride airplanes, own automobiles, use central heating and air conditioning, enjoy home refrigerators, and worry about high-cholesterol diets. How likely are we to change our ways?

For myself, I confess that I have been formed by a culture of high consumption and I like it. I value readily available water (hot and cold), an electrically wired home, comfortable indoor temperatures, plenty of food including meat. I want readily available hospitals, even though I hope to make little use of them. I enjoy books, newspapers, television, theater. I appreciate mobility, by auto and by air. If I could choose, I would prefer a world with fewer people who had more access to the things I like to use. If somebody then accuses me of elevating things above people, I reply that things can enhance personal life and that personal values are not directly proportional to numbers of people.

But if, looking at the present world, I see that my life-style is not a conceivable possibility for most of mankind, how willing am I to push for a social change that would require change of me? And how willing is the society, of which I am a part, to modify its life-style?

I expect little response to reasoned ethical argument. A rational argument, appealing to the world's interest or even the national interest, would now (I assume) call for reduced energy consumption throughout our society. But the national government does not dare to try seriously to do that. Immediate inflation and unemployment, not worldwide justice, dominate political controversy. The massive inertia of a high-consumption society will not readily be changed.

Major social changes come in response to pressures. I expect to see some mini-catastrophes: power blackouts, pollution crises, transportation difficulties. I expect to see society reluctantly face some hard choices: will we pay the high costs of pollution in medical bills or in restraints that either reduce production or make it more expensive?

I expect to see frustrating adjustments to shortages as the world withholds resources from us or makes them more costly. Ad hoc responses to necessity may add up to a considerable change in familiar ways of life or may prod our society into a more or less coherently planned pattern of change.

But such grudging acceptance of limitations will lead to a bitter, crabbed life unless they are joined to some animating vision of a future that is alluring, as well as threatening. The changes in racial patterns in this country have been painful enough to warn us that social change is not easy. But enough has happened to teach us some things. Moral idealism did not effect change until massive pressure moved against prejudiced privilege. But pressure alone brought stubborn and stupid resistance, except when it was joined to some vision of a more just society.

The pressure of circumstances, I suppose, will require new ways of coping with scarcity and new modifications of extravagant consumption. If some ethical commitment can meet the pressure, people may move with some grace.

That is why perceptive writers are urging reconsideration of the mythologies that embody human aspiration and adaptation. Arnold Toynbee has urged a return to something like premonotheistic animism. Others have urged something akin to Chinese Taoism, which bends humanity toward nature while bending nature toward humanity. Lynn White, Jr. has urged a recovery of Franciscan Christianity. Kenneth Boulding and Herman Daly, economists, have urged a recovery of the Sermon on the Mount—an idea so radical that a theologian would scarcely dare propose it. Robert Heilbroner has advocated a movement from the myth of Prometheus, with his ambitious daring, in the direction of the myth of Atlas, the enduring burden bearer.

My own hope is for a mythology that has more of fun and delighted contemplation, more of fervor for justice, more of joy in sharing than I see in Atlas. Although I expect limitations upon many freedoms that have been important in the past, I see possibilities for spontaneity that is minimized in contemporary industrialized cultures. As Margaret Mead has put it, "Prayer does not use any artificial energy, it doesn't burn up any fossil fuel, it doesn't pollute. Neither does song, neither does love, neither does the dance." [8]

To use terms that carry meaning for me, while encouraging others to find languages meaningful to them, I see possibilities in a revised version of Christian asceticism—what a World Council of Churches conference has called "a new asceticism":

An appropriate "asceticism" for our time will not be a punishing of the body for the sake of the soul or worldly denial for the sake of an otherworldly reward. It can become part of the voluntary life-style of committed people. [9]

Such a new asceticism draws on a tradition that celebrates the embodiment of divine spirit in flesh, that enjoys the gifts of creation while guarding against the temptations of affluence, that finds life enriched by a sharing of wealth and scarcity. Its precise programmatic meaning for the perilous future is yet largely unknown. Programs will require diligent empirical inquiry and accurate analysis of social processes. Such inquiry and analysis do not themselves produce ethical decisions, but they are a part of the process of ethical decision. Whether they serve the cause of oppression or liberation, of greed or generosity, of nations or wider communities depends upon their linkage with commitments and decisions.

The roots of such commitments and decisions are mysterious. Among the words that point to them are freedom, judgment, and grace.

12

The Technology of
Life and Death

WILLARD GAYLIN

We tend to become most interested in questions of values when we are unsure of the direction in which we are heading. Philosophy thrives on self-doubt, and self-doubt is a point every sensitive person will reach somewhere along that road from aspiration to fulfillment. When we are young and only aspiring we do not need doubts; we will shape our fantasies in perfect harmony. It is only after we begin to achieve our ends, when we realize that achievements and fulfillment are not necessarily congruent, that we must redefine our goals, guided by our doubts.

Perhaps I am merely saying that success leads to power, and power brings responsibility, and responsibility (to a decent human being) inevitably leads to anxiety. The urgent interest in medical ethics at this time is a symptom of the success of medicine.

It was not always this way. I look back somewhat enviously at the scientist of the late nineteenth century when it was, indeed, all promise and aspiration—when it seemed that science was on the threshold of achieving all of the solutions to man's needs.

This attitude persevered well into the present century, symbolized for me by a particular memory. When I first came to college, I was impressed by an unusual architectural detail: a small Victorian building of great charm, albeit somewhat rundown and decrepit, was almost completely surrounded by a functional, massive, ugly modern

Willard Gaylin is Clinical Professor of Psychiatry in the Columbia University College of Physicians and Surgeons and President of the Institute of Society, Ethics, and the Life Sciences.

building—red brick, as I remember it. The Victorian building, I was later to find out, was the Divinity School, and the red brick building was the Department of Biology. It was a symbol of the state of things then. Biology was about to swallow up theology.

But "swallowing up" is not necessarily destroying. The psychoanalytic concept of introjection describes how that which we devour to destroy, with poetic irony, may become a part of us. To absorb something, to take it into you, may also mean that you have unwillingly undertaken its burdens. I think that *has* happened. The scientist, particularly after Hiroshima, has often been the conscience of the nation—what little conscience there has been. It is he who has anguished over the moral implications of the technologies evolving from the scientific age.

There is certainly no lack of issues to grapple with. The rapid development of technology, permitting procedures once never contemplated, is now raising ethical problems of an ever more complex nature—and confusing some previously settled areas. If biomedical researchers are even partly correct in their predictions, the ethical issues we must soon face will require of us a considerably more subtle philosophy of life than we now command. To play the role of futurist by canvassing the more remote possibilities in biomedicine might be, in some perverse sense, more entertaining; but it is more useful to those who seek to examine the interface between science and values if I restrict my attention to technologies that are already upon us, or nearly so, and attempt to identify only the most obvious legal and ethical issues which they present and only the most rudimentary considerations of human values which they evoke. This I propose to do by addressing questions that relate broadly to two familiar categories: birth and death. In order to end with the cheerier subject, I will begin with the category of death, and begin with only one set of complications that arise from the seemingly simple task of redefining death.

I

Back in simpler days there was no medical need for a physician to consider the concept of death; the fact of death was sufficient. The difference between life and death was an infinite chasm breached in an infinitesimal moment. Life and death were ultimate, self-evident opposites.

With the advent of new techniques in medicine, those opposites have begun to converge. We are now capable of maintaining visceral functions without any semblance of the higher functions that define a person. We are, therefore, faced with the task of deciding whether that which we have kept alive is still a human being, or, to put it another way, whether that human being that we are maintaining should be considered "alive."

Until now we have avoided the problems of definition and reached the solutions in silence and secret. When the life sustained was unrewarding—by the standards of the physician in charge—it was discontinued. Over the years, physicians have practiced euthanasia on an ad hoc, casual, and perhaps irresponsible basis. They have withheld antibiotics or other simple treatments when it was felt that a life did not warrant sustaining, or pulled the plug on the respirator when they were convinced that what was being sustained no longer warranted the definition of life. Some of these acts are illegal and, if one wished to prosecute, could constitute a form of manslaughter, even though it is unlikely that any jury would convict. We prefer to handle all problems connected with death by denying their existence. But death and its dilemmas persist.

New urgencies for recognition of the problem arise from two conditions: the continuing march of technology, making the sustaining of vital processes possible for longer periods of time; and the increasing use of parts of the newly dead to sustain life for the truly living. The problem is assumed by many to be on its way to being resolved by what must have seemed a relatively simple and ingenious method— the issues of euthanasia would be evaded by redefining death.

In an earlier time, death was defined as the cessation of breathing. Any movie buff recalls at least one scene in which a mirror is held to the mouth of a dying man. The lack of fogging indicated that indeed he was dead. The spirit of man resided in his *spiritus* (breath). With increased knowledge of human physiology and the potential for reviving a nonbreathing man, the circulation, the pulsating heart, became the focus of the definition of life. This is the tradition with which most of us have been raised.

There is of course a relationship between circulation and respiration, and the linkage, not irrelevantly, is the brain. All body parts require the nourishment, including oxygen, carried by the circulating blood. Lack of blood supply leads to the death of an organ; the higher functions of the brain are particularly vulnerable. But if there is no respiration, there is no adequate exchange of oxygen, and this essential ingredient of the blood is no longer available for distribution. If a part of the heart loses its vascular supply, we may lose that part and still survive. If a part of the brain is deprived of oxygen, we may, depending on its location, lose it and survive. But here we pay a special price, for the functions lost are those we identify with the self, the soul, or humanness, i.e., memory, knowledge, feeling, thinking, perceiving, sensing, knowing, learning, and loving.

Most people are prepared to say that when all of the brain is destroyed the "person" no longer exists; with all due respect for the complexities of the mind/brain debate, the "person" (and personhood) is generally associated with the functioning part of the head—the brain. The higher functions of the brain that have been described are placed, for the most part, in the cortex. The brain stem (in many ways more closely allied to the spinal cord) controls primarily visceral functions. When the total brain is damaged, death in all forms will ensue because the lower brain centers that control the circulation and respiration are destroyed. With the development of modern respirators, however, it is possible to artificially maintain respiration and with it, often, the circulation with which it is linked. It is this situation that has allowed for the redefinition of death—a redefinition that is being precipitously embraced by both scientific and theological groups.

The movement toward redefining death received considerable impetus with the publication of a report sponsored by the Ad Hoc Committee of the Harvard Medical School in 1968. The committee offered an alternative definition of death based on the functioning of the brain. Its criteria stated that if an individual is unreceptive and unresponsive, i.e., in a state of irreversible coma; if he has no movements or breathing when the medical respirator is turned off; if he demonstrates no reflexes; and if he has a flat electroencephalogram for at least twenty-four hours, indicating no electrical brain activity (assuming that he has not been subjected to hypothermia or central nervous system depressants), he may then be declared dead.

What was originally offered as an optional definition of death is, however, progressively becoming the definition of death. In most states there is no specific legislation defining death; the ultimate responsibility here is assumed to reside in the general medical community. Recently, however, there has been a series of legal cases which seem to be establishing brain death as a judicial standard, and Kansas and Maryland have legislated approval for a brain definition of death.

This new definition, independent of the desire for transplant, now permits the physician to "pull the plug" without even committing an act of passive euthanasia. The patient will first be defined as dead; pulling the plug will merely be the harmless act of halting useless treatment on a cadaver. But while the new definition of death avoids one complex problem, euthanasia, it may create others equally difficult which have never been fully defined or visualized. For if it grants the right to pull the plug, it also implicitly grants the privilege *not* to pull the plug, and the potential and meaning of this has not at all been adequately examined.

These cadavers would have the legal status of the dead with none of the qualities one now associates with death. They would be warm, respiring, pulsating, evacuating, and excreting bodies requiring nursing, dietary, and general grooming attention—*and could probably be maintained so for a period of years.* If we chose to, we could, with the technology already at hand, legally avail ourselves of these new cadavers to serve science and mankind in dramatically useful ways. The autopsy, that most respectable of medical traditions, that last gift of the dying person to the living future, could be extended in principle beyond our current recognition. To save lives and relieve suffering—traditional motives for violating tradition—we could develop hospitals (an inappropriate word because it suggests the presence of living human beings), banks, or farms of cadavers which require feeding and maintenance, in order to be harvested.

To the uninitiated the "new cadavers" in their rows of respirators would seem indistinguishable from comatose patients now residing in wards of chronic neurological hospitals.

In the ensuing discussion, the word *cadaver* will retain its usual meaning, as opposed to the new cadaver, which will be referred to as a *neomort.* The "ward" or "hospital" in which it is maintained will be called a *bioemporium* (purists may prefer *bioemporion*).

Whatever is possible with the old embalmed cadaver is extended to an incredible degree with the neomort. What follows, therefore, is not a definitive list but merely the briefest of suggestions as to the spectrum of possibilities.

Training: Uneasy medical students could practice routine physical examinations—auscultation, percussion of the chest, examination of the retina, rectal and vaginal examinations, et cetera—indeed, everything except neurological examinations, since the neomort by definition has no functioning central nervous system.

Interns also could practice standard and more difficult diagnostic procedures, from spinal taps to pneumoencephalography and the making of arteriograms, and residents could practice almost all of their surgical skills—in other words, most of the procedures that are now normally taught with the indigent in wards of major city hospitals could be taught with neomorts. Further, students could practice more exotic procedures often not available in a typical residency—eye operations, skin grafts, plastic facial surgery, amputation of useless limbs, coronary surgery, etc.; they could also practice the actual removal of organs, whether they be kidneys, testicles, or what have you, for delivery to the transplant teams.

Testing: The neomort could be used for much of the testing of drugs and surgical procedures that we now normally perform on prisoners, mentally retarded children, and volunteers. The efficacy of a drug as well as its toxicity could be determined beyond limits we might not have dared approach when we were concerned about permanent damage to the testing vehicle, a living person. For example, operations for increased vascularization of the heart could be tested to determine whether they truly do reduce the incidence of future heart attacks before we perform them on patients. Experimental procedures that proved useless or harmful could be avoided; those that succeed could be available years before they might otherwise have been. Similarly, we could avoid the massive delays that keep some drugs from the marketplace while the dying clamor for them.

Neomorts would give us access to other forms of testing that are inconceivable with the living human being. We might test diagnostic instruments such as sophisticated electrocardiography by selectively damaging various parts of the heart to see how or whether the instrument could detect the damage.

Experimentation: Every new medical procedure demands a leap of faith. It is often referred to as an "act of courage," which seems to me an inappropriate terminology now that organized medicine rarely uses itself as the experimental body. Whenever a surgeon attempts a procedure for the first time, he is at best generalizing from experimentation with lower animals. Now we can protect the patient from too large a leap by using the neomort as an experimental bridge.

Obvious forms of experimentation would be cures for illnesses which would first be induced in the neomort. We could test antidotes by injecting poison, induce cancer or virus infections to validate and compare developing therapies.

Because they have an active hematopoietic system, neomorts would be particularly valuable for studying diseases of the blood. Many of the examples that I draw from that field were offered to me by Dr. John F. Bertles, a hematologist at St. Luke's Hospital Center in New York. One which interests him is the utilization of marrow transplants. Few human-to-human marrow transplants have been successful, since the kind of immunosuppression techniques that require research could most safely be performed on neomorts. Even such research as the recent experimentation at Willowbrook—where mentally retarded children were infected with hepatitis virus (which was not yet culturable outside of the human body) in an attempt to find a cure for this pernicious disease—could be done without risking the health of the subjects.

Banking: While certain essential blood antigens are readily storable (e.g., red cells can now be preserved in a frozen state), others are not, and there is increasing need for potential means of storage. Research on storage of platelets to be used in transfusion requires human recipients, and the data are only slowly and tediously gathered at great expense. Use of neomorts would permit intensive testing of platelet survival and probably would lead to a rapid development of a better storage technique. The same would be true for white cells.

As has been suggested, there is great wastage in the present system of using kidney donors from cadavers. Major organs are difficult to store. A population of neomorts maintained with body parts computerized and catalogued for compatibility would yield a much more efficient system. Just as we now have blood banks, we could have banks for all the major organs that may someday be transplantable—lungs, kidneys, hearts, ovaries. Beyond the obvious storage uses of the neomort, there are others not previously thought of because there was no adequate storage facility. Dr. Marc Lappe of the Hastings Center has suggested that a neomort whose own immunity system had first been severely repressed might be an ideal "culture" for growing and storing our lymphoid components. When we are threatened by malignancy or viral disease, we can go to the "bank" and withdraw our stored white cells to help defend us.

Harvesting: Obviously, a sizable population of neomorts will provide a steady supply of blood, since they can be drained periodically. When we consider the cost-benefit analysis of this system, we would have to evaluate it in the same way as the lumber industry evaluates sawdust—a product which in itself is not commercially feasible but which supplies a profitable dividend as a waste from a more useful harvest.

The blood would be a simultaneous source of platelets, leukocytes, and red cells. By attaching a neomort to an IBM cell separator, we would isolate cell types at relatively low cost. The neomort could also be tested for the presence of hepatitis in a way that would be impossible with commercial donors. Hepatitis as a transfusion scourge would be virtually eliminated.

Beyond the blood are rarer harvests. Neomorts offer a great potential source of bone marrow for transplant procedures, and I am assured that a bioemporium of modest size could be assembled to fit most transplantation antigen requirements. And skin would, of course, be harvested—similarly bone, corneas, cartilage, and so on.

Manufacturing: In addition to supplying components of the human body, some of which will be continually regenerated, the neomort can also serve as a manufacturing unit. Hormones are one obvious product, but there are others. By the injection of toxins, we have a source of antitoxin that does not have the complication of coming from another animal form. Antibodies for most of the major diseases can be manufactured merely by injecting the neomort with the viral or bacterial offenders.

Perhaps the most encouraging extension of the manufacturing process emerges from the new cancer research, in which immunology is coming to the fore. With certain blood cancers, great hope attaches to the use of antibodies. To take just one example, it is conceivable that leukemia could be generated in individual neomorts—not just to provide for *in vivo* (so to speak) testing of antileukemic modes of therapy but also to generate antibody immunity responses which could then be used in the living.

If seen only as the harvesting of products, the entire feasibility of such research would depend on intelligent cost-benefit analysis. Although certain products would not warrant the expense of maintaining a community of neomorts, the enormous expense of other products, such as red cells with unusual antigens, would certainly warrant it. Then, of course, the equation is shifted. As soon as one economically sound reason is found for the maintenance of the community, all of the other ingredients become gratuitous by-products, a familiar problem in manufacturing. There is no current research to indicate the maintenance cost of a bioemporium or even the potential duration of an average neomort. Since we do not at this point encourage sustaining life in the brain-dead, we do not know the limits to which it could be extended. This is the kind of technology, however, in which we have previously been quite successful.

Meanwhile, a further refinement of death might be proposed. At present we use total brain function to define brain death. The source of electroencephalogram activity is not known and cannot be used to distinguish between the activity of higher and lower brain centers. If, however, we are prepared to separate the concept of "aliveness" from "personhood" in the adult, as we have in the fetus, a good argument can be made that death should be defined not as cessation of total brain function but merely as cessation of cortical function. New tests may soon determine when cortical function is dead. With this proposed extension, one could then maintain neomorts without even the complication and expense of respirators. The entire population of decorticates residing in chronic hospitals and now classified among the incurably ill could be redefined as dead.

But even if we maintain the more rigid limitations of total brain death it would seem that a reasonable population could be maintained if the purposes warranted it. It is difficult to assess how many new neomorts would be available each year to satisfy the demand. There are roughly two million deaths a year in the United States. The most likely sources of intact bodies with destroyed brains would be accidents (about 113,000 per year), suicides (around 24,000 per year), homicides (18,000), and cerebrovascular accidents (some 210,000 per year). Obviously, in each of these categories a great many of the individuals would be useless—their bodies either shattered or scattered beyond value or repair.

And yet, after all the benefits are outlined with the lifesaving potential clear, the humanitarian purposes obvious, the technology ready, the motives pure, and the material costs justified—how are we to reconcile our emotions? Where in this debit-credit ledger of limbs and livers and kidneys and costs are we to weigh and enter the repugnance generated by the entire philanthropic endeavor?

Cost-benefit analysis is always least satisfactory when the costs must be measured in one realm and the benefits in another. The analysis is particularly skewed when the benefits are specific, material, apparent, and immediate, and the price to be paid is general, spiritual, abstract, and of the future. It is that which induces people to abandon freedom for security, pride for comfort, dignity for dollars.

William May, in a perceptive article, has defended the careful distinctions that have traditionally been drawn between the newly dead and the long dead.[1] "While the body retains its recognizable form, even in death, it commands a certain respect. No longer a human presence, it still reminds us of that presence which once was utterly inseparable from it." But those distinctions become obscured when, years later, a neomort will retain the appearance of the newly dead, indeed, more the appearance of that which was formerly described as living.

Philosophers tend to be particularly sensitive to the abstract needs of civilized man; it is they who have often been the guardians of values whose abandonment produces pains that are real, if not always quantifiable. Hans Jonas, in his *Philosophical Essays,* anticipated some of the possibilities outlined here, and defended what he felt to be the sanctity of the human body and the unknowability of the borderline between life and death when he insisted that "nothing less than the maximum definition of death will do—brain death plus heart death plus any other indication that may be pertinent—before final violence is allowed to be done." And even then Jonas was only contemplating temporary maintenance of life for the collection of organs.

The argument can be made on both sides. The unquestionable benefits to be gained are the promise of cures for leukemia and other diseases, the reduction of suffering, and the maintenance of life. The proponents of this view will be mobilized with a force that may seem irresistible. They will interpret our revulsion at the thought of a bioemporium as a bias of our education and experience, just as earlier societies were probably revolted by the startling notion of abdominal surgery, which we now take for granted. The proponents will argue that the revulsion, not the technology, is inappropriate.

Still there will be those, like May, who will defend that revulsion as a quintessentially human factor whose removal would diminish us all, and extract a price we cannot anticipate in ways yet unknown and times not yet determined. May feels that there is "a tinge of the inhuman in the humanitarianism of those who believe that the perception of social need easily overrides all other considerations and reduces the acts of implementation to the everyday, routine, and casual."

This is the kind of weighing of values for which the computer offers little help. Is the revulsion to the new technology simply the fear and horror of the ignorant in the face of the new, or is it one of those components of humanness that barely sustain us at the limited level of civility and decency that now exists, and whose removal is one more step in erasing the distinction between man and the lesser creatures—beyond that, the distinction between man and matter?

II

If we turn from the morbid subject of death to the more delightful subject of birth, we confront a range of prospects that are no less challenging, and a set of issues that are no less sticky. It has been suggested that the discussion of this subject can be organized into three categories: (1) helping people have a baby; (2) helping people have a normal baby; and (3) helping people have an optimal (ideal or special) baby. I find the categories useful, not simply because they form a rather convenient rhetorical triad, but because they allow us to see how the technologies possible under one rubric lead rather naturally to those of the next; how each category shades into the one that follows; and how from humanitarian ideals of the purest and most "natural" kind we come eventually to face the specter of infanticide and social engineering.

But all of this starts with the simple, humanitarian, and generous desire to help parents fulfill the normal longing to have a child.

The simplest interventions here are those procedures which facilitate ovulation in a woman—drugs, hormones, endoscopy, and the like. Further up the scale is the now accepted medical convention of artificial insemination with donor sperm when the prospective father is proved to be defective in the manufacture of viable sperm. It is one of the ironies in the field of bioethics that some of the most profound problems pass unnoticed because they use familiar language and familar techniques. Artificial insemination requires no advanced technology and little laboratory equipment; artificial insemination demands of the donor only that he do that which at an earlier age at least came quite naturally. As a result, the implications of the procedure have never fully been examined, while it has become an accepted institution of medicine.

Some very simple questions arise. Wouldn't a sperm bank with massive selection and computerized data represent a superior method of donor selection to the current technique of collaring a convenient medical student or intern who needs the money and finds the donation of sperm a more amiable procedure than the donation of blood? What then of criteria and controls? Should there be a free choice in the nature of the donor, and who should make the choice? In the past we have implicitly assumed our task to be an imitation of the natural process, and we have attempted to replicate as closely as possible the natural inheritance. The obstetrician does little scientific screening. The traditional donor is, therefore, a poorish (needs the money) medical student (available), healthy and intelligent enough to have survived to if not through graduate education, who looks something like the father who is to raise the child (satisfy paternal wants, maintain deception, quiet gossipy speculation).

Appalling as it may seem, this represents the traditional and typical

at the current time. One would think that common sense might indicate simple correctives. Why a student? A sixty-year-old grandfa-
ther who has, to use the veterinarian term, been at stud and proved a
successful sire would seem more logical. Certain late-developing diseases with hereditary tendencies could be eliminated, e.g., diabetes,
arteriosclerotic conditions, and the like.

This then would introduce some aspects of positive engineering: we
are moving from the problem of "having a child" to that of "having a
special child." Should such genetic variables as sex, size, I.Q., or race
be options? Or should they be left to chance? Student legislators at
Columbia University Law School, grappling with this problem, mandated matching two variables to the natural parental traits, I.Q. and
size—I.Q. under the assumption that a child with a high I.Q. might be
disadvantaged with parents of an 80 I.Q., and size presumably to
protect a small mother from a giant fetus. The question of race selection, beyond psychological and sociological considerations, raises
problems of constitutional law.

Similar possibilities, and questions, arise when the biological difficulty is not with the potential father, but with the potential mother.
Certain women fail to ovulate. It is now possible, however, to obtain
ripened ova directly from the ovaries of a woman, via the use of
microtechnologies. We know that once an ova is fertilized, whether in
a test tube or naturally, it can easily be implanted into the uterine
wall, where then it will continue to divide and progress as naturally as
if it had made its own way to that site. The crucial technology still
beyond us is the capacity to combine the egg and the sperm in the
test tube. We have done this with increasing success in lower forms,
but for some as yet unknown reasons the capacity to bring the
human sperm and ova together has eluded us. It may well be by the
time this manuscript is in publication, it will have been successfully
done—so close we are to this solution. Once we are capable of *in
vitro* fertilization, what would seem to be the more complicated process of encouraging division in a test tube moves spontaneously and
readily forward. And the artificial implantation procedures (from all
experimental data) produce as high, if not higher, incidence of normal children than the natural. Once the problems of *in vitro* fertilization are resolved, the use of borrowed or bought ova would seem to
reside in the same moral category as the use of borrowed or bought

sperm. In either case, one has the advantages and privileges of carrying one's own child, experiencing the birth process, and having control of the environment of the developing fetus and at least half of its genetic endowment.

Other possibilities occur when the woman ovulates, but due to blocked tubes, the ova has no way to make its way down into the uterus, nor does the sperm have the capacity to penetrate through to the ova. In this instance the woman's ripened ova can be removed surgically and fertilized *in vitro* with her husband's sperm, then—after allowed to develop to the four-to-eight-cell stage in the test tube—be implanted in her uterus, where it will grow naturally to term. The resulting child will be fully the couple's "own."

The frequency of this condition should not be underestimated. The chief cause of blocked tubes has traditionally been pelvic inflamatory disease (PID), the majority of cases being due to gonorrhea. This was a common problem in clinics during the period of my training, and seemed well on its way to resolution with the discovery of antibiotics. PID was a common diagnosis among the poor and uneducated. We have been most successful in limiting its spread, but now, ironically, we find a resurgence of PID, indeed of venereal disease in general, with a sociological shift toward the middle and upper classes, and better educated. It is definitely attributable to the new sexual freedom and new life-styles. Perhaps the desire for the "natural life" encompasses such of God's creatures as the gonococcus and the spirochete, in opposition to the nasty intrusions of modern technology.

Still further possibilities arise when a woman ovulates but has no uterus, either because of congenital defect or because a hysterectomy was necessary. Here we have a different set of problems. A mother and father can now have their genetically natural child; the mother's ova can be combined with the father's sperm in a test tube, and then be implanted in the uterus of another woman (or in an artificial placenta, once technology develops it). Whether this is indeed precisely their "natural" child is an intriguing question.

At a seminar in bioethics at Columbia University Law School, a contract was written for the hiring of a surrogate mother, i.e., planting a fertilized egg of one couple in the uterus of another woman to bear for nine months. The Surrogate Mother Act is in the service of an honorable old medical tradition, facilitating the desire of parents to have their own "natural," i.e., genetic child. But when the details are spelled out in the language of contracts the results are unnerving. The surrogate mother should have "rights of first refusal" in case the natural parents decide that they wish to abort. The surrogate mother must fulfill obligatory dietary commitments and interdictions during the gestation period. It is difficult for this to avoid sounding funny, yet renting womb space can be conceived as a respectable alternative, probably preferable to an artificial placenta, for solving that not uncommon problem of a woman who had her uterus removed, has perfectly functioning ovaries, produces ova, is married, and wants a "natural" child.

It should be obvious that once we have introduced these procedures—once they have become institutionalized and part of accepted technology—it will not be necessary, nor indeed possible, to limit their use to those conditions demanded by physiology alone. Psychological and sociological justifications will soon press their claims. A woman may be terrified of pregnancy and/or delivery and for psychological reasons choose to rent womb space. Or for reasons of career (as a movie star, executive, primary breadwinner of the family, and the like) a woman may choose not to bear her own child. Or her reasons may be purely those of vanity.

The permutations of the simple technologies for permitting one to have a baby raise incredible psychological, sociological, ethical, moral, and policy dilemmas. When we move on to the next two categories, they are further compounded.

The second category is facilitating the having of a *healthy* baby. The key procedure here, among technologies already developed, is intrauterine diagnosis by amniocentesis, which permits the diagnosis of certain congenital defects which can then be "prevented" by aborting the defective or unwanted child, thus permitting the parents an opportunity for another try at a healthy or normal child. Amniocentesis is a relatively simple and safe procedure. It involves the extraction, by hypodermic needle, of fluid from the amniotic sack which encases the developing fetus. This fluid contains cells shed from the body of the fetus. By microscopic examination of these cells, numbers of congenital abnormalities can be discovered. Interest in recent years has centered on such diseases as Tay-Sachs and Down's syndrome (or mongoloidism). But the number of diseases capable of being diagnosed has gone in the last few years from a mere dozen or two to literally hundreds.

It is not possible to even list the troubling legal and moral issues that evolve out of this procedure. What constitutes a "pathological condition" worthy of abortion? Who shall decide? Shall genetic screening be mandated? What happens when parents differ? What is the role of the genetic counselor?

Genetic counselors are now adopting the posture that psychiatrists once assumed. They presume to present the facts, eschewing value judgment. Genetic counselors, at this point in time, are geneticists first and counselors second, if at all. The sophisticated psychiatric counselor has abandoned the delusion of objectivity. He is aware that all counseling involves the conscious and unconscious intrusions of the values of the adviser. Since I have yet to meet a value-free psychiatrist, I am skeptical of the genetic counselor's ability to "only present the facts."

One of the offshoots of a typical screening pattern for congenital diseases in an amniocentesis is the revelation of the sex of the baby. Once this information is transmitted to the mother, she may exercise her right to abort because she is unhappy about the prospect of a son or daughter, whichever the case may be. This leads us directly into the third category: helping people have an ideal or special kind of child. Certainly it will be possible in the future, by more direct methods than amniocentesis, to select the sex of one's child. The implications of that alone are mind-boggling. It will also be possible to determine other genetic traits, partly by screening and selective abortion, but also by more advanced techniques.

Even selective screening raises monumental moral and political questions: the right of privacy; the relationship of the unborn to its parents; the implications of social attitudes about defective children for all living individuals with defects; the potential trivialization of the reasons for eliminating the defective; and on and on. These are the issues which will form a political battle for the future, shape the moral character of our state, and determine our value systems.

Beyond screening, however, is the current potential for direct intervention into the genes themselves. Recent work in molecular biology has demonstrated the feasibility, by the use of viruses and other transfer forms, of actually introducing genetic material from one organism into the protoplasm of another. By virtue of such technology we can combine genetic material from different species; we can, in other words, create new life forms. Such research, designated recombinant DNA, holds incredible promise for cure of disease and elimination of genetic defect—and, needless to say, incalculable risk. Great promise is held for such things as a cure for cancer or the introduction of an insulin-making capacity in diabetic individuals. The potential of the penetrable bacteria or viruses to escape and influence our genetic nature is a frightening and horrifying one, and has led to great debate in the scientific community as to how such research in molecular biology ought to be controlled.

We now have also the newly discovered ability to synthesize a gene! Who is to assay the risks and benefits of such research? Who will determine when high-risk/high-gain research is warranted and when not? What will be the criteria by which we will make our decision?

We are always happier in our justification of changing things when we can utilize a medical model—a therapeutic model, that is. We are happier when we are "normalizing" than when we are "idealizing." But it can be seen from some of the examples given that the borders between the two are inevitably fuzzy and indeterminant.

III

It is plain to see how easy it is to travel unwittingly from the safe role of comforter to the sick to the uneasy position of political advocate and social engineer. Yet, here we are. The issues are fused and cannot be avoided. Nor should they be. Even if the power of decision is shared by the physician, the public at large will always be an inevitable and important coparticipant. It is an authentic part of our role as citizens and we must be trained for it.

The physician has always been a value-maker, ethicist, social force, and political influence in the lives of his patients and beyond. He is whether he wishes to be or not. There is no such thing as a value-free medicine, nor has there ever been. To accept this as a fact of professional life is not arrogance but honesty—and the first step toward that redefinition of the physician's role that is necessitated by the coming of age of medicine. It is our success that demands a new sensitivity, a new humility, and a new training.

Nor must the lay person be seduced by some artificial extolling of the "natural" and fear of the "artificial." It is of our nature to change our nature. Technology is not the enemy of man, but rather the definer of man—for he is the only species capable of scientific pursuit. It is our glory, not our nemesis. It is in the uses of our technology that we will confront the questions that each generation has examined to determine the moral and political climate in which they choose to live. Anti-technology is anti-man, and self-hatred is a dangerous thing in any time. But particularly so in this technological age that has brought us together as never before—to share a common future or the lack of it. For such reasons science and technology are in the vanguard, demanding a rebirth of interest in the humanities and philosophy. We must ask the same questions that have been asked over the ages, and recast the answers in the metaphors of our day.

13

Biomedical Progress and the Limits of Human Health

DANIEL CALLAHAN

In thinking about the ideal of human progress and biomedical procedures, it is tempting to concentrate on the most dramatic possibilities:
genetic engineering and positive eugenics, *in vitro* fertilization and ex-
tracorporeal gestation, sex selection, and the like. All these possibilities have been celebrated as potential occasions for human progress,
primarily on the grounds that future people will be able to transcend
the mischief and vagaries of nature by devising a human being who
is brighter, more creative, more benign, and better adapted to modern life. An attractive vision . . . Yet even those with the weakest
degree of pessimism about human affairs can discern a few problems:
What assurance would we have that the humans thus produced
would actually be "better"? What does "better" mean? And who will
decide what it means? In short, whether these radical biomedical developments would represent "progress" is highly problematic. Worse
still, because of the inevitable time lags, it will not be our generation
which will have to live with any of the consequences. Our whole generation may be willing to bet that indeed it will represent progress,
but it will be our children or their children who will either collect or
have to pay up on the bet. We can guess whether it will represent
progress, but they will be the only ones who will *know*. Looked at in
that way, there is all the more reason to be wary of labeling something like genetic engineering as "progress."

Daniel Callahan is Director of the Institute of Society, Ethics, and the Life Sciences.

More pertinent at the present historical moment are those far more subtle, less dramatic developments which are already upon us: safe and simple abortion-on-request, prenatal diagnosis (via amniocentesis) of genetic defects, new surgical techniques for treating deformed newborns, increasingly sophisticated methods for sustaining the life of the aged and injured, the use of a wide range of drugs for increasingly focal effects on emotions and psychological predilections. The list could easily be extended. The importance of these developments, from a cultural and ethical point of view, lies less in their actual existence—they will help some people, harm others and create terrible dilemmas for still others—than in their cumulative and total effect upon the way people begin thinking about themselves, their children and other human beings.

I

If one can speak of a historical dialectic between technology and culture, it would probably amount to this: the advance of technology requires a set of affirmative cultural attitudes in the first place; and, in the second place, the technology once achieved brings some changes not only in the details of those sustaining attitudes but also in a host of other cultural attitudes and values, usually going well beyond anything envisioned in the first impulse to develop a particular set of technologies. The history of the automobile would sustain that description of the dialectic.

Our society has now become conscious of this dialectic at work in biomedicine, but I believe it has some special features which will sharply distinguish it from that brought about by the development, say, of machinery and industrialization in general. The latter have had the primary effect of changing the external conditions of life, the environment. While it would be foolish to underestimate the impact of that kind of change on human behavior and self-conception, biomedical change can have a very different kind of impact. Biomedical technology intervenes directly into the internal world, first into our bodies and then into our minds and emotions (with of course considerable overlap). The change is direct rather than indirect, unmediated rather than mediated. Whether the change is as simple as reducing infant mortality rates or as complex as psychosurgery, the intervention is straightforward and targeted, and the target is the human body and/or self in one or all of its features. Hence, one can understand both the sense of new power and, here and there at least, the kind of dejection or nervousness which can accompany it. For the dialectic in this case begins in the usual way, attempting to make use of scientific knowledge and technological skill to achieve some desired end; but then, as that end is gradually achieved, it comes to be seen that the changes in attitudes and values it brings with it are of a kind entirely different from what society has become accustomed to when the changes are environmentally introduced. One is working directly with human consciousness from the inside, whether the target happens to be the mind or the body, and that is a very formidable way of dealing with some ultimate values at their source, the mind and its bodily substance itself. One tinkers with a machine or tries to develop a better model in order to effect a different environmental outcome (knowing that this outcome may influence society and thus eventually bring about change and thereby affect the human mind). But in this case, the process is exactly the opposite: we begin with the mind and body and work out from there, changing very drastically the feedback circuitry between intervention and values.

Let me give a simple illustration. Spina bifida is a congenital disorder in which a child is born with an open spinal cord. Ten years ago the fatality rate was approximately 60 percent and even those who survived were ordinarily permanently and severely crippled and retarded. Here, then, was a perfect target for technological progress, in this instance an advancement in surgical technique; obviously, it was thought, if such a condition could be treated many children would be saved, many parents relieved, and the human condition bettered. Who could argue with that or fail to call it progress? And it happened. It is now possible to save the lives of 90 percent of those children born with spina bifida. Unfortunately, it has not been possible—nor may it ever be—to save those children from a crippled and mentally retarded life. Would they be better off dead? Do we really want to count that technological development as progress?

Those are the obvious questions at one level, and they merely symbolize one by now old and familiar story with technology, that of unintended and undesired consequences. But far more important is the way that intervention into the human body, together with the kind of troubled reflections it has engendered, has set in train (though not by itself, or course) a very different kind of logic of values. Put simply, the process of trying to deal with the dilemma of a procedure which saves lives only to create a new misery (a very crippled life) has pressed the value discussion back to a still more fundamental question: what is a life worth living and a life worth saving? The apparently emerging answer (judging only from scattered clues) moves in the direction of a narrower rather than broader view of a normative "quality of life." Fewer lives are now thought worth saving than was the case before attempts were made to devise surgical cures or ameliorations. Even those who under poorer surgical conditions would have had every effort expended to save them, however terrible their condition and however slight the relief offered, will now be allowed to die. This is increasingly true with spina bifida and even more strikingly with Down's syndrome (mongoloidism). Where it was once the case that routine simple surgery would have been customarily performed to save the life of a mongoloid, the trend now is not to perform the surgery even though there are more routine techniques available than before.

Paradoxically, then, the attempt to achieve what can be called "simple" progress—the treatment of a condition universally accepted as bad—led, in the end, to a very fundamental change of values toward defective newborns with the final result likely to be far fewer defective children being saved now than was the case before the advent of a medical "progress" meant to save more lives.

II

Now it is unlikely that this kind of change in values has come about in total isolation of other changes taking place in medicine and, no less importantly, in public attitudes toward health. In the case of spina bifida and mongoloidism it is not just physicians who are changing their thinking about saving the severely defective child, it is, even more, the parents of the children. Given a choice, they increasingly elect not to have lifesaving surgery performed, a striking and very rapid shift in attitude. It is a very delicate matter to interpret the origin and meaning of the shift, which at its extreme includes a sharp rise in the outright abandonment of defective children by their mothers in hospital nurseries, previously an exceedingly rare event.

At the risk of being wildly wrong, I believe the underlying cultural phenomenon here can be traced to some important changes in the concept of "health," in notions of what constitutes a "right to health," and, more generally, to the increased demands upon medicine and biology to solve human problems. The definition of *health* adopted by the World Health Organization in 1946 provides an insight into the historical origin of the changing concept of health. "Health," the WHO said, "is a state of complete physical, mental, and social well-being and not merely the absence of disease or infirmity." It is a definition worth meditating upon, for not only did it mean to encompass the burgeoning mental health movement, but it also meant—quite literally it seems from the record of early WHO discussion—to encompass everything that could be included under the heading of human welfare and happiness. It was an ambitious definition, giving medicine an enormous mandate and theoretically excluding no human problem from treatment by medical means or from interpretation in medical or biological categories. That much of the popular moral language of our culture is now cast in terms of deviants—criminals and radical politicians being "sick" rather than just wrong, stupid, or immoral—is a reflection of the power which animated the WHO definition of *health.*

Moreover, it is patently the case that the combination of an all-encompassing concept of health together with biomedical advances has created a situation in which nearly anything that anyone wants from medicine can find legitimation as a health need. The abortion reform movement, culminating in the Supreme Court decision to allow abortion-on-request for the first two trimesters of pregnancy, drew heavily on health and well-being arguments, and the wording of the Court's decision made much of putting the final choice jointly in the hands of women and their physicians, as if it were simply a medical matter. The emergence of genetic counseling as a major medical profession has drawn much of its force not only by concentrating on the genetic health of fetuses and neonates but also and simultaneously by concentrating on the mental health of parents faced with the possibility of producing a defective child. The word *suffering* is a common one in the literature of genetic counseling, but it is rarely clear whether the suffering referred to is that of the child, the parents and family, or the society as a whole; they are usually mixed together in ways which make them almost indistinguishable. *Health,* now seen as the alleviation of all suffering, provides a very handy basket into which can be thrown both physiological considerations and emotional responses. A similar analysis could be developed from an examination of much of the literature, now mountainous, about the dying patient.

To my mind, the most disturbing aspect of the way in which the concept of health is made amenable to any and all human demands is less the ultimate meaninglessness of the term when so employed but, instead, the way it so well serves a medicine and society which are highly individualistic in orientation. One major feature of individualism in our society is that it knows no limits: anything may be aspired to, anything may be hoped for, anything may be sought. A very loose concept of health, then, allows people to trade upon the fact that with minor ingenuity any desire can be seen as a health need, and any health need can be legitimated in the name of increased personal freedom. The political system cannot make all of the people happy all of the time, but it is just possible that physician-prescribed psychotropic drugs can. If all of us have a right to happiness, with an equal right to define happiness in our own private terms, why should we not call upon medicine to give us what other institutions cannot?

III

The central issue, I believe, is what our society will finally make of the "right to health." In that respect, I will present a threefold argument: (1) that the concept of a right to health can never be given a full and significant meaning unless, at the same time, we are clear about what the limits to that right are; (2) that the reason we have not managed politically to implement the notion of a right to health in our society in any sensible way is, paradoxically, because *we already* grant the right to health—but in a way which, because that right knows no limits, guarantees a maldistribution of health care; and (3) that for both practical and theoretical reasons, but mainly and crucially the latter, the right to health care is, intrinsically, a limited right.

There are some general premises upon which I will base these arguments:

1. The problem of the right to health care, and especially the limits of that right, currently reflects the more general tension between the tyranny of individualism and the tyranny of survival. Unless that more general problem is solved, it is hardly likely we can solve the right-to-health problem.

The unhealthy tension between the twin tyrannies can be thus described. Survival is a basic drive. In affluent technological societies, acceptable survival (ego- or self-survival) levels are raised very high, particularly under the influence of individualism, which leads people to place high, constantly escalating demands upon life and happiness. One outcome of this process is that such people can be as tyrannized by a perceived threat to survival as those whose survival-demand level is much lower. That the demands of the former are much higher than those of the latter is less a function of real need than it is of the high demand-level to which their individualism (and that of the culture) has led them. The result is a genuinely vicious circle: their individualism pushes them to a higher acceptable survival level, but because that level is higher, they must make more individual demands in order to achieve a sense of safety or security. The only real security lies in having a lower demand-level for survival, creating a situation where needs can effectively be met. But the individualism of the culture will not allow that; hence, people never feel secure, no matter how many of their wants are satisfied.

2. The language of "rights" ultimately becomes meaningless when that which is demanded in the name of rights is impossible of achievement in either a real or a plausible world.

In ordinary usage, the right to health care has come to mean, in its more modest sense, the right of individuals to equal access to available health care regardless of their ability to pay for that care. The pressure to establish the right to health care developed out of a perception that with laissez faire medicine, the rich could get what the poor could not. The principal value appealed to in that discussion is distributive justice, with an emphasis on the contention that, however medical resources are to be distributed, ability-to-pay provides a wrong, because inequitable, basis for that distribution.

One premise of that line of argument is that health care is, in many respects at least, a scarce resource. One cannot assume (and this the facts seem to support) that health care exists in such abundance that no problem of distribution exists; thus it is necessary to deal with the problem of equal access, and to deal with that means to invoke the value of distributive justice together with all the political and legal ramifications of that value. Even in its modest sense, then, the right to health has raised some difficult problems, and they become all the more difficult when one moves into the question of distributing such things as totally implantable artificial hearts, expensive treatment for hemophilia, and kidney dialysis machines.

However, that is not the end of the "right to health care" issue. As the phrase became more popular, a more extravagant sense began to appear. Not only ought there to be equal accessibility to *available* health resources, but it was asserted that it is also perfectly right and proper for individuals to demand, as a matter of right, that medicine *develop* such new resources as would be required to help them achieve whatever they may desire in the name of health (which in the end becomes indistinguishable from *whatever* they may desire). Thus a claim has been entered that infertile women have a right to *in vitro* fertilization (and the right to demand its development) in order to find relief from their infertility. Even more extravagantly, some women demand, also as a matter of right, that *in vitro* fertilization be developed in order that they may exercise the option not just to have a child but to have a child in the way they choose to have one. In the first instance, the assumption is that there exists a natural right *not* to be infertile—and that society, if that right is to be respected, has the duty to devise ways to implement that right. In the second instance, there is the assumption that the rights of women include the right to have a child in whatever way they may choose to do so—and again, that society has the duty to carry out such research as would be needed to implement that right.

These are, admittedly, rather lurid examples, but they do illustrate what seems to be happening as the concept of the right to health care is pressed beyond its original boundaries. It is used to legitimate a claim that there exists a right to have *all* bodily infirmities corrected if scientifically possible or conceivable (and a correlative duty on the part of society to take all necessary steps toward that end) and a right to demand that medical technology be brought to bear (well beyond the narrow boundaries of "illness") to satisfy all desires. Think of the following trends in the logic of the "right to amniocentesis" (even if it is now a subcurrent): if pregnant women have the right to amniocentesis in order to avoid bearing a Tay-Sachs baby or a Down's, why should they not have the right to avoid bearing a child with a cleft palate? And if they should have the latter right, why should they not have the right to avoid bearing a child with a crooked left little finger? But if they should have that choice, why then should they not also have the choice of using amniocentesis in order to have sex selection (if that is what they *want*)? And if they should have the last-mentioned right, does not society have the duty to make amniocentesis available to all women in order that they may gain their rights?

[I want to bracket here another development of the extravagant sense of the right to health: that of the argument—as presented by Bentley Glass et al.—that *every* child has the right to be born with a healthy mind and body (a variant on the notion that all humans have the right to be free of infirmities); and the correlative argument that, therefore, the way to protect the rights of such a child is to abort it—a really novel notion of a way in which rights can be protected.]

I V

With all that by way of preamble, let me return to my three central arguments.

1. The concept of the right to health care can never be given a full and significant meaning unless we become clear about what the limits to that right are. This argument can be defended on a number of grounds. First, the word *health* will be meaningless if we do not have a sense of what falls outside the realm of health; if *health* is used to bear on everything to do with human well-being, then the right to health care becomes tantamount to a right to happiness—a right, in effect, to everything anyone wants in the name of that which they believe will fulfill them. That is a route which guarantees that the right to health care will end up meaning nothing at all—because it will mean anything and everything. Second, even if *health* is given a more circumscribed meaning, the requirements of a manageable circumscription must include (by definition) limitations on the meaning of *health*.

Third, the history of discourse and discussion on human rights has always shown (and required) the development of some paradigm of a meaningful limitation of right (e.g., the right to freedom of speech: one cannot falsely cry "fire" in a crowded theater; the right to freedom of religion: one cannot burn heretics and pagans at the stake; the right to self-defense: one cannot torture little children to save his own life or that of others, and so on). The reason, of course, why such paradigms have proved necessary is that there is more than one human right. Eventually, rights come in conflict and, thus, if only to avoid logical absurdity, limitations upon any given right are needed in order that they may not cancel each other out. The right to health care has yet to go through that kind of historical process; until it does, there will be no way to keep the right to health from transgressing (at least in principle) all other rights and no way to establish the relative place of the right to health care in the constellation of other accepted rights. At best, this will mean that it will remain nothing but a slogan; at worst, it could become a monster.

2. One origin of the present maldistribution of health care is the fact that far from actually being denied, the right to health has been granted in practice. But it has been so extravagantly granted and with such an unlimited scope that it has been impossible to deal with the claims of distributive justice in its distribution. To make this case, one major assumption is needed: I will define as a (de facto) right any claim which any well-placed individual can successfully make without violating an extant law or ingrained social custom. In the case of health, it has been possible for any individual with money and/or clout to demand and frequently to get whatever he has wanted: his face lifted, his infertility treated, his psyche soothed, and so on. He cannot get an arm amputated without reason, but beyond that kind of (rare) exception, there is no one who will stand in his way if he calls upon medicine to take care of what ails him (or what he thinks ails him); all he needs is money. That is why I say there exists a *recognized de facto right* to health.

Precisely because that has been the situation, medicine has been prone to develop in a direction where money and/or power have determined how available resources have been distributed. Put another way, because of the recognized de facto right to health, there has been no principle available which would enable the society to say no to the person who wants something from medicine, proposes to break no laws, and has the money to pay for it. If (as a fantasy) every male and female in the country over the age of forty decided that having his or her face lifted was the greatest medical good, and every young doctor decided to become a plastic surgeon to meet the demand for face-lifting, I don't think there is any principle of limitation now available which could deny there was an intrinsic right to face-lifting for all or which could deny that all physicians would have the right to become plastic surgeons if they chose to do so. We might say that the males and females in question were acting absurdly and that the physicians were acting irresponsibly as a group. But we could put neither group in jail—no law would be broken—and it is hard to see how we could say that any given *individual* physician was acting irresponsibly (if we don't consider it irresponsible for any given physician now to become a plastic surgeon, why would it be irresponsible for all to become plastic surgeons—particularly if they could show, as a group, that sufficient demand for their services existed?).

In short, we are faced with a classic situation. The de facto right to health has been based on the principle of freedom: anyone can demand anything in the name of health and, if he can pay for it, can have it. When freedom has that much scope, it is hardly surprising that justice can hardly get its foot in the door. Justice is possible only when freedom is limited in some way. But the right to health has (again, de facto) been premised on total freedom; hence, justice has been excluded. If there is to be a just distribution of medical resources, there must be some limitations placed upon the present de facto freedom of individuals (mainly the wealthy and powerful) to get what they want. That, in turn, means that the right to health care must eventually be seen as a limited right. Justice is incompatible with absolute individual rights, and a just medical care system is incompatible with unlimited individual health rights.

3. The right to health care will always be to some extent limited by practical considerations, even in the most affluent societies. But more importantly, there are intrinsic limits to the right. These limits stem from the fact that the human body is a finite material organism, subject to decay and eventual dissolution. Illness is one manifestation of human finiteness. The demand for an absolute right to health care is, in the end, a demand to be free of bodily finiteness. One could, I

suppose, speculate that medicine may eventually find ways to transcend that finiteness. But I don't think it will ever happen, probably because I believe that finiteness is an intrinsic part of matter and organicity; finiteness may be ameliorated, but it won't be cured. In any event, finiteness is surely a present fact, and our generation and the next (at least) will have to live with that fact. Therefore, in our present situation, the right to health care is intrinsically limited.

14

Technology and
the Structuring of Cities

DAVID P. BILLINGTON

In this essay, I shall try to sharpen the definition of technology, to develop a framework for the comparative critical analysis of major public structures, and to use that framework to evaluate a specific pair of examples in urban technology. My central concern here is that there is almost no reflective, critical analysis of the large-scale structures built in American cities during the twentieth century and hence that public policy in this area has been denied the perspectives of history, philosophy, and art that are essential to humane urban life.

I

Structures and machines are related by contrast.[1] Structures are roads, bridges, terminals, dams, harbors, waterworks, power plants, office towers, and public housing blocks whereas machines are cars, trains, trucks, turbines, ships, pumps, motors, television sets, computers, and window air-conditioners. But such contrast also implies a close interdependence, for structures are built by machines and machines have structure to hold them together. Technology, consequently, has two sides—that of structures and that of machines—each having a peculiar nature and requiring a distinct form of criticism.

A *structural view of technology* may be outlined as follows:

David P. Billington is Professor of Civil Engineering at Princeton University.

1. Structures are static. If a bridge moves visibly, something is wrong. Structures are usually large-scale, which means that failure will involve widespread danger, often to both life and property. Collapse of a bridge, a dyke, or a dam can kill hundreds of people and destroy entire cities. Structures are more permanent than machines; in particular, they must be designed to withstand environmental threats from their immediate physical location and from weather conditions over a long period of time. Designed to be static, of large scale, and permanent, structures can be said to possess ideally the attribute of durability, which means that they will stand up.

2. In addition to being large-scale, structures are custom-made. This implies high cost, which means in turn that structures are usually public works whose funding involves political action. Moreover, they are normally designed for a single site. Attempts to avoid individual designs by mass-producing bridges, for example, lead to the thoughtless uglification characteristic of much turnpike bridging. Being public, custom-made works meant for one locale, structures are characterized by their convenience, which means that they will be useful.

3. Individual structures—large-scale, custom-made works—are thus unique. Being unique does not require a novel shape for every structure; it means rather that the physical setting and social context need to be reflected in a design that respects both economy and environment. Moreover, structural possibilities change very slowly over time because they are so closely governed by the changeless forces of gravity. Although new possibilities arose in the midnineteenth century with the development of cheap steel and reinforced concrete, they were worked only slowly into engineering practice. Thus, development in structural principles involves slow-paced change rather than the so-called breakthroughs of technology. A study of an Eiffel bridge from the 1880s can raise all the technical questions essential to a study of a well-designed metal-arch bridge of the 1970s. Since such works take a long time to build and are difficult to remove, they demand long-range planning or patience in design. Finally, because of their permanence and uniqueness to a locale, structures become principal documents for historical studies. Being unique, changing slowly, and documenting the past, structures are singled out for study primarily on the basis of their beauty or their lack of it; that is, they impress themselves on a culture mainly through their appearance.

A *machine view of technology* may be similarly outlined:

1. While structures are static, machines are dynamic. If a car does not move, there is trouble. Machines are usually small-scale, which means that failure will involve private danger but normally not wide destruction—though some machines, of course, are designed to be destructive. While a structure is built to last, a machine is designed for a relatively short life-span—a fact emphasized by the tendency toward building disposable machines.

2. In addition to being small-scale, machines are generally designed to be mass-produced. This implies low cost, which means in turn that individual machines are usually private works funded by private citizens. They are made to be independent of the environment and hence are useful almost anywhere; one may call them universal. A particular human setting and a local physical environment have no meaning for mass-produced machines, which are made to be used anywhere from Maine to California.

3. Individual machines—small-scale, mass-produced works—are thus reproducible. Machines are not only mass-produced and built by standardized techniques, but they are also designed to perform in a precise and predictable manner, characterized by standardized movement. Moreover, new discoveries in science along with new research in marketing lead to new machines, such that old ones become obsolete and rarely survive. Thus, machine technology involves fast-paced change. A study of Eiffel's water elevator of 1889 would add little of value to the technical education of tomorrow's mechanical engineers. Since machines are quickly built and readily disposed of, they encourage short-range planning or restlessness in design. Finally, being disposable and thus always subject to renewal, machines have come to be seen as the precursors of the future.

Just as the virtue of a machine, like that of a scientific formula, is its universality—it is made to perform the same function everywhere at all times—the virtue of a structure is its particularity. It reflects a specific and unique situation—like an artist's painting—and indeed it is much closer to a work of art than a work of science, although science is necessarily involved in its construction.

Machine form is closely analogous to a scientific formula: abstract, universal, capable of being used anywhere, it is as immediately clear to all the scientifically trained as it is consistently mysterious to all those not so trained. Structural form, on the other hand, is closely analogous to art form even though most structures, like most paintings, are very far from being great works of art.

Structures are characterized by a single personality—an Eiffel, Roebling, Nervi, or a Maillart—who creates very large works in very small offices whereas machines are often very small works created by teams of technicians in very large offices. As machines become smaller or as they are mass-produced in greater numbers (an individual piece being then a smaller part of a larger enterprise), careful calculation, precision, and the development of detailed standard formulas all become highly significant and force the design team to draw more and more on the abstractions of mathematical science. As machine technology advances, it becomes more and more scientific, and judgments about it and about its control need analysis having a strongly scientific basis.

Analysis and criticism of machine technology is essentially scientific, whether the assessment be that of natural science evaluating the performance of the machine as such (its power consumption, noise level, emissions, vibrations, and the like) or that of social science weighing a machine's performance in broader human contexts (such as those of environmental pollution and the energy crisis). In either case, criticism must deal initially with quantification, statistics, and models, though its end result will be political formulas for control, taxation, fines, and the like.

Criticism of structural technology, on the other hand, is essentially artistic or humanistic. Its initial concern must be the qualitative evaluation of individual works and designs, though it will lead to the identification of artistic forms as examples of high quality and human enrichment or of thoughtless design and degradation.

Both forms of criticism converge on public life but from different perspectives. Political formulas are intended to control machine excesses from the perspective of science, but the choice of formula will be determined by beliefs, ideals, and prejudices that transcend science. Analogously, artistic forms are intended to control structural excesses from the perspective of the humanities, but the choice of form will be determined by quantification and rationality that transcend the humanities. In short, social criticism of machines, beginning in science, ends by including judgments of value whereas artistic criticism of structures, beginning in value judgments, ends by including science.

I I

By reconsidering the three points definitive of the structural view of technology, we can identify three characteristics or meanings of structures which will provide a framework for later criticism.

1. The *scientific meaning* of a structure can be divided into three relatively precise areas, each characterized by separate, quantifiable analyses; each of these in turn relates directly to one of the categories in the structural view of technology outlined above:

The *efficiency* of a structure, measured by static analysis, is determined by considering the influence of gravity and pressure loadings on the development of a structural form. The design objective is a slender form using a minimum of materials to provide a required strength.

Because of a structure's large scale, its *safety* is determined by considering the influence of wind gusts, water waves, earthquake shocks, or even static loads on various structural forms. The design objective is a stable form using materials and connections that provide a required stiffness.

Because of the permanence of a structure, its *endurance* is determined by considering the influence of temperature, humidity, erosion, and pollution on various structural forms and materials. The design objective is an intact form using materials and connections that require a minimum of maintenance while providing a required durability.

2. The *social meaning* of a structure can be divided into three relatively distinct areas, each characterized by separate, partially quantifiable analyses:

Because large-scale structures are custom-made, the *economy* of a structure is determined by considering the influence of various modes of construction on the cost of a completed work. The design objective is a useful work for a minimum of money.

Because of the public nature of a structure, its *political value* is determined by considering the need for the work in relation to the paying public. The design objective is to satisfy a need felt by the largest number of voters.

Because of the local effects a structure has, its *community value* is determined by considering the influence of the change in the local environment on the life of the surrounding community. The design objective is to make the changed environment acceptable to a maximum number of local residents.

3. The *symbolic meaning* of a structure may also be broken into three categories, each characterized by a different visual image: unity, daring, and complexity. Many structures will exhibit all three characteristics—as indeed others not mentioned. For symbolic meaning cannot be exhaustively categorized, and the distinctions I offer are meant to be merely suggestive.

Unity of form, bound up as it is with the uniqueness of an impressive structure, often derives from a designer's efforts to harmonize the conflicting requirements of efficiency and economy, i.e., to avoid wasting either materials or money.

Given the slow-paced change typical of structural technology, a *daring* form results frequently from a designer's attempts to harmonize the conflicting requirements of safety and politics, i.e., to prevent structural collapse while avoiding a tasteless, if popular, massiveness.

Complexity of form is a matter of historical perspective, a feature of the symbolic nature of unique and daring structures deriving from the differing responses that successive generations have to the same form as it ages. To achieve complexity, the designer must ensure the endurance of his structure while respecting the viewpoints of the local community; he must induce community change while not rigidly determining community behavior.

This attempt to distinguish three kinds of meaning in structure is a very old one. It dates at least from the time of the Romans, whose principal structural spokesman, Vitruvius, considered the meaning of buildings in terms of their durability, convenience, and beauty—terms which directly correspond to the scientific, social, and symbolic meanings I have outlined.[2] These I shall now illustrate by reference to two major modern urban structures.

III

As an example of comparative critical analysis of structure, I shall consider within the framework just sketched the John Hancock Tower in Chicago and the World Trade Center in New York, two of the world's largest structures. My aim is to set forth some ways in which structures should be evaluated in order for public policy to have the benefit of humanistic perspectives.

The table, which describes significant features of the nation's four highest city towers, will set the stage. In it, l is the height and c is the base width of these essentially square-plan towers; *psf* means pounds of structural steel per square foot of floor area, and *Msf* means million square feet of usable floor area.[3]

	l ft.	l/c	Stories	Steel psf	Area Msf	Date
(a) Sears, Roebuck, & Co., Chicago	1,454	6.4	109	33.0	4.4	1974
(b) World Trade Center, N.Y.C.	1,350	6.9	110	37.0	4.5*	1972
(c) Empire State, N.Y.C.	1,250	9.3	102	42.2	2.75	1930
(d) Hancock Tower, Chicago	1,107	7.9	100	29.7	2.8	1968

   * 4.5 million square feet of gross floor area for each of the two twin towers.

The John Hancock Tower

1. The *scientific meaning* of the tower: The tower's efficiency derives from its tube form. In such a form, the external walls are load-bearing, carrying both vertical and wind loadings. The form is especially efficient in the tower because diagonals tie together the otherwise widely spaced columns, thus distributing the vertical forces evenly among them (Fig. 14.1). "What is so special about a new 100 story building?" asks Fazlur Khan, the building's designer.

The answer lies in three basic characteristics of the high-rise buildings built in the thirties. First, a 20-ft. column spacing was considered adequate for office spaces. Today a minumum of about 40 ft. is considered adequate. In fact, the longer the spacing, the better the office space. Second, the partitions used were generally made of solid masonry from floor to floor, adding considerably to the rigidity of the entire building. Today most partitions are removable, therefore very low in weight and stiffness. Third, the exterior wall detail was generally made of solid masonry or stone, and the window opening consisted of a small percentage of the total wall surface. Today the glass curtain wall is generally attached to the frame as a non-rigid skin.[4]

By making a tube, Khan took maximum advantage of structural changes introduced in tall-building construction over the last forty

Figure 14.1 Elevation of John Hancock Tower
(photo D. P. Billington).

years—namely, larger column-free interiors, lighter removable interior partitions, and more open exterior walls. Thus his tube structure provides, with relatively little material, a relatively high strength and stiffness. As Khan says,

Taking advantage of this bearing-wall characteristic of the system, all exterior columns on each face at any floor were made of the same size irrespective of their nominal tributary areas. This resulted in a considerable reduction in construction fabrication and erection time as well as cost.[5]

The safety of the tower also derives from its tube form. Since the dead load in a tube form is carried primarily by the outermost columns, they are sufficiently heavy to carry the wind load without significant overstress. The dead load column forces are made high by bringing most of the dead load on the exterior columns, whereas the wind load column forces are made low by having those columns on the exterior. This ensures safety against wind loads. Stiffened by the floors, which form diaphragms at every level, the tube form also provides safety against excessive vibrations.

The endurance of the tower is aided by having the exposed steel frame covered by aluminum with black anodic coating.

2. The *social meaning* of the tower: Economy of construction was achieved in three ways: in the design process, since the owner, developer, architect, engineer, and builder all worked on the original plan; in the chosen structural form, which permitted smooth upward construction by standard means; and in the construction procedure, which consisted of the standard cantilever-type construction of steel erection.

The politics of the tower were relatively simple because the owner, a private corporation, having once satisfied city ordinances, needed to be accountable only to itself. In the future, however, the tower may prove politically significant since some people of high talent who would have lived in the suburbs are now living in the city and thus provide it with a potential source of new leadership.

The preexisting local community appears not to have expressed strong feelings about the tower before its construction. Since its completion, however, there has been substantial traffic-jamming in the neighborhood, partly because of the tower's inhabitants and partly because of visitors. The most significant community feature of the tower is its dual use: the lower forty-five stories contain offices and the upper fifty, apartments. Thus, the tower is a city within a city and has brought a new community into the Chicago business district.

3. The *symbolic meaning* of the tower: By using a tubular structure made up of few columns with huge diagonals, Fazlur Khan was able to visually express an efficiency of structure which is probably more easily understood by the general public than has been the case for any other major skyscraper. That it was built economically in comparison with similar buildings lends further meaning to this highly visible and unique structural form.

The designers clearly saw themselves standing in the long tradition of the Chicago school of architecture begun during the time of Louis Sullivan, for whom structural engineering was a powerful stimulus to tall-building design. Thus, the idea of daring new forms that express structure might be said to come more naturally to designers living and working in Chicago. Moreover, it is not insignificant that the designers can see their own works from their own offices, and indeed some designers now live in the Hancock Tower themselves.

Finally, the idea of a city within a city has been realized in this tower on its largest scale to date. Especially significant is its height because no commercial apartments have ever before been available at such immense elevations. The project was originally planned as a forty-five-story office building and a separate seventy-story apartment building, but to reduce expenses and to use less land, the designers decided to put one building on top of the other.[6] The meaning of this new cloud-level community is not yet well established. Newspaper accounts tend to portray the tenants as pleased. Clearly, the rents on the offices below tend to subsidize the apartments above which would otherwise need higher rents to be commercially successful. The tower city has many people who live and work in the building, and their style of living presents new possibilities not yet well observed.

The World Trade Center [7]

1. The *scientific meaning* of the center: The center's twin towers are tubes like the Hancock building except that each consists of a closely spaced steel grid instead of widely spaced columns and a few

large diagonals (Fig. 14.2). The grid is formed from columns spaced 3 ft. 4 in. on center crossed by horizontal spandrel beams 4 ft. 4 in. deep at each floor level (each level is 10 ft. 10 in. on centers). This arrangement requires more steel than in the Hancock and partly accounts for their different relative efficiencies: 37 psf compared to 29.7 psf on the Hancock. Another major difference is the use of different grades of steel at different levels of the World Trade Center, with higher-strength steel used at lower levels. By contrast, ordinary steel was used nearly everywhere on the Hancock.

For safety, the towers were founded on bedrock, and, as with the Hancock, their tube design permits wind to be carried easily. In the World Trade Center, moreover, vibrations are counteracted by 10,000 damping units, 100 per floor, to reduce wind sway that might be uncomfortable to building occupants. As in the Hancock, the floors serve as horizontal stiffeners.

The endurance of the towers is aided by an aluminum cover over the exterior columns.

2. The *social meaning* of the center: Comparison of the economy of the two projects can be based on technical sources. For the World Trade Center, *Civil Engineering* in June of 1971 stated that "the $650 million project will include some 9 million square feet of office space"; [8] this yields a ratio of $73 psf. For the Hancock Tower, the *Architectural Forum* of July/August 1970 reported a construction cost of $91 million for 2.8 million square feet of usable space [9] (1.0 million for offices, 1.0 million for apartments, and 0.8 million for parking and commercial uses [10]). The *Forum* stated that fees were $9.0 million, but it is not clear whether or not this is included in the $91 million. Even if it is not, the total cost is $100 million, yielding a ratio of $35.60 psf—less than half the value for the World Trade Center. One important aspect of economy probably arises from the connection between designers and builders. After rejecting a presumably very high bid for detailing, furnishing fabricating, and delivering 180,000 tons of steel, the Port of New York Authority decided to break the one contract into fifteen separate contract packages—thirteen for fabrication, one for erection, and one for detailing. This great complexity in construction contrasts with that of the Hancock Tower for which all the steel work was virtually done by one contractor.

Politics in New York were much more complicated than in Chicago

Figure 14.2 Elevation of the World Trade Center
(photo D. P. Billington).

if only because the owner of the twin towers is the Port Authority, a public agency. After the announcement of the World Trade Center in early 1960, the choice of architects, Minoru Yamasaki and Associates and Emery Roth and Sons, in late 1962, and the development of plans for a $270 million complex, the appellate division of the state supreme court declared the legislation for the project passed in both New York and New Jersey unconstitutional. The plaintiffs, a group of businessmen who would be displaced by the new center, eventually lost their case: the Court of Appeals overruled the decision of the lower court, and in November 1963, the U.S. Supreme Court refused to hear the case.

The local business and residential community objected to the center as did people interested in the Empire State Building. The former objected to being physically displaced; the latter, to having their building displaced as the world's highest. The objections of urban planners were well summarized by Lewis Mumford:

The Port of New York Authority's World Trade Center, 100 stories high, is a characteristic example of the purposeless giantism and technological exhibitionism that are now eviscerating the living tissue of every great city. . . . This policy has resulted in mounting traffic congestion, economic waste, and human deterioration—though with a constant rise in land values and speculative profits.[11]

3. The *symbolic meaning* of the center: The smooth-faced, constant-section towers of the World Trade Center contrast with the visual expression of structure in the Hancock Tower. Since the twin towers are essentially columns, it is ambiguous to see the base of these immense towers formed by a visually thinner set of slender tall columns (Fig. 14.3). On the Hancock the vertical columns get visually heavier as they get lower, and they continue right down to the ground (Fig. 14.4). It is not easy to see the visual transition between the dense wall and light columns of the World Trade Center.

The daring nature of these towers lies not in their structural form but in their scale and in their being two. The form of both towers is entirely regular, a simple rectangular block; therefore, scale dominates. For this reason, critics like Mumford consider the Trade Center to be a Homage to Giantism, and even those who praise it do so essentially for its daring scale:

Figure 14.3 Base of one tower of the World Trade Center
(photo D. P. Billington).

Figure 14.4 Base of the John Hancock Tower
(photo D. P. Billington).

Construction of the record-breaking skyscrapers will be a prodigious under-taking. Each will require about 86,000 tons of structural steel; the whole project about 200,000 tons. (The Empire State Building took 60,000 tons, the huge Pan Am Building 45,000 tons.) Elevators will be the world's fast-est, at 1700 ft. per minute, and have by far the largest high-speed cabs ever installed. The project's air-conditioning system will require 40,000 tons of refrigeration and 80,000 gal. per min. of river water. Electrical needs are es-timated to total 60,000 kw, equivalent to that of a city with 400,000 popula-tion, such as Syracuse, N.Y. Pressure in water pipes may exceed 500 psi. And these are only a few of the enormous requirements of this colossal project. If the Empire State Building's 33-year-old height record is to be exceeded, it will pass to a worthy successor in this huge World Trade Center.[12]

Partly because of a different history of building in New York, the de-signers, Yamasaki of Detroit and Skilling of Seattle, did not feel ob-liged to maintain any particular tradition.

Finally, the towers also characterize their owner, the Port Author-ity, which, in the words of the editorial just quoted,

as an agency of the states of New York and New Jersey, can exert the power of condemnation to assemble the plots needed for the site. Also, while the Trade Center is expected to be self-supporting, the Authority's credit is backed by tolls from its bridges and tunnels. And PNYA holdings are tax exempt, though it usually makes payments in lieu of taxes to local ju-risdictions. Thus, the PNYA has prestige, power, resources and privileges that ordinary owners lack, and these can be a powerful force in promising success for the record-breaking skyscrapers.[13]

Thus, these huge-scale structures also symbolize the relationship be-tween the local community that opposed construction and the much wider communities, the two states, that had "prestige, power, re-sources, and privileges." Unlike the city within a city of the Hancock Tower, the twin towers are pure office buildings; thus they are novel in neither form nor content. Reaction to these two skyscraper projects will certainly change as the structures age, as their use develops new patterns of behavior, and as the perceptions of articulate critics are in-fluenced by changing cultural attitudes. It is certainly clear, for ex-ample, that events connected with United States policy in the 1960s have influenced Lewis Mumford.

IV

One can observe from this analysis that *scientifically,* the Hancock Tower is superior to the Trade Center because less material was used per square foot of floor area; that *socially,* the two structures represent different ideas: one novel—a self-contained city within a city; the other traditional—a pure office building; and that *symbolically,* the Hancock Tower presents a strong visual statement of technology, whereas the World Trade Center has a form that in one major way denies technology, i.e., in a visually reduced base wall which occurs just where the forces are greatest.

Beyond these obvious conclusions lie more central issues regarding the influence the two structures are having on their respective cities and thus the possible consequences of such works on future proposals. By reflecting on the recent past, by analyzing structural technology according to its three central meanings (scientific, social, and symbolic), and by appropriating the kind of critical attitude proper to the best commentators on the arts—the humanist and the humanistically alive citizen have powerful tools to affect the future course of public policy as it relates to the basic restructuring of American urban life.

15

The Aesthetics of Technology

In Response to David P. Billington

MARIO G. SALVADORI

Since David Billington and I belong to the elite club of structural engineers who have been infected by the virus of aesthetics and humanism, I know he will take my comments as offering additional perspectives on the subject of his paper, even if I must at times refute what he writes.

What I wish to say first is so obvious that I feel almost ashamed to mention it. I want to remind Billington that, according to Le Corbusier, "la maison c'est une machine à habiter." And this brings us to a first difficulty: what is a machine and what is a structure? Easy distinctions are hard to come by, especially in the case of the large buildings Billington considers in his essay; in such buildings, the structure accounts for only 20 percent of their total cost and the machinery in it for 50 to 60 percent. At least from one important point of view, then, the building's structure is merely a shell to house a lot of machines. I will return to this point.

Another equally basic question I wish to ask is whether a structure should have what Billington calls the "right" form or whether instead it should have the "wrong" form. What if one believes, as I do, that whatever man has done that was truly worth doing, has been done, so to say, against nature or God? God did not decide that men should speak in poetry, but some of us, thank God, have written poetry. God did not decide that the Eiffel Tower should be represented

Mario G. Salvadori is Renwick Professor Emeritus of Civil Engineering at Columbia University and partner at Weidlinger Associates.

the way Delauney painted it, but that's what he did, and we are richer for it. To realize this truth takes us to the center of the problem of a "right" or "wrong" form. Is the self-evident form of a wrestler who stands with bulging muscles, ready to fight, more "right" than the elusive form of a ballerina who is ready to jump and seems to defy the laws of nature? Her limbs are so graceful that they hide the strength needed to leap into the air. Now, I would suggest that when we design what may be considered a good structure, we often try to do just that: to defy the laws of nature, at least visually.

It so happens that the firm in which I am a partner participated in the structural analysis of the Hancock building—we did one of many wind analyses of the building—and I am in a position to have some knowledge of this particular building. It is a very good building, a well-conceived building (although, contrary to Billington's statement, having offices and apartments in the same building is not a new idea: it was done in the Marina Towers many years ago). The idea of bracing its four sides with big X's and of making it into a tube is a very sound structural idea. But those diagonals are about six feet wide, and if someone happens to have a window in a section of the tower crossed by the diagonals, he's presented with a problem. There is a story that the problem was cleverly handled in Chicago by the renting agent for the building: he made the diagonals into such a status symbol that if someone has a window blocked by the X-bracing, he pays more rent than someone who doesn't. If true, this is a brilliant example of American marketing genius, but I'm not sure that it shows whether that particular structure is more "right" or more "wrong" than the structure of the World Trade Center.

As for the columns on the World Trade Center which come together on the facade so that the building opens up at the bottom . . . Don't we have here exactly the same point of view that Brancusi used in designing his beautiful piece of sculpture, *Bird in Flight*? The same point of view that Nervi had in designing his world-shaking Turin Exhibition Hall? And, after all, since one has to get into a building, one must open it up at the bottom.

It is fairly obvious to anyone with a little aesthetic sensibility that the Hancock Tower is more "beautiful" than the World Trade Center. But the more important issue is why such tremendous towers are built at all. (And they are built everywhere, not only in this country.) I think it's because, in the present stage of our culture, fighting a difficult battle against the forces of nature is one of the expressions of the human spirit. These towers are not pure works of art, or pure social instruments, or pure technological breakthroughs. They are an expression of *our* culture. They were never built before and may never be built again, but they must be respected for their total symbolical value.

Beside stating this basic semiotic viewpoint, I would like to make one or two purely technical comments. It is *not* true that the Hancock Tower, as Billington states, is "scientifically superior" to the World Trade Center towers because the latter used more steel than the former. One needs only to point out some well-known structural facts and do some simple arithmetic for this to be understood. The amount of steel needed to construct a floor has no relation to where the floor is: it doesn't matter whether it is the second or the hundredth floor. Once the loads to be carried by a floor are taken into account, its steel weight is whatever it comes out to be. For this type of building, one needs about eight pounds of steel per square foot of floor. The weight of steel for the columns, however, grows linearly with the building height, and at great heights, as Billington notes, the wind becomes an extremely important factor in steel requirements: steel needed for wind bracing grows as the *square* of the building height as against the *linear* growth of the column steel and the *independence* from height of the floor steel. (And at this stage, Billington may take the viewpoint of the wrestler that it's better to express the kind of structure required by the wind while I may take the viewpoint of the ballet dancer and not want to show how the wind interferes with structural appearance: and I will spend days, months, thinking and figuring and adjusting in order to bring this about.)

The World Trade Center towers are 1,350 feet tall, and the Hancock tower is 1,107 feet tall. Therefore, the ratio of their heights is 1.22. Since the weight of steel in the Hancock Tower is 29.7 pounds per square foot, subtracting the 8 pounds per square foot needed for the floors leaves 21.7 pounds per square foot; on an average in very tall buildings half of this weight, or here 10.85 pounds, is required for the columns and half, or 10.85 pounds, for the wind bracing.

To adjust these figures to account for the height of the World Trade Center, the 10.85 pounds for the columns must be multiplied by 1.22, giving 13.2 pounds, and the 10.85 for wind bracing by the square of 1.22, i.e., by 1.48, giving 16.1 pounds. The total steel in the World Trade Center is the sum of the steel in the floors, that in the columns, and that required for wind bracing; it should weigh, therefore, approximately $8 + 13.2 + 16.1 = 37.3$ pounds per square foot. It weights 37! The two buildings have an *identical* structural efficiency. And this must be true because if the engineers who designed the World Trade Center had put 23 percent more steel in those towers than strictly needed, they would have been fired. I mention this to show that one must be very careful in judging whether one building is better than another, even from a purely technological point of view.

Another technical matter brings me back to the point I raised at the beginning of these comments. Billington wants to keep a clear distinction between structures and machines: he's concerned that the World Trade Center uses 10,000 dashpots. Now, a dashpot is simply a device used to dampen a building's oscillation: if under the impact of the wind the building starts oscillating, these devices are activated and the oscillations stop much more rapidly than they would otherwise. I have reason to believe that the top of the towers of the World Trade Center may move laterally as much as three feet under a hurricane wind: one must counteract and stop these high-amplitude oscillations either by using more steel or by using dashpots.

The use of dampers (or dashpots) is one more indication that today even a structure can become in part a machine. And it would seem that if the engineer can, by means of 10,000 dampers (which are very simple and inexpensive devices), save a considerable amount of money in steel, without harming the social, the symbolical, or any other meaning of the building, he should certainly do so.

We ran into a problem of this sort when my office did the wind and thermal design for the Place Victoria tower in Montreal, a building designed by Pier Luigi Nervi. Nervi asked us to check the thermal conditions which, simply stated, are these: inside the air-conditioned building, the temperature is constant; on the outside, temperatures vary and can get as high as 120°. At this temperature, the outer columns (it is now very fashionable to show the skeleton in architecture) become longer than the inner columns, and the building tends to come apart. There are many ways to avoid this dangerous tendency, but one we suggested consists in circulating a liquid inside the outer columns in order to prevent them from becoming too hot. But

Nervi disagreed: "I will never allow one of my structures," he said, "to depend for its safety on a mechanism." But there is now a building in Pittsburgh, the Alcoa Building, which is built exactly on this principle.

I was anxious to add these technical comments to my remarks because I believe that if a culture produces solutions in one field to problems which exist in another field—here, the fields of structures and machines—and if the solution is properly transferred—in this case, saving money and making for better buildings—all of us, technicians and artists alike, are bound to gain. Why should we maintain distinctions between fields, particularly in a culture like ours in which boundaries are getting more blurred by the day?

By now it should be clear that I am extremely interested in the work of engineers like Billington, for they are introducing an important new dimension to our trade. They remind us that if the architect provides us with the symbols of our society, the engineer makes the symbology possible. They help us answer Lewis Mumford and others who believe engineering civilizations to be evil and to undermine the human value system.

Let me emphasize by means of a concluding story how important it is to clarify the humanistic role of the engineer. A few years ago, the head of a large architectural firm in New York City summoned me to his office to ask whether I would design a roof for a certain hall: "because I understand," he said, "that you can design imaginative roofs." I asked him what the roof was for and was told it was for the dining hall of a large hotel. I asked where the building was, and was told it was in Ponce, Puerto Rico. "Fine," I said, "and what do I see if I look out in this direction?" "The Caribbean," he said. "And if I look out in this other direction?" "Hills." "Brown, arid hills or green hills?" I asked. The architect did not answer my question, but asked: "And what do you care?" "I care," I said, "because, although you want the roof of a dining hall to provide protection, you want at the same time, if there is a beautiful view of the Caribbean, to have the whole landscape come in. And therefore I would like to give you a roof that seems to go up to the sky in one direction and just embraces you in the other. It is called a conoid." The man paused for a moment, and then said, "Are you sure you're an engineer?"

And this, in a nutshell, is our problem.

War and the Social Order

What inducement has the farmer, while following the plough, to lay aside his peaceful pursuits, and go to war with the farmer of another country? or what inducement has the manufacturer? What is dominion to them, or to any class of men in a nation? Does it add an acre to any man's estate, or raise its value? Are not conquest and defeat each of the same price, and taxes the never-failing consequence?

Thomas Paine
The Rights of Man, 1792

16

On National Frontiers

Ethnic Homogeneity and Pluralism

WILLIAM H. McNEILL

What can one say about national frontiers? First and foremost: it seems obvious that frontiers delimiting political power structures change through time within limits set largely by geography and prevailing technologies of transport and communication, of administration, and of military organization. Some frontiers have reasserted themselves frequently: the best examples I can think of are the Himalayan barrier delimiting the northern frontier of India and the desert barrier delimiting China's northwest abutment upon the steppe lands of Asia. But geographical-cultural zones of transition are seldom so sharp as in these two regions; and where strong and compelling geographical obstacles do not exist, most political frontiers have proven themselves vastly pliable across time—man-made, in short, and capable of being remade as human conditions alter.

This, however, scarcely needs elaboration. It is more interesting to observe that civilized history seems to demonstrate alternations between two types of political frontier: either an imperial structure, uniting different peoples under one administrative, tax-collecting roof; or smaller and more nearly homogeneous (characteristically also more barbarous) national units. We are often misled by the very unusual and profoundly atypical pattern of modern Europe into assuming that the national alternative is normal. Yet this is not so. Modern Europe, in fact, offers the only instance I know of in which instead of creating a civilized polyethnic empire at a reasonably early stage of develop-

William H. McNeill is Richard A. Milikan Distinguished Service Professor of History at the University of Chicago.

ment—as once seemed possible in the days of Charles V—the atavistic, barbarian notion of nation survived as the dominant political form while in other respects men advanced in civilization. Elsewhere and in other ages, civilization always came to be encapsulated within one or more polyethnic imperial structures; and of course it is not clear that modern Europe will not belatedly undergo a similar evolution, if not in our time then in some future age.

I

There is no great mystery as to why polyethnic empire has been the dominant, normal political form for civilized mankind. It is merely an exemplification of the general advantage professionalized skills commonly enjoy in competition with amateurs. Professionalization is, indeed, the principal hallmark of civilization: and when the principle is applied to military activity, one gets standing armies, bureaucratic caging of organized violence, and, in most cases, such superiority on the field of battle that he who commands such an army is in a position to subdue neighbors who have not succeeded in creating comparably effective organizations. The result is to create a polyethnic, and at least potentially, civilized empire.

How swiftly and irresistibly the dynamic of warfare pushed men along this path in ancient times can be appreciated by reflecting upon the biblical account of David and his mighty men. For David, who began life as champion of national self-determination for the Israelites against Philistine aggression, ended up as king of a polyethnic, if still comparatively miniature, empire, ruling from Jerusalem—a foreign capital and stronghold—by means of professionalized fighting men, including foreigners like the ill-fated Uriah the Hittite. Moreover, David's kingdom was modeled on older imperial patterns, as worked out among the Hittite, Egyptian, and Babylonian rulers of the bronze age; and they in turn were elaborating patterns of administration that started with Sumer and Akkad, at the very dawn of civilized history.

The Middle East remained the theater for a succession of empires from that distant day until almost yesterday, for the breakup of the Ottoman polyethnic empire that had united most of the ancient heartlands of Middle Eastern civilization in the sixteenth century is an affair of the twentieth. In all essentials, the patterns and techniques of imperial civilized administration were perfected by the ancient Assyrians (who seem to have created an army with regular ranks and promotion according to demonstrated efficiency) and Persians (who added a navy, traveling inspectorate, and a universalistic religiopolitical ideology—Zoroastrianism—to the older structures of law, market, bureaucratic office-holding, roads, professionalized military forces, and the like that held ancient civilized empires together). The only subsequent improvement I can think of was the establishment of formalized education of a kind that systematically prepared individuals for a career in public administration. This was pioneered in China in the time of the Han dynasty and became formalized under the Sung, to the great advantage of subsequent rulers of China down to our own day.

I need not argue at any length the fact that the imperial pattern of political management was characteristic not merely of the Middle East and China but of nearly all other civilized regions of the earth as well. The succession of empires in the Mediterranean and eastern Europe is thoroughly familiar. The Macedonian, Roman, Byzantine, Arab, Ottoman, Russian, and Habsburg imperial monarchies all exhibit a strong family resemblance. In India, imperial consolidation was never as stable as in lands north of the Himalayas. Empire always retained a perceptible foreign flavor, from the age of the Mauryas, who modeled their court in some details at least on Persian and Macedonian exemplars, to the Moslem and European conquerors who ruled India in more recent times.

Presumably, Indian recalcitrance to empire rests upon the strength of the competing institution of caste, which performed some of the functions reserved for public authority in other lands and thereby diminished the scope of political-military government, making it in principle an affair of only one caste among a multiplicity of other castes. Yet for all the fragility and superficiality of empire in India as compared to similar structures elsewhere in the civilized world, the caste principle itself guaranteed that any territorial state—even one of quite modest extent—had to embrace a plurality of diverse groups, some of whom could trace descent from "forest peoples" and other ethnic exotics, and whose diversity was protected and perpetuated by caste rules and behaviors despite prolonged proximity and complex interactions across caste lines. Hence even when India was not politically united into a single vast imperial state structure, the constituent political units were themselves of imperial complexity.

As for China, the weight and continuity of the imperial principle there needs no elaboration from me. It is more interesting to note that in the Americas, where all possibility of contact with Old World patterns of government and administration can safely be excluded, the evidences for the development of an "Imperial Age" in both Mexico and Peru by about A.D. 1000 is clear. Professionalized warriors and administrators, just as in ancient Mesopotamia, set themselves up as masters over a mixed bag of taxpayers and subjects of different language and with different rituals from those the rulers employed. In other words, the dynamism of power and professionalization that found such broad scope in the Old World from the third millennium B.C. onward also asserted itself among the Amerindians, although belatedly in comparison with the Old World patterns that gave Spaniards a model for their post-Columbian imperial administration of both the great centers of native American civilization.

I must confess that the Aztec imperial structure, which rested upon the idea that a massive supply of human hearts was required to keep the Sun God nourished—a supply provided by annual raids against militarily less well organized neighboring and semi-subject populations—represents an extreme case of imperial oppression. But taxes that took away crops required for peasant survival, at least in a bad season, were quite normal in civilized imperial history; and the effect in such cases was not, perhaps, so very different from the Aztecs' murderous system. In both instances, that is, macroparasitism assumed such proportions as to limit peasant numbers and affect individual survival in thousands, or hundreds of thousands, of cases.

Nevertheless, this points to a fundamental weakness of civilized imperial structures. How can the depradations of the ruling element be held to tolerable proportions when, by definition, overwhelming force rests in its hands? Too heavy a tax and rental load will provoke heavy peasant die-off and will eventually weaken the parasitic imperial structure itself. On the other hand, too small a rent and tax income—a far rarer circumstance—implies an ineffective guard against attack from without, leading, characteristically, to raiding, civil war, and heavy peasant die-off too. What is needed, obviously, is just the right *via media,* according to which enough but not too much food and other resources are devoted to the maintenance of a ruling element. But in the absence of clear and reciprocal community of interest and shared values between master and servant, such a *via media* is supremely difficult to achieve or maintain. And it was and is even more difficult to create or maintain a vivacious sense of community among all elements of an imperial structure whose nature is such that some take while others have to hand over the product of their labor as taxes and perform various other services that are (or appear to be) entirely unrequited.

The result, therefore, was a persistent and recurrent weakness in civilized imperial structures when confronted by barbarian attack. For a barbarian people, even if less well armed and less expertly trained than the professional fighting men of civilized armies, were far more likely to enjoy a solid morale. "One for all and all for one" is a plausible slogan among barbarians whose pattern of life does not involve much occupational differentiation; it is all but impossible to believe in a society based upon harsh taxation of some to support the luxury of others. Yet this was the situation in all traditional civilized empires; and particularly when members of the professionalized armed forces were drawn from the lower classes—a not infrequent situation—the possibility of disaffection within civilized ranks drastically reduced efficiency. As a result, rebellion in the rear coinciding with assault from without constituted a recurrent cause for imperial collapse.

In addition, one should also recognize that the ruling element in and of itself was difficult to keep together. Disputed successions and rivalries for recognition, status, or wealth were liable to flare into organized struggles, since the attitudes and habits of life appropriate for a military ruling clique—aggressive assertiveness, ready resort to force, sensitivity to punctilios of personal honor, status, deference expectation, and the like—could just as well be turned inward against fellow members of the ruling element as against outsiders, despite obvious and often disastrous results for the stability of the imperial political structure as a whole. Finally, the decay of military habit resulting from the enjoyment of wealth and soft living in urban communities is not a myth. Many a conquering horde, having won previously unimagined wealth and luxury by dint of hard riding and hard fighting, has found the delights of urban life irresistible, with the result that grandsons or great grandsons of the conquerors were no longer capable of enduring the fatigues or exertng the force their forefathers had been capable of—even if, as was also likely to occur, diseases and infections indigenous to cities did not diminish their numbers to a critical point before they had to face the dual test of rebellion in the rear and assault from without.

The consequence of these factors interacting with the persistent advantage attainable by the sort of professionalized fighting forces that could be sustained by tax and rental income was to create an irregular pattern of imperial breakdown, barbarian invasion, and subsequent reconsolidation, often at the hands of some newcomers from the barbarian fringes of the previous imperial structure. Examples are innumerable: Macedon *vis-à-vis* Greece and Persia, Rome *vis-à-vis* the Hellenistic world, Isaurians in the eighth-century Byzantine state and Austrasian Franks in the Merovingian empire of the same century are examples from European history. Indo-European, Turkish, Mongol, and Tungus conquerors issuing from the Eurasian steppes in an unending sequence, beginning with the third millennium B.C., and climaxing in the sixteenth and seventeenth centuries A.D. with the Mughal conquest of India and the Manchu conquest of China, constitute by far the most significant series of barbarian conquerors who from time to time overran civilized lands and set up new (often short-lived) imperial structures. A similar though geographically less extensive reservoir of barbarian prowess and solidarity existed in the grazing lands of Arabia and the adjacent savanna lands of Africa. As a result, the political history of North and West Africa, and of the Middle East, has been punctuated by recurrent invasions from the southern grasslands that recapitulate in all essentials the history of similar conquests issuing from the northern steppelands of Eurasia. And in the Americas, too, imperial peoples like the Incas and Aztecs seem to have originated as barbarian conquerors coming from the fringes of older civilized society.

As this pattern of periodic breakdown and barbarian conquest followed swiftly by reconstitution of an imperial structure became familiar to civilized communities—Ibn Khaldun calculated that the cycle took no more than three generations in North Africa—attitudes and devices for regularizing the transfer of authority from one hand to another tended to shorten the periods of disorder and to accelerate the next imperial consolidation. Thus the readiness of tax administrators to serve a new conqueror became standard in Eurasia from early Christian centuries so that a new-sprung ruler no longer needed, as in King David's day, to reconstitute a workable tax system from scratch. Thus Ostrogoths in Italy, Vandals in North Africa, and Arabs in Egypt, Syria, Mesopotamia, and Iran were able to call upon the willing services of preexisting administrative experts; and the continuity of civilized structures correspondingly increased, despite continuing (but increasingly superficial) conquest from without. Chinese success in accommodating invaders by offering them the services of a well-trained administrative elite is well known, but it is properly viewed as only the most efficient and longest standing such administrative system.

II

By comparison, the principle of national self-determination and maintenance of an effective coincidence between ethnically homogeneous populations and sovereign political-administrative structures of public life must be described as trifling and evanescent on the canvas of world history. Such arrangements were, of course, characteristic of barbarian tribes wandering around in grasslands and forests, whose poverty, ignorance, and brutality sustained hardihood, homogeneity, and heroism. Such peoples could, as Tacitus' *Germania* proves, sometimes win admiration among civilized populations because they embodied, or seemed to embody, a style of life reminiscent of the civilized rulers' own distant and admired past. But efforts to combine the virtues and simplicities of barbarian existence with the power, wealth, and beauty attainable through professionalization, occupational diversification, and urban concentration of specialisms were intrinsically unstable and—save in a very few instances—very short-lived.

Yet those few instances have bulked very large in the European literary inheritance and in our historical imagination. Fifty years of Athens—or, to be generous, one hundred and fifty Athenian summers—matter more to us than four thousand years of normal imperial civilized existence in China, India, or the Middle East. Similarly, Romans like Livy and Tacitus set the tone for modern Europeans by idealizing the semibarbaric virtues of the Roman republic and deploring the corruption of their own imperial days. The idea that history records the heroic struggles of free men pitted against the slavish subjects of imperial dynasts entered European historiography with Herodotus. It was reiterated by the Roman historians, picked up by Italian publicists in the Renaissance, believed by American patriots of the eighteenth century, and made canonical by national historians of the nineteenth in each of the leading countries of the Western world.

Yet the ideal of freedom, according to which individuals cooperate in public matters of their own volition because of common recognition of dangers from without and of the costs of civil strife within, stands and has always stood in persistent conflict with experience. Athens' victory of 479 B.C., reverently recorded by Herodotus, immediately projected Athens toward empire, whose ambiguities were ably analyzed by Thucydides a scant generation later. That is to say, any locally successful realization of the ideal of freedom, allowing substantial numbers of skilled men to cooperate effectively and voluntarily across the spectrum of civilized occupational diversity, generated such compelling force that barriers to territorial expansion crumbled. The free city thus transmogrified itself—as it were, willynilly—into the tyrant city that Pericles discerned after the early setbacks of the Peloponnesian War. Athens found herself ruling over unwilling subjects in a fashion not fundamentally different from that in which kings and emperors ruled over their unwilling subjects. Moreover, abdication or defeat simply called a new tyrant into being, as the Greeks discovered when Athens' overthrow, in 404 B.C., substituted first a Spartan, then a Theban, next a Macedonian, and eventually a Roman master for the vanished Athenian hegemony.

The city-states of Italy that rose to wealth and power in the thirteenth century also experienced the ambiguity of freedom early in their development. One city swiftly subdued others until only five "great powers" remained to divide the soil of northern Italy. Simultaneously, internal divisiveness opened a path for usurping princes to command the obedience that could no longer be secured voluntarily. And when the local sovereignty of the Italian city-states in turn yielded before the superior power of larger territorial sovereignties, chiefly France and Spain, the drama was reenacted north of the Alps within less than five centuries.

For in our own age as much as in the fifth century B.C., any vigorous and successful pursuit of national greatness and security verges immediately upon empire, whether the dominant people and their leaders deliberately aim at empire or not. In recent centuries, the way in which the ideal of freedom, blown up to white heat by ideologues of the French Revolution, translated itself swiftly into the Napoleonic empire is almost as "classical" an example of the dilemma we are exploring as is the Athenian example itself. Similarly, we readily discern in the two German wars of the twentieth century the germs of the same evolution. Had World War I led to German victory, it seems clear that a German *imperium* would have been fixed upon Europe; and it already appears probable that Hitler's most enduring monument may turn out to be the massive mingling of peoples that Speer's war economy inaugurated, which in turn provided the principal base for the more lasting success of the European Common Market since 1949.

The obvious decay of nationalistic sentiment in Europe since 1945 is matched by the no less obvious polyethnic imperialism of the Soviet Union and of the United States, which was such a prominent feature of the immediate post-World War II scene. Both our own and the Russian government already preside over ethnically diverse populations as a result of the patterns of frontier expansion, migration, and englobement of weaker peoples that shaped both polities from the seventeenth through the nineteenth century. Reassertion of this tradition of frontier expansiveness seemed likely in the 1940s, obvious in the 1950s, and though the prospect has faded in the 1960s and 1970s, no one can assume that the tradition, so central to both the Russian and American pasts, will not reappear if circumstances provoke the possibility again, as they did in the immediate post-World War II years.

Hence it is far from obvious that ancient imperial patterns of government have lost their relevance for modern times even though the dominant ideal and widespread practice of recent decades has emphasized national self-determination and disrupted older polyethnic imperial structures in the name of freedom and liberty. Old dilemmas that led repeatedly in times past to the establishment of empires embracing divergent and mutually distrustful elements into a single administrative structure have not been lastingly resolved; and the possibility of reconstituting such structures on a territorially enlarged scale, perhaps even embracing the entire globe, remains and will remain alive as long as the technological possibility of world-girdling administration remains as strong as is now the case.

III

Yet this is not the only nor perhaps the most interesting way in which ancient polyethnic patterns of political association may be relevant to our age. What I have in mind results from the radically changed limits of the possible in our time. This in turn arises from the multiplication of wealth we commonly refer to as the industrial revolution. Until comparatively recently, it took about four persons working in the fields to raise food to support one person who did something else. Today it is possible for a single farmer to support nineteen others who do not raise their own food but depend instead on the enormously magnified productivity that power machinery, fertilizers, seed selection, and systematic livestock breeding and feeding permit. This is a very new situation indeed, whereby the majority of mankind are potentially (not yet actually in any but the more advanced industrial regions of the earth) emancipated from age-old rhythms and patterns of agricultural labor and introduced instead to the perils and potentialities of urban living.

What may come of this radical tipping of numerical balance between town and country remains the biggest question of modern history—or so I think. So far, the major result has been a multiplication of layers of bureaucracy, both private and public, of a luxuriance that would have amazed rulers and subjects in any earlier age. Multiplication of managerial personnel has numerous effects, not least being a distancing between "people"—whoever they may be—and rulers, while rulers find themselves increasingly ineffective in deflecting the behavior of the administrative bureaucracy they nominally command.

What may be happening can be conceived of as an alteration of old patterns of ethnicity. What I mean is this. In older times, individuals learned the customs and behaviors appropriate to their status in life as defined by birth and, as part of this educational process, acquired membership in some kind of ongoing *ethnos* or *natio*—depending on whether one prefers the Greek or the Latin form. But insofar as learning itself becomes institutionalized and therefore separable from accidents of birth, individuals may through formal training and apprenticeship assimilate themselves to new groups to which their parents and remoter forebears had been entirely alien. Such shifts have always occurred since occupational differentiation began among men; but in our own age the scale of social mobility has enormously enlarged, and the variety of groupings into which an individual may aspire to fit himself has undergone a no less extraordinary parallel process of elaboration.

The result, of course, is a society much more flexible overall than one dominated by simple inherited ethnicity; at the same time, the multiplicity of—as it were, artificial—*ethnoi* to which a person may assimilate himself raises unresolved problems of rank and precedence amongst the various corporate, bureaucratic groupings into which our society seems more and more to be dividing. There is, after all, no compelling reason why one sort of occupation should receive higher wages than another that better organization, more determined disruption through strikes, and skilled bargaining with paymasters cannot alter. The fact that white collar and professional groups have not hurried to organize does not mean that they will not do so in time to come; and even at a time when an absolute majority of the population remains aloof from formally constituted corporate organizations, it seems obvious that as such entities increase in number and power, adjudication of rights and precedence amongst them will become increasingly political, in the sense that patterns of conduct formerly associated with dealings between ethnic and territorial human groupings will have to come into play to define workable patterns of behavior.

In such competition and conciliation, it seems entirely probable that something very like the age-old imperial pattern of politics might creep in through the back door. That is to say, the armed *ethnos,* comprising professional military people, can expect to find themselves in a position to act as ultimate arbiters between rival and competing organized groups. To them belongs the exercise of ultimate force: and when all else fails, force remains as a way of settling disputes. In such a case, the self-interest of the military would soon reduce other occupational *ethnoi* to tributary status. This, indeed, has already happened in nearly all the so-called new nations where military personnel have assumed public rule.

The dependence such ruling cliques have upon suppliers of modern arms, who remain relatively few in the world, is an interesting echo of far older forms of clientage and dependence. For instead of realizing national independence, as current slogans require all self-respecting governments to do, these regimes institutionalize a new form of dependence. Ruling juntas of majors and colonels as much as repudiated generals and generalissimos need tank and airplane parts to keep their own self-respect and the support of the armed forces from whence they come. Often they need rifle ammunition as well. The result is to create interrelationships between the armed *ethnos* of one supplying country and its counterparts in a number of others. Once upon a time, kings and warriors distributed gold rings and other plunder as sign and seal of patron-client relationships; today, military hardware plays a similar role. And the role assigned to the rest of society while these transactions are going on—to observe and applaud or at least acquiesce—is roughly equivalent, too.

Nor is military usurpation of sovereignty necessarily limited to new nations, inexperienced in self-government. Current policy clearly opens the door for a domestication of military rule in our own country by isolating the military from the rest of society as much as possible and thereby increasing the likelihood of bringing conflicts of interest between civil and military elements of the population into the open. Why, for instance, should not every private become a tycoon with an income of $100,000 per annum? Why should the rest of us not be assigned the honor and responsibility of tending to his and his leaders' comforts and conveniences? This is what happened in older and far poorer civilized societies. What prevents members of the modern military from realizing the potential of their position and learning to enjoy the standard of living modern technology allows them to aspire to?

To be sure, there are powerful and lively countervailing forces amongst us still: not least a lack of imagination among our military personnel. Yet it seems to me plain foolish to overlook the possibility that the corporate self-interest of our professional military element might some day assert itself at the expense of the rest of society. Insofar as citizen solidarity decays and our primary identity becomes membership in some corporate or semicorporate occupational grouping whose self-interest in maximizing income and minimizing effort is not fundamentally different from the self-interest of any other group, including the military: insofar as this occurs, it seems clear, at least to me, that a way is open for a sort of backdoor reassertion of the validity of ancient imperial patterns of social ordering, despite our inherited ideals of free men in a free state, sharing the burdens as well as the benefits of political participation.

This may seem far from my topic, yet I hope I have not wandered off the subject entirely. For the burden of my remarks is this: national frontiers and political organization into ethnically homogeneous nations whose members all share meaningfully in a common public enterprise are rare and fleeting structures in world history, whereas polyethnic imperial states predominated in all civilized lands. In addition: in our own time and country, the political ideal and practice of citizenship that was handed down to us by our forefathers is as fragile as ever.

When most citizens were farmers, the shared burdens and benefits of a commonwealth meant one thing—seldom attained, yet often striven after, at least in the Western world. Now that few of us are farmers and most of us belong to occupational groups for which education and training rather than inherited status qualify us, the character of the *ethnoi* to which we belong has altered in some important respects; yet ancient patterns of interaction between corporate and self-conscious groupings have not thereby been canceled; and old, very, very old ways of settling quarrels and stubborn differences of interest and opinion as between such organized groups are likely to challenge the idea and practice of citizenship in the nation, conceived as one and indivisible, supreme over all alternative groupings among mankind and at the same time open to the influence and active participation of each and every individual citizen.

Historical perspective on national frontiers, in short, suggests not only how much they have altered through time but how permeable they are in our own age to ancient alternative ways of ordering human life in which voluntary acquiescence plays a lesser role and the direct exercise or threat of force plays a greater one than we are accustomed to think or wish to believe.

17

The Lower Middle Class as Historical Problem

ARNO J. MAYER

The lower middle class is a complex and unstable social, political, and cultural compound that deserves close and systematic analysis. No doubt, it is more than the mere sum of its many and changeable elements; and it appears to be far more polymorphous and tangled than the landed aristocracy, the peasantry, the bourgeoisie, or the working class.

Too frequently, scholars who have wrestled with the lower middle class have done so not in its own terms but with a view to determining which of two directions it is fated to move in: toward proletarianization or toward embourgeoisement. In the one perspective the petite bourgeoisie is condemned to become proletarianized, the old artisans being dissolved or forced into the industrial working class, the new white-collars losing whatever economic and social advantages momentarily differentiated them from manual workers. In the other perspective the concern is with the swelling ranks of clerks, technicians, and professionals: these are viewed as the spearhead for the embourgeoisement of would-be classless postindustrial societies.

Perhaps the time has come to stop thinking about the petite bourgeoisie primarily or even exclusively in terms of other classes. The lower middle class has been and continues to be of sufficient historical and political moment to warrant study in its own right: its economic roots, its inner structure and life, its place and role in society at

Arno J. Mayer is Professor of History at Princeton University.

large, its political behavior. Instead of ignoring, disparaging, or dismissing the world of the petite bourgeoisie as transient, insipid, and counterfeit, intellectuals should examine and understand that enigmatic universe for what it has been, for what it is, and for what it is constantly becoming. For, like the laboring class, the lower middle class generates and keeps generating a separate culture, ethos, lifestyle, and world view. Even if at first sight it seems to have no apparent class consciousness—except in times of acute crisis—it does have a sharply defined class awareness. In addition, members of the petite bourgeoisie have a distinctive taste in dress, ornament, furnishing, and food, as well as a distinctive family structure, home life, marriage pattern, consumption budget, and social deportment. Moreover, they have leisure habits, speech forms, social values, associational activities, and political attitudes that are typical of their class.

I

The lower middle class differs from place to place, and within any given country it may vary with changing local conditions as well as with historical time periods. The petite bourgeoisie of independent peasants, artisans, and shopkeepers of precapitalist and preindustrial times had a significantly different character—in terms of its internal structure as well as of its relationship to society as a whole—from that of the lower middle class of predominantly dependent clerks, technicians, and professionals of mature or late capitalist industrial societies. Needless to say, earlier layers—those that Bukharin characterized as "transition classes"—survive into later periods, usually invigorated by mentalities, attitudes, and values that retain greater vitality than the economic and social conditions that originally shaped them.

Even so, the notion of petite bourgeoisie or lower middle class, however ill defined, has long been in use, and it has signified the existence of a real and enduring social phenomenon. But the changing signification of this concept over time cannot be understood except in historical terms.

There is no need to comment on what either the Greeks or the Romans understood by middle class, there being no genetic link, either conceptual or historical, between the middle class, high or low, of ancient times and the nascent middle class of the later Middle Ages. It was in late medieval and early modern times, whose social contours remain vague, that "new men" proceeded to found and settle in commerical, manufacturing, ecclesiastic, or bureaucratic towns. According to Pirenne, the incipient urban merchant capitalists did not originate among the noble landlords, the knights, the clergy, or the villeins of medieval Europe. Rather, their ancestors were "the poor men, the landless men, the nomadic folk who wandered about the country, working for hire at harvest time." These indigent men were self-reliant and ingenious "adventurers" who "knew their way about," had seen "many countries," and spoke "many languages." From "this floating scum" came those "poor devils" who made "their capital out of nothing," thus becoming the heroic "creators of personal property." In sum, "a group of outlaws, vagabonds, and poverty-stricken wretches furnished the first artisans of the new wealth, which was detached from the soil" and was implanted in the cities.[1]

No doubt Pirenne overstated his case: traders and merchants never disappeared all that completely from the medieval scene. Moreover, in some regions landed aristocrats and *ministeriales* invested in mercantile ventures or sold their surplus produce in nearby cities, a certain number of them even taking up residence there. But in spite of these and other qualifications, Pirenne's thesis about the essentially lowborn origins and marginal place of much of the mercantile stratum in the towns ought not to be lost from view. Perhaps his perspective could also be adopted to chart the beginnings of artisanal production in the cities; it would seem that, even though there were some artisanal workers within the manorial economy, and some of these might have gone from villages to towns, most of them are likely to have come from other sectors of what Pirenne portrayed as the "floating scum." For whatever the origins of these mercantile and artisanal strata, they were composed of *menu peuple*. Admittedly, a few big merchants, shippers, financiers, traders, and manufacturers eventually emerged, and some of these magnates certainly were of rural origin and noble lineage. But these notables, though powerful, were at best a tiny fraction of the population of the new or renascent cities. The bulk of the population of those early towns consisted of small tradesmen, artisans, and craftsmen, of petty clergy, professionals, and officials, and of journeymen, apprentices, semiskilled day workers, and menial laborers.

At the same time as the towns fought for their corporate rights and privileges against the surrounding territorial feudal lords, within towns the different groups formed an indigenous system of stratification. The economic basis of this evolving social and political structure, which was distinctly separate from the manorial and feudal one, was preeminently trade, finance, and manufacture, rather than land. By sidestepping the feudal order the merchant and manufacturing patriciate came to occupy first position in the cities, leaving an intermediate lower or petite bourgeoisie, both independent and dependent, between itself and the underlying population of plebeian journeymen, apprentices, and ordinary laborers. Quite clearly, then, the patrician bourgeoisie was not a middle class, for it was in a manifestly superordinate, rather than intermediate, position in a new as well as autonomous structure of class, status, and power.

The size of the city affected its political culture. Only medium and large cities, not small towns, developed separate and self-perpetuating castes in which there was a commingling of wealthy patricians and high bureaucrats with an occasional petit bourgeois who had made his fortune. The grand patriciate of commerce, finance, trade, and administration owned or controlled capital resources rather than the means of production, which in the postfeudal but preindustrial period were in the hands of small artisans, traders, and merchants.

Compared to the disenfranchised, propertyless, and unskilled underclass, the petty independent producers and tradesmen unquestionably were better off, in that they owned their homes, shops, tools, inventories, and land either outright or with the help of credit. But even those guildsmen who eventually managed to admix their own labor with that of a few apprentices, journeymen, or navvies rarely acquired great wealth and standing: their profits and savings remained moderate, their standard and mode of life modest, and their political power marginal. Whatever political influence they had was a function of their numbers as well as of their contingent potential and propensity for joint action.

This middling class of city dwellers was the backbone of the municipal militia, which the patriciate needed not only to ward off external dangers but also to control or repress internal disturbances. Although the armed civic guards normally could be depended upon to suppress rioters, the possibility that they might become disaffected and throw in with the sinister underclass kept the patriciate from inordinately ignoring, contravening, or burdening the lower middle class.

In turn, the petty craftsmen and retailers were dependent on the governing, financial, and mercantile patriciate for credit, subsidies, contracts, and protection against competition from other cities. Especially in times of economic contraction or stagnation, when the ruling oligarchy felt threatened by popular eruptions, the artisans managed to extract additional economic safeguards. Quite apart from manning the city militia, artisans and tradesmen were bonded in guilds for combined action. In some ultimate sense these guilds were instruments of economic, technological, social, and cultural preservation—and obstruction—rather than of innovation, adaptation, and mobility: their regulations for admission and training and their standards for quality were designed to protect a limited and privileged membership against competition from outside.

While these monopolistic protective associations were relatively permeable in periods of prosperity, in periods of economic adversity, both structural and conjunctural, they became intensely rigid. Under the pressure of bad times, guildsmen accentuated their exclusiveness by giving ultrapreferential access to their apprentice sons. In addition to exercising their prerogatives of self-governance, the guilds pressured the ruling magistrates into enacting ordinances to further curtail the influx of labor and finished goods from nearby or distant cities. They also prevailed on individual magnates to provide credit and contracts on preferential terms. Such aid was the price or bribe for which, in moments of social crisis, the lower middle class stood with the patriciate and also agreed that this patriciate should retain its virtual monopoly of political power and authority.

This bent toward tight economic preclusiveness by the petty burghers was coupled with an unpliant and hidebound personal as well as social morality, a localist mental set, a rudimentary literacy, and a factitious sense of status. Members of the "guilded" lower middle class may never have had more than a moderate income, a comfortable life, and a low drive for infinite self-enrichment. Even so, they felt themselves superior not only to unguilded artisans but, above all, to the much larger pool of menial workers in the urban labor force whose existence was impoverished (not to say destitute), benighted, and insecure.

Of course, urban and rural societies, even at the highest echelons, were never totally disconnected. Just as the vertical stratification within cities was far from rigid, so also was there lateral fluidity, overlapping, and commingling between city and countryside. Still, the substantial bourgeoisie's struggle with the landed and bureaucratic elites for influence and power in the nascent national polities and unitary social structures never ceased to be significantly conditioned by this group's relations with the petite bourgeoisie, which was itself torn by competition between petty tradesmen and craftsmen.

The changing relations between the Independents and the Levelers in the English Revolution and between the Jacobins and the sansculottes in the French Revolution ought perhaps to be reviewed from this perspective, which underscores the crucial swing position of the lower middle class in the battle for control of the modern state. In the meantime, the revolts of 1848, in which the lower middle class is generally agreed to have been a pivotal *masse de manoeuvre,* may be considered paradigmatic.

In a first stage of the upheavals of 1848, the independent lower middle class had a conspicuous place in the constellation of forces that challenged the nondemocratic, aristocratically ingrained authority systems. This lower middle class was the principal social base and popular force that the liberals and democrats used to stage their initial assault. From the preindustrial, urban crowds came the insurgent force and the specter of revolution which gave political thrust to the substantial liberal bourgeoisie of business (manufacture, trade, finance) and of the professions (law, medicine, teaching). The advanced democrats, the ambitious but frustrated lawyers, journalists, teachers, and university students—themselves a relatively new segment of the incessantly mutable lower middle class—would also have been too inconsequential to furnish a credible popular force.

But in a later stage of each of the multiphased upheavals of 1848, when the circumstantial coalition of defiance broke up, most of the lower middle class swung around to support the existing economic, social, and cultural institutions: it fell in behind old-guard rulers and energetic new elites who presently unified against the more radical challenge, real or imagined, not only of traditional journeymen, apprentices, and casual laborers but of new-fledged factory workers as well. The small master artisans and the petty shopkeepers were, or felt, economically and socially threatened by the machine, the factory, and the enlarging retail store. Consequently, along with the substantial bourgeoisie and most of the cramped professionals and civil servants, they put up with—even championed—a conservative restabilization. The independent lower middle class did so in the expectation that the old order would prolong and shore up its vulnerable economic and social position with protective measures; and it readily paid the price of continuing political disenfranchisement for this new lease on life.

In 1848 the independent lower middle class still faced *both* ways: it formed the swing sector of the urban mass that first propelled the liberating thrust of the substantial bourgeoisie, but it then fueled the conservative recoil and retrogression. Thereafter, and most certainly after 1871, the petite bourgeoisie increasingly looked in one direction only: backward.

Between them, Pirenne and the founders of Marxism provide a suggestive chart of the long trajectory of the "old," independent lower middle class from the late Middle Ages to the middle of the nineteenth century, Pirenne being interested in its point of entry, Marx and Engels in its point of exit. According to Pirenne, the petite bourgeoisie sprang from among the "floating scum" of late feudal society, and it was to be celebrated as the glorious progenitor of capitalism. Marx and Engels, however, saw the small merchants, traders, shopkeepers, and handicraftsmen of the midnineteenth century as a "transition class" [2] that was "tossed about between the hope of entering the ranks of the wealthy class, and the fear of being reduced to the state of proletarians or even paupers." [3] Instead of glorifying a petite bourgeoisie that was suspended between hope and fear, Marx and Engels warned that, in the crunch of a revolutionary crisis, it would seek to save its special social position that had an obsolete material base, joining a coalition against the claimant underclass.

II

The lower middle class was expected to shrink drastically in importance, perhaps even to sink altogether into the proletariat; but this expectation, which was so widely shared, has not been fulfilled. Even the old lower middle class of petty independents has displayed an extraordinary longevity. To be sure, since the midnineteenth century the economic and numerical weight of small craftsmen, tradesmen, and peasants has declined strikingly. But this decline has been less rapid than commonly assumed. In fact, now and then the number of small shopkeepers and service operatives actually increased, especially under conditions of acute economic distress.

In addition, a "new" lower middle class of dependent clerks, technicians, and professionals expanded quickly and, in terms of numbers, more than compensated for the gradual withering of the "old" strata of independent small producers, tradesmen, and peasants. This *second* "making of the lower middle class," with the new dependent strata in ascendance and potentially predominant, was a by-product of the rapid development of industrial, commercial, and financial capitalism, which also stimulated the forced-draft expansion of the working class. By the last third of the nineteenth century, this second birth of the petite bourgeoisie coincided with and was stimulated by the swift growth of government bureaucracies, schools, hospitals, and armies.

Accelerated urbanization was another characteristic feature of this transformation: the growing cities, both old and new, became the natural habitat for the burgeoning white-collar sector of mere clerks, technicians, and professionals, many of them civil servants. But in a very fundamental sense, the old proverb *Stadtluft macht frei* lost its resonance: the sprawling urban milieu became a particularly fertile ground for culturally mediated enslavement. Because of their high literacy the new-fledged petits bourgeois, together with the traditional *Kleinbürger*—rather than the less educated industrial workers— became the first avid consumers of commercialized popular culture, made possible and profitable by ongoing advances in the technology of mass communications. Media and audience found each other. The penny press, followed by the illustrated periodical, the six-shilling book, the lantern slide, the radio, then the film, and finally television, catered to a mass audience of literate but essentially uncultivated petits bourgeois. The substantial bourgeoisie, especially its women, is likely to have been the prime, not to say exclusive, audience for the novels of Flaubert and Zola, as well as for the fiction of second-rank writers. But it seems to have been the lower middle class, both old and new, that consumed the incipient popular culture that, in contrast to a spontaneous folk culture, was contrived, standardized, and disseminated from on high. The modern opiate-dispensers provided a heady mix of entertainment, distraction, diversion, and fantasy for a composite lower middle class that was characteristically insecure. Whereas its old segments were increasingly fearful of economic and social decline, the new segments were forever anxious about their uncertain chances for future economic and social advancement.

Although the two segments of the petite bourgeoisie may have had rather different attitudes toward advanced capitalist industrialism and commercialism, for both segments the manual worker became the indispensable negative reference point. It may be worth exploring the extent to which the purveyors of popular culture—its business entrepreneurs and government administrators as well as its creative intellectuals, artists, and scholars—contributed to reinforcing the petite bourgeoisie's sense of superior dignity, place, and milieu over the blue-collar proletariat by manipulating symbols, stereotypes, and norms through words, sounds, and images. For it was this sense of honorable status that gave unity to an otherwise disconcertingly heterogeneous lower middle class. Moreover, while enacting protective legislation for the benefit of the petite bourgeoisie, the ruling class played upon this same craving for recognition when rallying the petite bourgeoisie into a mass movement to counter the socialist challenge.

Clearly, then, in the history of Europe the lower middle class has differed and changed with time and place. But whatever these variations, it has always been a critical mass and, in moments of crisis, a *critical swing group*. Politics may never have been carried on *for* its benefit, and certainly it was never a class *for* itself. Even so, the lower middle class has been a class *in* itself: it has demonstrated its proclivity and capacity for fitful insurgency although it has never managed to mobilize and organize itself for independent and sustained political action. Actually, without the support of the lower middle class the bourgeoisie could not have conquered, consolidated, and defended its prime or coequal position in modernized and modernizing societies.

III

The place of the lower middle class in pre-Nazi Germany, Lenin's Russia, and Middle America suggests some of the larger contemporary concerns that might conceivably stimulate scholars, intellectuals, and politicians to define and examine the lower-middle-class problem.

Between 1866 and 1914 Germany experienced strikingly rapid and intensive industrialization, urbanization, and alphabetization. Even so, at the outbreak of the First World War, the Continent's farthest outpost of industrialization was far from polarized between a minute class of *grands bourgeois* and a vast mass of proletarians.

Admittedly, there were some thirteen million industrial workers in Germany, although at the very minimum many labor aristocrats in big industry and workers in small enterprises were of petit bourgeois disposition. In any event, clearly separated from this vast proletariat in the labor force were 2 million white-collar employees, 2 million lower and middle civil servants, 1.5 million traditional craftsmen and artisans (of whom close to 500,000 still were guild members), 700,000 retailers, and around 2 million small peasants.

It is generally agreed that, by the turn of the century, Germany's white-collar sector was expanding rapidly: between 1883 and 1925 the number of white-collars increased more than fivefold while the number of industrial workers less than doubled. There was now one employee for every five workers. But perhaps it should also be noted that, at this same time, the number of small retailers, which is so often assumed to have declined significantly, actually grew faster than either the population or the national product. During the first quarter of the twentieth century the number of retail stores rose from roughly 700,000 to 850,000, an increase of some 20 percent. By the time of the big crash in 1929 the small sector had expanded by an additional 3 percent.

German economy, society, and polity were severely strained precisely because the prodigious development of industrial, commercial, and financial capitalism was matched by an equally powerful and complex resistance to it. Consequently, the intermediate strata, old and new, which became the popular carriers of this resistance, assumed major political importance. From within the establishment, embattled conservatives sought to enlist the composite lower middle class against the forces of modernity. In the socialist movement, Eduard Bernstein, in his iconoclastic *Evolutionary Socialism* (1899), warned against losing sight of the *Kleinbürgertum* and throwing it into the arms of the ruling classes. Among intellectuals, Werner Sombart, Johann Wernicke, Emil Lederer, and Theodor Geiger sought to provide a conceptual frame with which to deal theoretically as well as politically with this refractory and fragmented class.

At the same time as the composite lower middle class attracted increasing political and scientific attention in late imperial and early republican Germany, toward the end of the Russian civil war Lenin examined it from the vantage point of bolshevism.

In *"Left Wing" Communism: An Infantile Disorder,* written in March and April 1920, Lenin noted, above all, that "very, very much of small scale production still remains in the world, and small scale production gives birth to capitalism and the bourgeoisie continuously, daily, hourly, spontaneously, and on a mass scale." [4] In his preplanning cast of mind, Lenin insisted that even after the revolution the proletariat would "for a long time remain *weaker* than the bourgeoisie" not least because "the small commodity producers in the land which has overthrown the bourgeoisie spontaneously and continuously restore and regenerate capitalism and the bourgeoisie." He saw quite clearly that it was "a thousand times easier to vanquish the centralized big bourgeoisie than to 'vanquish' millions and millions of small proprietors, who by their everyday, imperceptible, elusive, demoralizing activity achieve the *very* results desired by the bourgeoisie and which restore the bourgeoisie."

Using a historical frame, Lenin noted that in the past, in times of crisis, such fledgling capitalists—agrarian, commercial, industrial—had at best been won for a *"petty-bourgeois revolutionism,* which smacks of, or borrows something from anarchism, and which differs in all essentials from the conditions and requirements of the sustained proletarian class struggle." But even this commitment was mercurial and ephemeral. The "small proprietor . . . who, under capitalism, suffers constant oppression and very often an incredibly sharp and rapid worsening of conditions of life and even ruin, easily becomes extremely revolutionary, but is incapable of displaying perseverance, ability to organize, discipline, and firmness." From a

Marxist perspective, as interpreted by Lenin, "the petty bourgeois, 'furious' over the horrors of capitalism, is a social phenomenon which, like anarchism, is characteristic of all capitalist countries." True revolutionaries needed to be on their guard concerning this petty bourgeois revolutionism, which was unstable, barren, and prone to swift transformation "into submission, apathy, fantasy, and even into a 'mad' infatuation with one or another bourgeois 'fad.' "

For Lenin, outside the *grande bourgeoisie* and nobility, whoever was not a genuine and full-fledged proletarian was a narrow-minded and fickle petit bourgeois; craftsmen, artisans, shopkeepers, small and middle peasants, low- and middle-level civil servants, and labor aristocrats all fell into this category. In other words, his definition was so sweeping that all except select sections of the intelligentsia were covered by it. After the seizure of power, it would be easy enough to dislodge decisively the big landlords, big capitalists, and upper government bureaucrats. Moreover, with the *grande bourgeoisie* and high civil servants replaced by Bolshevik party cadres, the dependent low-echelon government officials as well as labor aristocrats would be easy enough to bring into and keep in line.

But it would be far more difficult to assimilate and control those economic and social classes that, though part of the same "petite bourgeoisie," broadly defined, had elements of economic independence in the form of capital, land, or tools. Although these independent strata, notably the "small commodity producers," would have to be gotten "rid of," they could not be *"driven out* or crushed." The revolutionary proletariat would have "to live *in harmony"* with producers who "encircle" it on "every side with a petty-bourgeois atmosphere which impregnates and corrupts the proletariat and causes constant relapses among the proletariat into petty-bourgeois spinelessness, disintegration, individualism, and alternate moods of exaltation and dejection." These independents, including the peasantry that had just seized the land of the large landholders, "can (and must) be remoulded and re-educated, but this can be done only by very prolonged, slow, cautious organizational work" by the centralized, disciplined, and vanguard party of the class-conscious proletariat.

In short, after the seizure of power and the end of the civil war, Lenin conceded that in Russia as elsewhere, there would be a large, economically productive but politically unreliable petite bourgeoisie to contend with. Marxists and liberals had long since taken to using the term "petit bourgeois" as one of denigration, abuse, and contempt. Although Lenin had contributed to this defamation of the petite bourgeoisie, he also realized that this classless class or half-class of quasi workers and quasi bourgeois would continue to be more than an economic, social, and political anachronism or relic.

In the first half of the nineteenth century, the United States may well have been the closest thing there has ever been to a country of independent small producers and property owners. The United States was what might be considered an exemplary *lower-middle-class* nation, society, and culture. In terms of its heritage and early beginnings, however, it began to live by the myth and dream of *middle-*class America. This spurious vision of a *middle-*class rather than *lower-*middle-class society and culture persists to this day: it has survived a century of capitalist development in which the petty independent proprietors yielded first place to captains of industry, factory workers, and white-collars.

Presently, the myth of middle-class fulfillment and promise is being reconsecrated by America's emergence as the first modern nation in which nonmanual workers distinctly outnumber manual workers in the labor force, due notice being taken of the preponderant place of women among clerical and sales employees. While other advanced industrialized countries have not crossed this threshold as yet, they nevertheless are approaching that critical transition from a wage-earning to a salary-earning society. But the Middle America that is the unexamined prototype for so many other industrialized or industrializing societies is anything but middle-class or bourgeois. It is, rather, a uniquely lower-middle-class nation whose labor force is preeminently nonmanual, modestly salaried, and totally dependent. The fast growing manual and wage-earning labor aristocracy is another critical segment of this Middle America in terms of income, property, life style, aspirations, and self-perception—and so are the small independent retailers, building contractors, and service operatives.

This lower middle class has a compliant but also strained relationship with the upper establishment, which it both aspires to and resents. Its relations with those "beneath" it—the underclass of essentially unskilled and ethnically disadvantaged and trapped manual workers—are becoming increasingly strained as well. In what may well be a chronically unsettled economy, the rising income, standard of living, and aspirations of the petite bourgeoisie are in jeopardy. More important, its prestige—its status and self-esteem—is being undermined, for two important reasons. To be lower middle class is no longer a conspicuous badge of social achievement in a society that is predominantly nonmanual and nonproletarian and that is typically Middle American. Moreover, the equalization of economic rewards, life chances, and modes of living between the *mere* white-collars and the skilled blue-collars undercuts yet another source for real and felt superiority.

In other words, in the United States, the presumed differences between large sectors of the lower middle class and large sectors of manual labor are being eroded. As a result, those who have climbed the first rungs of the mobility ladder become apprehensive lest they or their children be blocked from rising higher or slip downward.

Under such constricting conditions, the lower middle class loses self-confidence and becomes prey to anxieties and fears which, in the event of sharp economic adversity, may well predispose it to rally to a politics of anger, scapegoating, and atavistic millenarianism.

I V

By the late nineteenth century the old, independent lower middle class was neither fossilized, nor extinct, nor pressed or merged into the proletariat. For one thing, it braced its failing position by taking measures of collective self-help and by prevailing upon governments to provide aid and protection. More important, however, the same economic, technological, and bureaucratic transformation that devitalized key segments of the old lower middle class quickened the growth of a new lower middle class of dependents and independents in major cities and provincial towns. These declining and rising segments formed a syncretic lower middle class, a heterogeneous and often incompatible occupational, economic, social, and ideational mixture.

Nominalists argue that this constantly evolving lower middle class is too heterogeneous, indeterminate, and polymorphous to be given any single label or to fit into any rigorous conceptual definition. But the danger of creating a self-fulfilling sociological abstraction can be minimized by taking account of both the complex multiformity of the lower middle class and the concrete structural conditions which frame the political response of its principal components in times of crisis.

Any rigid conceptual definition of the lower middle class certainly risks being too general and static to be of much use to the historian. Even so, a preliminary and tentative definitional statement is an indispensable aid in any inquest of this nature. In these provisional terms, the lower middle class can be said to be composed of individuals (1) who earn their living by work that is not preeminently manual labor requiring steady physical exertion and that demands a minimum of alphabetization; (2) who by objective criteria (of income, wealth, education, residence, etc.) are neither upper nor lower class; (3) who are singularly self-conscious about being neither one nor the other, but aspire upward; (4) who are inclined to be highly individualistic in their pursuit of upward mobility; (5) who consider private property to be sacrosanct; (6) who are glaringly susceptible to personal cooptation and patronage; (7) who are bent on protecting or improving the life chances of their children; (8) who ultimately and particularly in situations of stress are more fearful of sinking down or back into dishonorable trades or manual labor than eager to rise into the absolute middle class; and (9) who coalesce for concerted political action only in such times of severe stress.

The lower middle class presents an exceptionally difficult challenge to social scientists, whatever their world view. Internally, it is divided into a multiplicity of components, many of which have incompatible economic interests, especially in times of acute economic contraction. But whereas the economic and social coherence of the lower middle class is fragile, its belief system is singularly cohesive; a thick ideational cement fills the cracks in the structure of common economic interests. Particularly in moments of intense social and political tremors, when these cracks widen, it becomes clear that the lower middle class is a class in and of itself, with significant elements of autonomous class consciousness. Even so, in such unsettled periods it has difficulties jelling as a class for itself, with the result that in the end it remains a political auxiliary of superordinate elites.

The Communist Manifesto proclaims that "the small manufacturer, shopkeeper, the artisan, and the peasant" are doomed to "finally disappear in the face of modern industry." But this political pamphlet also insists that rather than meekly resign themselves to their fate, these *Mittelstände* seek to stave off or slow down their economic superannuation and social degradation. In this quest for survival the lower middle class, determined to "roll back the wheel of history," readily goes beyond political conservatism to outright reaction.

Engels and Marx stress the homologies of the economic interests, mentality, and political conduct of the lower middle class: the *"mesquin* character . . . [and] short-sighted, pusillanimous, wavering spirit" of its business practices and operations are the key to its political vacillation.[5] Especially in the press of revolutionary turmoil, when it is threatened from below by the urban or rural proletariat, or by both, economic vulnerability and social insecurity predispose the lower middle class to cease wavering and avert further risk. In such conjunctures it backs up whoever promises to guard "property, family, religion, order" and to provide cheap credit as well as reduced tax rates.[6]

Admittedly, Marx and Engels, in their logical rather than historical projection, overestimate the pace and scope of the decay of this independent lower middle class, and fail to anticipate the vast expansion of the independent and economically "unproductive" lower middle class of clerks, technicians, and professions. Even so, they provide a valuable frame for examining the connections between the unstable economic and social position, the jejune world view, and the convulsive politics of the composite petite bourgeoisie.

V

What, then, are the principal occupational groups that can be classified as being lower middle class? Perhaps the most fundamental division is between the petty independent producers, merchants, and service operatives, on the one hand, and petty dependent clerks, managers, and technicians, on the other. As for the lesser or lower professionals, including teachers and professors, they are located at different points along the spectrum that runs from independence to dependence. There are also the small rentiers, such as retired persons and widows.

More is known about the master artisans and petty manufacturers (weavers, tailors, shoemakers, etc.) than about all the other independents combined. They have attracted the attention of historians because of their contribution to the growth of capitalism, their periodic revolts, their unsteady behavior in revolutionary situations, and their alleged assimilation into the insurgent industrial proletariat. Of course, until recently they also had the force of numbers, the more so since in the cities their workshops and homes were clustered in special districts, in separate neighborhoods.

But although they are less numerous and more dispersed, there are all the other independents as well: bakers, grocers, butchers, brewers; haberdashers, druggists, stationers; barbers, hairdressers, cleaners; tobacconists, cafe owners, restaurateurs, hotel keepers; repair operatives (for carriages, cars, appliances), haulers, taxi operators, truckers; and specialty retailers and franchised concessionaires. The vast majority of these independents operate enterprises that are "family" owned, controlled, and managed. Moreover, the principal owner works—with his own hands—either by himself, alongside one or more members of his immediate family (wife, children), or side by side with a small number of hired employees. It is characteristic of these independent proprietors and operatives that they deal directly with the consumer and that they commit most if not all of their limited capital to their enterprises. As a result these enterprises are singularly vulnerable in the event of the prolonged illness or death of the principal owner and in times of severe economic adversity. Rather than use their small stake as a vehicle for the maximization of profits, these petty proprietors avidly strain to preserve their modest holdings for the benefit of their descendents.

Small-scale shopkeepers, contractors, caterers, and service operatives occupy a much more prominent, influential, and dependable position in local society and politics in small and provincial towns than they do in large cities and urban centers. In the big cities they are geographically dispersed according to the social level of their clientele and the neighborhood in which their shops are located. Compared to their counterparts who are members of the ruling elites in small towns, the petty independents of large cities have minimal status, exercise little political influence, and are volatile in crisis.

As for the dependent segments of the lower middle class, they, too, are concentrated in towns, cities, and urban centers. Although they have a long history, the ranks of petty clerks, salespersons, administrators, and managers have grown massively since 1870. At the lowest levels their tasks are completely routinized, with at best limited scope for personal initiative, decision, and creativity. The operators of keyboard machines do work that is manual, though not menial, labor. In fact, in terms of dependency, of routinization, subordination, and (probably) income, the vast battalions of ordinary white-collars in sales, distribution, accounting, banking, and insurance are not unlike industrial workers.

But there are those important differences that set them apart from the blue-collars: the clean work environment, in which they soil neither their hands nor their clothing; the relaxed tempo and discipline as well as the shorter work week of the office and salesroom as compared to the industrial plant; the opportunity to deal with people or symbols instead of material objects; the chance to associate with individuals of higher status, income, and influence both on the job and after it; and the prospects for job security, promotion, and fringe benefits. Particularly on the score of social contacts, job security, and advancement, the mixture of illusion and reality tends to be weighted in favor of the former. Even so, at different times and places, when hard-pressed economically, segments of these "nonmanual workers" organize for collective bargaining. In particular, the clerks of the civil service, while normally forswearing strikes and partisan politics, quite often combine first to secure and then to preserve a special status of tenure, retirement, and economic benefits.

The higher grades of clerical workers, though no less dependent, identify most intensely with owners and top management. This white-collar aristocracy has as ambiguous and unreliable a relationship with the rank-and-file clerks as the labor aristocracy has with the rank-and-file workers. Department heads, administrative executives, and branch managers have higher salaries than ordinary clerks. Furthermore, in their middling administrative, sales, or service posts they have functions that, while largely routine, nevertheless involve varying degrees of predefined initiative, responsibility, and authority. Rather than provide leadership for the main body of nonmanual workers, this salaried and careerist aristocracy of dependent clerks sees itself as the anointed associate as well as the postulant member of the establishment.

Within the professions themselves there is also a considerable division between the lower and the higher ones. In addition, there has been a vast proliferation of specialties and numbers in recent times, notably in the low-status technical fields. Among these professions that have the oldest ancestry and remain the most numerous are the lower clergy (particularly in Catholic areas), petty military officers, teachers, and solicitors. Professionals of newer vintage are nurses, social workers, laboratory technicians, librarians, curators, bookkeepers, and draftsmen. The sons and daughters of the small peasantry, the urban petite bourgeoisie, and the labor aristocracy are the ones to enter these unexalted professions, not least because the necessary training is accessible in terms of both cost and time.

Veterinarians, pharmacists, opticians, accountants, engineers, dentists, architects, lawyers, and physicians, might be ranked in that order on a crudely ascending scale of social origins, time and cost of study, income, and status. With notable exceptions (among lawyers, doctors, and perhaps teachers), all these professionals are defined by their specialized technical or scientific training and practice and have little of that general culture which was the hallmark of the notables of the prestigious liberal professions in their infancy. Most of them are subject either to government certification or to the corporate standards and ethical codes of their respective professional associations, not to say guilds, or to both.

As for income, the overwhelming majority of these professionals, whether of high or low status, earn it in the form of a salary. Moreover, an ever increasing number of lawyers, doctors, and architects, who by tradition charge fees, honoraria, or retainers for their services, are becoming dependent, salaried experts with private and public corporations or government agencies and as such should perhaps be considered paraprofessionals.

Among the professionals, the intelligentsia, including the university professoriat, is particularly difficult to locate. Gramsci's notion of a hierarchy of intellectuals can help to define the petit bourgeois elements among them. According to Gramsci, there is a sharp distinction to be made between those few men of ideas at the summit, who create the new knowledge and art that nourish and inform the reigning hegemony, and the constantly growing ranks of intermediate and lower intellectuals and artists who interpret, diffuse, teach, administer, and apply the "authorized" knowledge and culture. These expanding subsidiary or auxiliary cadres form a special preserve in the self-perpetuating petite bourgeoisie in terms of recruitment, dependence, income, deferential attitudes, and world view.

In other words, in a taxonomy of the lower middle class there is no place either for the free-floating intelligentsia of independent wealth or for the artists of the unfettered avant-garde and self-abnegating *bohème,* whatever their social origins may be. Rather, the main body of the petit-bourgeois intelligentsia consists of intellectuals, scholars, experts, and artists who either are permanently integrated in existing institutions or seek temporary contracts, commissions, or assignments from established individuals, corporations, institutions, and governments. The two groups are equally dependent for their income, the one on steady salaries, the other on variable fees and retainers.

To be sure, the public image and tone for the entire intelligentsia is set by its commanding figures, who transcend their social origin and situation. But this tone and image do not convey any sense of the lower-middle-class realities which condition and perhaps govern the world of the intelligentsia. Even the professoriat of the universities and professional schools is predominantly petit-bourgeois. This is less the case in the humanities, most notably at the most prestigious universities and *grandes écoles.* But there can be no denying the lower-middle-class origin, outlook, income, life-style, and political posture of the rank-and-file experts and scholars in the more recently matured fields of the natural, technical, social, and behavioral sciences, both within and outside the academies and major cultural centers. The full-time students in these newer fields, as well as increasingly in the more traditional ones, are part of this same lower-middle-class universe.

Not unlike the newer and lower professional occupations, these modern sciences depend heavily on government programs and funding for survival and expansion. The same is true of writers and artists—actors, dancers, musicians, painters, sculptors. The rank-and-file and lesser officers of the creative and performing arts have long since abandoned the narrow and hazardous route of erratic commissions for the broad highway of fixed employment by government-financed or -subsidized institutions and by the mass media, including publishing.

One further sector of the work force must be considered for inclusion in the lower middle class: the labor aristocracy. Technical professionals, such as engineers, are formally trained in higher educational institutions and command knowledge that lies beyond the comprehension of the average workers. Labor aristocrats—foremen, supervisors, ultraskilled workmen—start as workers and acquire their special training on the job. Compared with the unskilled and semiskilled industrial worker, the labor aristocrat earns higher wages, has greater employment security, has more job satisfaction, exercises authority over fellow workers, and has closer contact with management. Moreover, he has better prospects for mobility either into low management or into the ranks of petty independent entrepreneurs. Compared with the ordinary worker, the labor aristocrat also harbors higher aspirations for his children, whom he encourages to study for one or the other new nonmanual professions.

VI

In periods of normalcy the superordinate elites see the lower middle class in a rather negative light. From above, the petit bourgeois is disdained for being mediocre, provincial, conformist, unambitious, parasitic, selfish, rigid, resentful, prudish, and moralistic. In recent times this contemptuous attitude has been further reinforced by the ease with which clerks, technicians, and professionals have turned—and have been turned—into zealous consumers of standardized merchandise, depersonalized services, yellow journalism, and commercialized popular culture.

Even in times of stress, incumbent elites have little to fear from the lower middle class. The *Kleinbürger* is languid by habit. Moreover, because he has a tangible as well as a psychic stake in the established order, he is disinclined to press for radical changes that are incompatible with the maintenance of the existing system. Besides, at the first sign of a serious challenge to their rule, incumbent elites remember or conjure up a more positive image of the lower middle class: they celebrate technicians and engineers for exercising delegated authority, foremen and supervisors for enforcing control and discipline, sales and clerical personnel for assuring the orderly distribution of goods and services, and civil servants for providing an essential link between government and citizen by delivering vital public services.

In social terms, the lower middle class is valued for being the shock absorber that helps brake the eruptions of the underlying strata. A buffer between capital and labor, or between landlord and peasant, it also serves as a bridge and mediator between them. Moreover, the petite bourgeoisie is the preeminent channel for social mobility: skilled manual workers can and do move into it from below, while from within its bulging ranks it raises its own spiralists to higher rungs on the income and status ladder. This lower middle class also serves as a net that cushions the fall of the skidders and the superannuated of both the higher middle class and the *grande bourgeoisie.*

Most notably, though, in certain conjunctures segments of the elite shift from snubbing or derogating the petite bourgeoisie for its philistine commonplaceness to glorifying it for its sterling virtues. Not unlike the virtuous peasant, the sober petit bourgeois is hailed for embodying those cardinal personal and social qualities that are being tarnished by capitalist industrial society: self-reliance, thriftiness, diligence, and moderation. The *Kleinbürger* is honored and flattered for being the principal standard-bearer of property, family, the "work instinct," religion, and patriotism.

Besieged elites also prize the petite bourgeoisie as a reserve army of political support. A class of but not for itself, it seems available for mobilization through the organization and manipulation of appeals, signs, and symbols grounded in the petit-bourgeois belief system.

While from above the power elite insouciantly disdains the lower middle class in good times and speciously flatters it in bad times, from below many workers look to the lower middle class as an escape route from the blue-collar world. Precisely because the aspiration to achieve petty independence or to secure white-collar posts is so common among workers, socialist ideologues have been bent on staining its image: they denounce the individual petit bourgeois for being pretentious and parasitic, and the class as a whole for having a false consciousness that entraps it into being the "bribed tool of reactionary intrigue."

As regards the way members of the lower middle class perceive themselves, the central point is that they are ostentatiously proud of being or having risen above the level of sheer physical toil. In other words, they see themselves as superior to those workers who have not crossed the line of strictly manual labor, which for them is the all-important divide. Needless to say, although the artisan and craftsman also works with his hands, he mixes personal labor with personal capital, and he also makes administrative and commercial decisions. In fact, the work ethic, independence, and integrality of this independent stratum of the lower middle class continued to be central to the ethos of the entire petite bourgeoisie until petty clerks, technicians, and professionals appeared in critical numbers. Until then, compared with the manual worker, the petit bourgeois, whether independent or dependent, was better off. He earned a higher income, owned more property, had more security, had greater possibilities for advancement, worked in more pleasant surroundings, and had a better chance for contact with higher-placed individuals both at work and socially.

VII

Perceptions aside, objectively the lower middle class until recently absorbed not only old, independent *Kleinbürger* who were becoming superannuated but also those elite workers who were inclined to break out or be lifted out of the proletariat. Furthermore, the lower middle class performed this dual absorptive function more frequently than it served as an escalator into the *grande bourgeoisie* that owns the means of production, exercises concentrated political influence, commands high social prestige, and has a cosmopolitan rather than parochial outlook.

On balance, however, the lower middle class is the up-and-down escalator par excellence of societies that are in motion. But because the petite bourgeoisie is so massive and populous, the movement into, within, and out of it is difficult to determine and measure. Even so, the effort should be made to study the different channels of economic, occupational, and social advancement, the extent of intra- and intergenerational ascent, and the role of education in the improvement of the life chances of children. Particular care should be taken to specify the larger societal context that conditions the means, processes, and rates of lower-middle-class mobility: account needs to be taken not only of the curve of economic activity and growth but also of the nature and rate of those technological, scientific, and bureaucratic developments that open up new career opportunities.

The lower middle class is generally assumed and said to be

uniquely gripped by *individual anxiety* and *collective fear* of downward mobility. This assumption calls for verification, care being taken to compare perceptions with the realities of the nature, pattern, and extent of social decline. Any such probe should be designed to distinguish between those petits bourgeois who are afraid of sinking back into the ranks of manual labor from which they have risen and those who are frightened of falling from the lower-middle-class stations in which they started.

No doubt there is a customary blend of objective fact and subjective misinterpretation in this petit bourgeois anxiety and fear about being deposited on the escalator that moves downward, into degrading occupations or the proletariat. But whatever the blend of fact and distortion that feeds this *grande peur,* it seems to be a reality, and one that invites closer investigation than it has received to date. Based on adverse economic or technological trends and stimulated by cyclical contractions, this panic is the root of the eratic and intermittently frenzied politicization of the lower middle class. Just as the uppermost classes do not make a habit of either lying down to die or committing suicide, so also the infirm segments of the lower-middle-class fight to reverse or slow down their superannuation and decline. In their struggle for survival they establish or bolster economic and professional organizations; they form political lobbies, pressure groups, or factions; they join or support political organizations, such as leagues, parties, or movements; and, ultimately, when cornered, they take to the streets and to the piazzas.

It is in such moments of extreme crisis that the vague sense of negative commonality—of being neither bourgeois nor worker—is transformed into a politicized awareness or consciousness of economic, social, and cultural identity. The economic incompatibilities and the status incongruities between the major occupational and professional segments of the lower middle class do not disappear. But in moments of soaring disequilibration, which bring existential issues to the fore, these *internal* strains and tensions become of lesser importance than the immediate conflicts with *external* economic and social forces. Although the frontiers with the other classes remain ambiguous and labile, these frontiers nevertheless continue to be lines of demarcation from other classes rather than bridges for fusion with them.

In other words, in spite of the fact that a grave crisis stimulates tensions and rivalries between and among the principal segments of the lower middle class, it also heightens their consciousness of a common social condition and destiny. Their occupational and professional self-awareness and their lingering sense of social separateness are sharpened to such a degree that they assume qualities of class consciousness. This transcendence from occupational and status awareness to political consciousness is only partly spontaneous. To be more than ephemeral it requires articulation, mobilization, and direction by political agents from within as well as outside the lower middle class. It is with the help of such agents that the petite bourgeoisie acquires elements of ideological specificity and direction. This ideological stock-taking and mobilization serves to reduce but not completely eliminate the internal divisions which are conditioned by the economic and social relations of each of the principal segments of the lower middle class.

A predisposition *against* revolutionary confrontation may well be the basic common denominator of the strikingly polymorphous petite bourgeoisie: few of its factions are likely to mount or join a frontal assault on the existing economic and social system either on the eve or in the midst of a crisis. But beyond this negative unity in defense of the status quo against a revolutionary challenge, the lower middle class provides support for each of the three major antirevolutionary positions: some segments of the lower middle class rally to conservatism, others to reaction, and still others to counterrevolution.[7] In brief, although the petite bourgeoisie is united in its rejection of radical revolutionism, it tends to remain divided over the means and objectives for the restabilization of unsettled polities, economies, and societies.

It may, however, be as misleading to stress the lack of tactical and strategic agreement in the lower middle class as to emphasize its underlying unity. To be sure, the petty producers of goods and services have different material interests and social relations from those of the petty clerks, civil servants, technicians, and professionals. But all alike have a tangible stake in the existing economic and political system that is the essential prop for their precarious survival in the economic, social, and cultural stations in which they also have such a heavy psychic investment. This being the case, the defense of the traditional rights of old, petty independents can quite easily be reconciled and combined with the advocacy of a square deal for new white-collars.

The bulk of the lower middle class can quite comfortably rally around a program, even an ideology, that advocates respect for order, property, individual effort and talent, open careers (for their children), and even-handed government; that denounces economic and governmental bigness, excessive economic and social inequalities, cultural innovation, sexual liberation, and cosmopolitanism; and that proposed reforms which, though radical in appearance, are compatible with the continuance of the existing economic, social, and cultural system. Without being too specific, such a reform platform can hold out the promise of a fairer distribution of income and wealth, of a restoration of self-determination, and of new legislation—economic, social, educational—favorable to the besieged intermediate class. There are, of course, some segments of the lower middle class—those for whom a conjunctural threat of economic and social decline aggravates latent inner anxieties—that are particularly prone to metapolitical appeals of a xenophobic and conspiratorial nature.

In sum, however amorphous the petite bourgeoisie may be in time of normalcy, in times of severe crisis it develops considerable internal coherence. In an early phase of destabilization it may join other groups in an antiestablishment radicalism. But the lower middle class stops short of participating in coalitions and confrontations that threaten to topple the economic, social, and political edifice in which it occupies a definite, though insecure, place. Eventually it even becomes an important social and political carrier of a restabilization that benefits not the petite bourgeoisie but those higher social classes and governing elites on whom it never ceases to be dependent and for whom it feels envy exacerbated by resentment. Although it may waver along the way, in the final analysis the lower middle class resolves its ambiguous and strained class, status, and power relations with the power elite above and the underclass below in favor of the ruling class. It is this inner core of conservatism, ultimately revealed in moments of acute social and political conflict, that is common to all segments of the lower middle class and that justifies treating them as a coherent phenomenon and as a significant historical problem.

18

Reflections on War, Utopias, and Temporary Systems

ELISABETH HANSOT

Utopias may be understood in at least two ways: as tools for the criticism of present society or as visions of a future good society within the reach of the present. As critical tools that give us both distance and perspective on the present, utopias are mental constructs or "thought experiments"; they are "temporary systems" wherein values which differ sharply from those of the authors' own societies shape the institutions and the lives of men. But when social change is distinguished from natural change and is perceived to be subject to men's control, the temporary, heuristic character of utopias diminishes; then utopias become "permanent systems," serious blueprints for a future end-state to which societies should aspire.

As temporary or permanent systems, however, utopias almost uniformly postulate societies in which military institutions have either been eliminated or so transformed that their activities are merely protective and reactive. And war, considered as a state of armed conflict or preparedness for conflict, either never occurs or is only remotely possible. Let us, then, take a look at utopian societies and their explanations of war; and since we will consider utopias as both temporary and permanent systems, we will ask not only whether their remedial prescriptions for the eradication of war are plausible but also what would be the cost of living in a society in which war does not occur. The latter half of this question may appear perverse—who,

Elisabeth Hansot is a senior associate with the National Institute of Education in Washington, D.C.

after all, desires more frequent and extensive warfare?—nonetheless, the question warrants attention.

I

Plato is highly skeptical that pointing out the causes of dissension within society will make any difference. For him, conflict occurs because men are ignorant: they have no single mark at which to aim. They are content with the shifting, kaleidoscopic form of knowledge which is based only on opinion (*doxa*); the frenzied lawlessness of the multitude, expressed in limitless appetites and unbounded ambition, is totally at variance with the norms of permanence, intelligibility, and completeness that obtain when right opinion (*ortha doxa*) or reason (*epistēmē*) govern men's conduct.

Plato is unique among utopian writers in providing for a warrior class within his utopia. Men whose dispositions and training suit them for warfare, the warriors exist to protect the polis against threats of aggression issuing from "outside." The communism under which the utopians live precludes the unsettling influence of extreme wealth and poverty and prevents them from falling prey to the luxury, idleness, and avarice typical of wealthier states.

Such a regime creates stable character traits that prevent the warriors from being tempted to subvert their own city. And Plato comments that in case of war, the warriors' task will not be difficult: any state with which the Republic might engage will in fact be made up of two states—the rich and the poor. Such states are easily subverted from within by exploiting the rancor and disaffection between classes. Moreover, the Guardians or philosopher-kings are, in Plato's words, primarily preservers and protectors: they have no incentive for initiating war so long as the Republic is not threatened from the nonutopian states that surround it.

There is no special warrior class in Thomas More's Utopia; instead, all the inhabitants bear arms in a citizens' militia when required. But as in the Republic, care is taken to shield the Utopians from the habit of violence. Utopians are forbidden to slaughter their own animals for market lest the activity encroach upon that "finest feeling of our human nature," mercy. They use foreign mercenaries to fight their wars, and the vast treasure kept in Utopia is used solely for the hire of mercenaries. The Utopians primarily fight defensive wars against incursions into their own territory or that of their allies although war is also undertaken to free an oppressed people from the yoke of a tyrant (it was in this fashion that Utopia initially acquired friendly neighbors).

Like Plato's warriors, the Utopians prefer to conquer their enemies by fomenting dissent and encouraging subversion from within; only when such stratagems fail do they reluctantly engage in battle. They celebrate the victories won by strength of intellect as peculiarly befitting humans; in comparison, the victories won by physical prowess are disdained because animals can always excel humans in fierceness and physical strength.

Thomas More attributes the causes of war in society to pride in all its forms: avarice, ambition, conspicuous consumption, vainglory. Essentially based on spurious distinctions, pride measures where a man is *in relation to another*—wealth by another's poverty, dominance by another's subjugation.

Both Plato and More, then, locate the causes of conflict in certain tendencies of human nature, tendencies which may be curbed by social arrangements, but not extirpated. Nonutopian societies, by contrast, reveal what happens when social institutions encourage rather than curb men's depravity. The resultant unchecked expansion of human appetites leads by implication to the expansion of the states which are their instruments. But within utopia, the belligerent tendencies of human nature are made to serve the common good. Plato's and More's utopians are armed but not belligerent, with enough imaginary distance between themselves and the nonutopian world to offer a buffer to conflict.

Both Plato's and More's utopias are spatially limited societies, contemporaneous with their own. Most eighteenth- and nineteenth-century utopias, by contrast, are global societies situated in the future.* When utopias become future-oriented, they also tend to be portrayed as universal states. This tendency reflects technological devel-

---

* The terms "spatially limited" and "future-oriented" align for the most part with utopias written before and after the sixteenth and seventeenth centuries. Spatial utopias like Howells' and Skinner's continue to be written, but they are basically like future-oriented utopias in that they locate the causes of human conflict in social institutions and consequently view human nature as capable of being bettered, not merely curbed, by utopian institutions.

opments as well as the absence of unoccupied geographical space that might act as a shield for the utopias. A state that wishes to preserve itself inviolate in the world must either be shielded by distance from nonutopian states or must have the power to control other states; such control implies the responsibility of ruling over the subject states, and this leads to the establishment of a universal state. Moreover, since a future-oriented utopia is *attainable* by the conscious volition of right-minded men, it is open to the charge of moral obtuseness if it does not try to help less fortunate societies progress toward utopia. A future-oriented utopia, therefore, tends to be universal in scope in order to avoid charges of imperialism or self-satisfied indifference to the presence of the less fortunate.

Locating utopia in the future emphasizes the possibility and the need for moving present-day society in the direction of utopia. Men are not perfect in Plato's and More's utopias; they are simply fortunate to live in a society that curbs the bestial parts of man's nature (pride, the passions) and provides stable support for those elements which are distinctively human. The assumption of future-oriented utopias is that flaws in present social arrangements, not human nature, are the cause of conflict in society: as Theodor Hertzka puts it in his *Freeland,* "Thoughtlessness and inaction are, in truth, at present the only props of the existing economic and social order." Men's ignorance thus becomes remediable, and deliberate social change is the solvent powerful enough to dissolve the habits that sustain obsolescent social arrangements. The image by which morality is understood in recent utopias is clearly movement, not vision; as Iris Murdoch notes, the good in modern philosophy, instead of being an object of contemplation, is an attribute of the will.[1]

Universal utopias contain a broad range of explanations for conflict in contemporary society—economic inequity, leading to unbridled competition and unfair distribution of goods in Edward Bellamy and W. D. Howells; men's tendency to form aggregates based on occupation, class, religion, nation, or race in H. G. Wells. If the future-oriented utopian author's diagnosis of the causes of social malaise seems to us too sweeping, it is rivaled in breadth by an optimistic estimate of men's abilities to work deliberately and consciously to change the offending social and economic arrangements. A primary emphasis of many nineteenth-century utopian authors is on the redistribution of wealth, and the societies they designed are ones in which economic production and distribution are regulated in such a way that men have the freedom to develop their talents and capacities. Wars result from economic anarchy or racial or national parochialism, and their remedy is the deliberate, often evolutionary, redesign of the institutions which support such chaos.

It is precisely here that a curious paradox arises. When utopias become universal, they avoid the problem of what to do with non-utopian neighbors only to see it emerge in another form: what attitude should utopia take to its own history? To ask what should be the utopians' attitude toward nonutopians is to question what function, if any, a knowledge of preutopian misfortune, egoism, or evil will have in utopia.

The answer is, not much. Spatially limited utopias, if literally thought into existence, have the surrounding nonutopian world as a reminder of what life is like outside utopia. Lacking nonutopian contemporaries to reinforce a sense of their difference, future-oriented utopias must retain a vivid understanding of the history from which they emerge. That, however, is extraordinarily difficult to do. One example will suffice.

In Bellamy's utopia (located in Boston in the year 2000), a historian, "moved by a certain crabbed sense of justice," decides to make a record of past events called "Kenloe's Book of the Blind." But as the heroine of *Looking Backward* admits, the "Book of the Blind" has become irrelevant not only because the utopians have sight but also because they have diagnosed the cause of the few remaining cases of misbehavior among them to be atavism—and indeed, what other explanation can there be if human depravity is understood exclusively as a symptom of institutional malfunctions?

The example is telling: Bellamy's utopians are ahistorical. Utopians have no reason to interest themselves in a history that has no bearing on the present and of which they have had no experience; to do so would be to indulge a degree of morbid curiosity of which utopians are innocent. Utopians live in an eternal present in which human nature is at one with itself, neither changing nor undergoing change. The paradox of a future-oriented utopia is that its raison d'être is extrinsic to itself and derives from pre-utopian problems that utopians can no longer understand.

Even in nineteenth-century spatial utopias, such as W. D. Howells', the problem of coping with the past appears intractable. The mutinous crew of a trading ship docks in Altruria. When the Altrurians hear of the squalid conditions in which the sailors live, they feel horror and pity. But as the incident is further described, it becomes clear that the pity and horror the Altrurians experience is uncomprehending; it does not work in the direction of greater understanding or enlarged sympathy because there is nothing in the utopian nature or experience in which to anchor such emotions.

Let us, for a moment, refer back to the two issues with which we started this discussion: the utopian diagnoses of the causes of war and the costs of living in a peaceful society. While utopian authors do identify a wide variety of ills which cause conflict in their own societies, they tend to divide into two broad camps by locating the underlying causes either in human nature or in societal arrangements. The responses of utopias (imagined to be in existence) to the problem of depravity in nonutopian societies divide accordingly. When the cause is judged to be in human nature, the utopia maintains a defensive posture, primarily concerned to protect the relative innocence or rectitude of its own society. When the cause is judged to be primarily societal, the problem is avoided by making utopia a universal state, or the problem remains unsolved to the degree that the inhabitants have lost touch with the experience, feelings, or habits that maintain nonutopian institutions.

When we launched this brief excursion into utopian thought, we also asked what might be the price of living in a society in which war was no longer a possibility. The question of course must be modified. In Plato's and More's utopias defensive wars are certainly possible, and their citizens retain a very vivid and operational sense of the causes of aggression, if only to better exploit the weaknesses of the aggressor. More recent future-oriented utopias have no need of defenses. But in the process of achieving utopia they seem to have paid a high price, a price alluded to by Howells when he calls Altruria a civilization so fundamentally alien to the American one as to leave "a sort of misgiving as to the reality of the thing seen and heard." That is, a sort of vitality, a restlessness, is lacking in these societies, and this is due to the congruence of human nature and social institutions.[2] In earlier spatial utopias, tension is generated and maintained by an ambivalent concept of human nature, a concept that is reflected in the defensive posture of these states to nonutopian societies. This tension is missing in future-oriented utopias where for want of any causes or purposes common to men, utopians drift off into the privacy of individual self-development and self-expression. Even politics, the art of maintaining the arrangements of society, disappears in all but name. Utopian achievements may be real, but they do not signify. Despite Wells's hope that utopia would unleash the creative energies of men, such energies are harnessed to no common needs and stand for no enduring qualities.

II

The price of utopia, were such a state ever to come into existence, may be the loss of significance, the loss of belief in that combination of qualities and accidents that give hope to some people that what they do may make a difference. In *The Warriors,* Glenn Gray describes his World War II experiences on the Italian front in language that captures brilliantly that sense of significance. Gray speaks of the "secret attractions of war" as the delight in seeing, the delight in companionship, and the delight in destruction.[3] War is a spectacle and there is in all of us what the Bible calls "the lust of the eye." The eye is lustful because it seeks out the spectacular and novel; it is this passion (which precedes in most of us the urge to participate) that moves people to crowd around the scene of an accident or fire. In war the fascination that power and its sheer magnitude hold for us, as in aerial combat, permits the spectator to lose himself in the majesty of the spectacle.

Another appeal of war is the communal experience, the sense of comradeship. The excitement generated by assembling people to achieve a common goal is heightened by the increased vitality afforded by the presence of danger; together, the experience produces feelings of release and liberation. Individual freedom requires reflection and choice; it is hemmed in by uncertainty and experienced as a burden. By comparison, the communal feeling of fellowship is experienced as a liberation from individual impotence; the "I" is capable of losing itself in a superordinate "we" that becomes the carrier of meaning and purpose.

A third and more sinister impulse Gray describes as the delight in destruction. This lust to kill is the impulse within us that works for the dissolution of everything living and united. Gray calls the satisfaction in destroying a peculiarly human impulse—or peculiarly devilish in a way that animals can never be. The delight in destruction has an ecstatic character, stemming not from the loss of ego but from a heightened sense of the self and its power.

In Gray's careful, concrete descriptions of the secret attractions of war, we recognize parts of ourselves—indeed, we recognize just those capacities, with what is both fine and degrading in them, that are absent from the texture of utopian life. And in his chapter on "Play and War" in *Homo Ludens,* Huizinga describes yet another element found in play and archaic combat, the psychological and social significance of which remains undiminished in modern society.[4] That element is honor. A person's or a group's honorable qualities must be manifest to all, and, if that recognition is endangered, it must be reasserted and vindicated by agonistic action in public. The point, Huizinga notes, is that honor is not founded on truthfulness, righteousness, or any other ethical principle: at stake is social appreciation as such. Again, the description evokes our recognition, but not in a way that allows us to locate it or its analogue in utopian society. Peace is not only the absence of war, it appears to be the absence of certain human capacities or passions which have come to be partially associated with combat.

At the close of his meditation on war, Gray quotes from Nietzsche's *The Wanderer and His Shadow:*

Rendering oneself unarmed when one has been the best armed, out of a height of feeling—that is the means to real peace, which must always rest on a peace of mind; whereas the so-called armed peace, as it now exists in all countries, is the absence of peace of mind. One trusts neither oneself nor one's neighbor and, half from hatred, half from fear, does not lay down arms. Rather perish than hate and fear, and *twice rather perish than make oneself hated and feared*—this must someday become the highest maxim for every single commonwealth too.[5]

The issue Gray raises is again the one with which we began this discussion: what are the costs of living in a peaceful society? The utopian answer, at least in its literary form, appears inadequate because it is grounded in an oversimplified view of human nature.[6] Peace and private contentment in utopia appear to produce a one-dimensional man incapable of purposes and projects beyond himself. We are not convinced by utopias, most frequently because we do not sufficiently recognize ourselves in their citizens.

There is no easy answer to the utopian dilemma, nor to Glenn Gray's. How does one picture a world without conflict in which, for instance, the capacity and need for camaraderie is given adequate purchase? In our repugnance for war, do we deliberately overlook the opportunities it affords for spectacle, for ceremony, the sheer delights of contest and the opportunities for honor? One response to the dilemma simply states: the game is not worth the candle. Human life is too high a price to pay for such opportunities. Another response argues that the human animal is infinitely ingenious, and when he stops waging war he will discover other outlets or vehicles for the qualities war now occasions. I am not sure that "substitutes" for war are possible or that, like members of the animal kingdom, we can limit intraspecies conflict to ritual feinting and symbolic submission.[7] If being human is to be halfway between angel and beast, it would seem that movement both up and down that chain of being is mysterious.

War is a complex social institution, a vehicle by which the great range of psychological, economic, and political motives of a society will, at times, express themselves. I am not at all clear as to the significance of institutions within our culture which seem to afford opportunities for some of the experiences associated with war. Nonetheless, I would like briefly to allude to several activities which take place in peaceful societies and which afford experiences similar to those provided by war.

Huizinga published his seminal anthropological study of play in 1938; there he describes play as "a free activity standing quite consciously outside ordinary life . . . [and] being 'not serious'. . . . It is an activity connected with no material interest. . . . It proceeds within its own proper boundaries . . . according to fixed rules."[8] Play is make-believe in that it requires us to assume a role, knowing that it is a role, for no useful end; or, as Paul Weiss puts it, play invites improvisation and imagination because it is serious with respect to the domain it bounds but not with respect to what is done in it.[9] Play takes place in a heightened and concentrated present, in one dimension of time. It issues from no prior events in the past and has no consequences for the future. Nothing changes as the result of play; the rest of the world is left intact. In contrast to the complex contingency of the work world where it is difficult and often impossible to trace effects to their causes or to link actions with their consequences, play takes place in a realm in which the consequences of actions are usually immediately apparent and the success or failure of a stratagem is measured by how it contributes to the outcome of the game.

Games and contests often begin and end with ceremonies that serve to set apart the distinctive time, space, and causality in which games occur.[10] The ceremony helps the spectator make the transition and mark the difference between one set of events and another. It also signals, in Eastern cultures, the intent to respect the rights and dignity of the opponent and thereby the nature of the spectacle about to unfold.

Like war, many games aim at victory and ideally end when one side has demonstrated superiority. But the differences are important. War often takes place without respect for the rules or conventions which are meant to limit it. The objective of war may be to annihilate the opponent, not to best him in a fair contest. As Plato and More remind us, in war improvisation is limited by human ingenuity, not by any pacts or rules agreed upon ahead of time. In war the contestants seek to discover how much power they have; in sports, how effectively to use that power to win under established rules.[11]

The differences are profound: Gray's delight in destruction can be accommodated by war but not by games. Nonetheless, it is useful when we consider play as a part of our culture to be aware of the ways it may respond to such human traits as the capacity for disinterested activity (this despite the avowed objective of winning), camaraderie in the pursuit of a single goal, the need for spectacle and for ceremony.

Play is an activity set apart from ordinary life, from past causality, and from future consequences. It is a temporary system—and a fragile one, in that it requires a willingness to be serious about nonweighty matters. From a different perspective, Matthew Miles has drawn our attention to the role of temporary systems within the more permanent and durable structures of our social life.[12] Some such systems are "total institutions" such as prisons, hospitals, and perhaps war itself, where the participants' departure is contingent upon the existence of a general state of affairs (rehabilitation, health, victory). Other temporary systems, like games and "task forces," end when a specific product is attained or when a specific event occurs (nine innings, a report). The temporary system phenomenon occurs in its purest form when the termination point of the system is explicitly timebound and when all members enter and leave the system at the same time, as in the case of a carnival or conference.

Some temporary systems appear to serve the purpose of maintaining persons in the surrounding social system by affording temporary breaks and expressive release from their normal routine activity. Vacations, games, recognition banquets, retreats, conventions, and carnivals can be so classified. They are primarily restorative—sometimes symbolic or ceremonial. But the systems that interest Miles the most are those set up to bring about change in a permanent system, necessitated because the latter's energies are often absorbed in carrying out routine operations and in maintaining existing relationships within the system. Such systems (workshops, conferences, training sessions, and the like) afford the opportunity to stand back from the usual pressures of work and focus on a particular issue or problem. There is a stepping out of real life into a temporary sphere of activity with a focus of its own and a commensurate release of energy which Miles compares with Huizinga's understanding of play activity.

But in the context in which Miles studies it, this isolation or temporary change of cultures serves another function: it reduces barriers to change by shearing away a person's (or group's) allegiance to things as they are. In addition, the isolation serves a protective function. Since life in the system is temporary, the penalties for making mistakes are reduced. The existence of limited time in which to achieve goals causes stress and tends to narrow the time perspective, forcing the person to live more in the psychological present with a commensurate single-minded attention to the goal or objective. To quote Miles:

The existence of superordinate goals, which one has personally helped to formulate, combined with the knowledge that the goals must be reached here and now, during the life of the temporary system, appears to cause an associated phenomenon—heightened *significance and meaningfulness* in the activity.

. . . the participant often reports that he is working at the height of his powers on the immediate, focused, short-run goals which he had helped redefine.[13]

There are dangers to the temporary system because the outcomes are uncertain and at best difficult to measure. Change for its own sake can become the norm. Alternatively, the "as if" character of temporary systems can invite feelings of infinite possibility, a tendency to support grandiose and unattainable goals. And often the engrossment in the temporary system leads its members to ignore the difficulties they will encounter when they take their product (and themselves) back to ordinary life with its routine conflicts, pressures, and vested interests.

Finally, Miles observes that a temporary system, like a utopia (conceived *not* as a permanent but as a temporary state of affairs), can be precisely designed to evoke the best possible contribution from its participants. The decline in the number of literary utopias in the twentieth century and their replacement by antiutopias which satirize or inveigh against the uniformity of a mass or mindless society has served to obscure the great increase of short-term quasi utopias—such as meetings, conferences, commissions, and training exercises. It is, Miles observes, "as if we had traded the grand visions of social life for miniature societies, to which one can become committed intensively, meaningfully, satisfyingly—and impermanently." [14]

III

Where, then, have these excursions into play, utopias, and temporary systems left us? When utopias are seen as visions of the future within the reach of the present, they give us a series of snapshots of what the warless society might look like. In the future-oriented, universal utopia the snapshots are hardly recognizable because some of the motives that fuel war—self-interest, greed, honor—seem inexplicably (and unconvincingly) banished. In the earlier, spatially limited utopia such motives are acknowledged, and those judged harmful are effectively curbed by carefully designed, interlocking utopian institutions.

War is a complex cultural institution and no single explanation of it remains plausible for very long. But one of the motives that has led men into war from biblical times is the will to preserve an environment—a society and its institutions—that is believed to afford leeway for men's values and nourishment for their capabilities. Like future-oriented utopias, this motive reflects a belief in the critical importance of institutions, as essential prerequisites if not as full determinants of human development. Paradoxically, the belief in adequate social institutions as the necessary matrix for human development often has not entailed a parallel belief in the potency of such institutions to curb unwanted human traits—Plato's avarice or More's pride.

War likewise has its paradoxes. It is often the occasion for experiences we value highly—honor, camaraderie, ceremony, and even spectacle occur in it in peculiarly heightened and poignant form. War is, it seems, frequently the carrier of just those forms of human experience found wanting in utopia.

The cost, however, is high, and war is not the only temporary system that offers the occasion for heightened and valued experiences. Play activity, carnivals, conventions, ceremonies are all in different ways vehicles for the sorts of intense and peculiarly absorbing human experiences we have described.

Perhaps it is the character of temporary systems, rather than war as a uniquely disruptive instance of them, to which we should attend. War is too interlocked with our culture and traditions to offer any easy alternatives: while there may be analogues to war in our society, these analogues are not substitutes. But because historians and philosophers have reached no agreement on the complex causality of war does not mean that we are without resources. If we are alert to the range and variety of institutions we wish to preserve, if we are aware of the lust of the eye to which ceremony or spectacle appeal, and the requirements of honor or sense of a heightened present which play activity answers to, we might over time come to view war with some of Nietzsche's contempt. Utopian explanations of how to get from here to there leave us unsatisfied. But they do suggest that once there, old necessities lose some of their binding force and we experience familiar values differently.

Certainly temporary systems are no remedy for war. The combination of luck and accident, of human will and social necessity that might make war an anachronism defies our imagination, much less our rational faculty. But if we are now frustrated because our institutions are not more effective curbs for man's malevolence (crime statistics are but one example), perhaps we can usefully concentrate our energies on developing, preserving, and extending the rich array of institutions that elicit his more notable qualities.

Education and the Good Society

I do not believe that a private education can work the wonders which some sanguine writers have attributed to it. Men and women must be educated, in a great degree, by the opinions and manners of the society they live in. In every age there has been a stream of popular opinion that has carried all before it, and given a family character, as it were, to the century. It may then fairly be inferred, that, till society be differently constituted, much cannot be expected from education.

Mary Wollstonecraft
Vindication of the Rights of Women, 1792

19

The University
and American Society

GEORGE W. PIERSON

My topic is "The University and American Society" as one aspect of "Education and the Good Society," with special concern for "the role of humanistic learning in public policy formulation." The subject is immense and complex, but so terribly important that I think it deserves our gravest concern. Obviously in a brief introduction one cannot even hope to do it justice. But perhaps I can make a useful beginning by offering three cautionary remarks and later, one cautionary tale.

I

In trying to relate the university to the good society and the humanities to public policy issues, it may be that we have undertaken a next to impossible task. For we can't agree on what constitutes a *good* society. We have "a thing" about issues or problems. And we don't know what the word *university* means. That is, we are trying to connect two unknowns by a very dubious link. Let me explain.

1. What is the *good society?* We don't know because we don't agree, we haven't agreed, we're not likely to agree. Within the "American consensus" we have made an uneasy compromise, a sort of unstable blend of certain powerful and sometimes antagonistic beliefs or visions of society (and we have reconciled them chiefly by

George W. Pierson is Larned Professor Emeritus of History and Historian of the University at Yale University.

ignoring them or by applying them one at a time). I will say more about these rather contradictory visions later; here I merely wish to observe that if we are fighting about what constitutes a good society, then we can't decide what would be "a good education" for that society, and the proper university becomes impossible. Or rather it becomes possible only if we apply the word *university* to everything or anything that people want, and so also change the word *good* to whatever suits the ambitions of our many social, economic, religious, or ethnic interest groups.

2. In the second place the suggestion is that we try to link the intellectual and humanistic university to the secular society by focusing on current issues and public problem-solving. It's an ingenious and engaging idea, and I think there's a great deal to be said for it. It's an awful lot better than trying to sell the humanities like a patent medicine. However . . .

We Americans have a problem with problems and issues. We overdo them. We're addicted to "problems": they've become a cultural habit. Anything and everything that bothers us we try to formulate into a problem, then we go to work on it and solve it. All "problems," in the American definition of course, are solvable. So we keep reminding ourselves that nothing's impossible and keep hoping that even the greatest "issues" can be resolved if only they can be stated and studied. So we attack crime through the environment and devise game plans against poverty and injustice. Pretty soon things change a little, and we've achieved a "breakthrough." We have a complex, too, about "reform." We almost worship the "new." And *every* change or novelty, of course, is good. It can also be dramatized. Americans love "innovations" and "breakthroughs." One could say that we cross watersheds at every subway station.

Problem-solving seems rational; it can sometimes be very useful; and it's good sport. Game plans are fun, after all, and they have the further advantage that there will always be new games. Happily so, because we've got this drug habit, this addiction to problem-solving, so bad it seems impossible to shake. Have you ever stopped to wonder what's happened to sin in America? We've packaged it . . . and feel much more cheerful on that account.

What I am trying to suggest is that not all evils or controversies or public discomforts can be shaped into problems; not all novelties are excellent; and not all "issues" are real. They may be superficial or narrowly selfish or obsessional or disguises for self-serving schemers. Thus, nine-tenths of the party issues in American politics have perhaps never been issues at all but strategic ploys to appeal to the emotional fears or prejudices of particular blocs of voters. Certainly the university should recognize that not all difficulties are "problems"; not all "problems" are solvable; and not all "issues" are urgent, or even issues at all. Above all, the university must not let passing fads or public clamor or neighborhood bullies determine which are the important issues or what problems it can most usefully study and try to solve. But . . .

3. *What is the "university"?* We in the academy are not agreed, and if we were, the nation might not be willing to accept our definition. What is the university? Perhaps at bottom that is the principal question before us. And I'm not neutral: I have a decided position. I am the partisan of a particular kind of university. My ideal is the liberal and independent university: that is, a great tree-like institution with its feet in the younger generation of ability and conscience and its crown in the skies. Its great trunk is a strong liberal arts college with a humanistic emphasis. Its branches are the graduate and professional schools of the most learned and intellectual professions, reaching for the highest learning. As a self-governing organism it exists, or it should exist, primarily to shelter teachers and scholars. And in return these privileged participants have certain great obligations: (a) to preserve and study the experience of mankind; (b) to pass on that cultural heritage to the next generation, i.e., to educate the young; and (c) to seek new truths and so enlarge or improve our civilization. To remember, to teach, to explore: these are the three commandments. And no doubt we should add: (d) to train the future teachers and explorers; and (e) to perpetuate and refine the skills of the great professions.

So I am old-fashioned in that I would first educate and then train the problem-solvers. And as for the problems, I would rather study and try to understand them than take a hand in applying solutions. I believe that individual professors and students should be entirely free as individual citizens to take a hand in problem-solving for society. But I do not believe that the university—as an institution—should undertake to solve such problems for the nation. And I have grave doubt how far the classrooms or the processes of instruction in the humanities should be bent toward the resolution or compromising of issues.

I also emphasize a liberal education, hopefully to be carried up into the top branches. To put it all metaphorically, "Nations without universities are treeless plains—universities without liberal colleges are hollow trees" (and even hollow trees cannot be reshaped overnight). Which brings us to the history of American colleges and a look at the rather curious road we have come. How were we shaped?

II

We were shaped by the three great visions of America. In the beginning, and indeed for almost three hundred years, there came and kept coming three kinds of people, or rather, people with three major ideas or visions of what America was or should be.

First, there were the Reformation Puritans who wanted to make a holy experiment or create a good society (we still sometimes use the phrase "good society" among ourselves). The Puritans wanted to build "a city on a hill." There were not many of them, but they were determined.

Then, in considerable numbers, America drew the freedom-seekers, those who wanted to get away from the lord or the bishop, the law or the tax collector. In the main their great desire was to stand on their own two feet, as free citizens.

Finally, there were the vast multitudes who wanted to better themselves economically: to get out of poverty, to own land, to eat-dress-live better, i.e., to prosper. All three groups obviously looked to the future, but they spelled the American future in differing ways.

There came also a few disappointed aristocrats, but otherwise hardly anyone looking for power. There were no great judges or doctors or churchmen, no scientists or artists or skilled craftsmen. And no scholars, no seekers of knowledge (except for the explorers, those curious about the land or the minerals or adventure).

So we started the Great Experiment with three great ambitions, a decapitated society, and a decapitated university—or one unit of what the English perhaps mistakenly called the university (for the English university had lost the schools of law, medicine, and divinity immemorially maintained for the learned professions by the Continental universities). The American "university" was therefore one college, not even a collection of colleges, and of course it had no professional schools. And what did the three colonial peoples do with their college—the "Saints," the "Independents," the "People of Plenty"?

I am speaking here, of course, in parable: I know there were few if any complete saints, or pure anarchists, or complete materialists, and I know that the pure virus of selfishness or idealism could not be passed down in the blood stream. The man on the make also hoped to be more free; the city on the hill did not wish to starve. We were and are of mixed motives. But for some Goodness came first; for others, Freedom; and for the masses Plenty mattered more than either. And so, again: what did these peoples, each with their dominant vision, do with the colonial college?

The Puritans believed in learning, took good and evil very seriously, and practiced a rigorous self-discipline. So they used the college to make a learned clergy and lettered citizens by means of the liberal arts, a reformed theology, and a disciplinary moral code for character training. With these materials they built no fewer than seven colleges to help make the New World a good society. But they built no universities; for they believed chiefly in religious learning and moral law, and there the great truths had already been revealed.

The Freedom Folk, the fugitives from oppression or justice: they built no colleges.

And the Property-minded weren't interested either. Or they weren't until Benjamin Franklin thought a business school would be a good idea and founded the University of Pennsylvania to foster that idea—prematurely, as it turned out. (The ninth institution, King's College, seems to have been a conservative Church of England venture, with some commercial ingredients added.)

Then the continent opened, and what did each of these groups try to do with the West?

The first little group, in little town-size swarms, tried to civilize the howling emptiness; they came across the continent, in the words of Henry Ward Beecher, "driving before them their lowing herds of churches and schools and libraries and lyceums," and also planting new denominational colleges, so many nurseries of faith and good citizenship, everywhere as they came.

The People of Plenty used the continent to produce, but very gradually they began to realize that their farming was using up the land and their new factories and railroads needed steam engine experts. So in the nineteenth century finally rose a demand for experimental farms and polytechnic schools and land-grant colleges for agriculture and mechanic arts. The early state universities were not true universities of the highest learning, either, or even just colleges of the liberal arts, but mainly clusters of technical schools for the utilitarian needs of society (for the People of Plenty).

The independent Republicans, the frontier freedom-riders and champions of democracy, saw no need of a college education or of instruction from the churches or of discipline by the classics. They did begin to notice some of the social advantages of going to college but were willing to study only what they wanted to study. So began the cry that all studies (like all men) should be equal and the demand for the so-called Elective System (another lineal descendant of this freedom faith is the Open University today). It seems to me that we can see these three great visions still struggling for the soul of America— and for control of our universities.

Perhaps we ought also to notice one other great lesson of the nineteenth century: the agony of disappointment and postponement. Again and again men saw what was needed long before the public could be brought to support it. I cite my own college for illustration: as early as 1777 at Yale, Ezra Stiles proposed a genuine university with three professional schools, but all he got was one professorship of ecclesiastical history which he had to fill himself without extra pay. Then in 1810–24 (some forty years later), Yale managed to open schools of medicine, law, and divinity, but few students came to take the higher training. Again in the age of Andrew Jackson, some professors saw the need of European science, but all the public clamor was for a little applied engineering. . .

Which they then wouldn't support. Witness what happened to Wayland at Brown or to the Sheffield Scientific School and Land Grant College at Yale, where the professors offered real engineering or agricultural chemistry, but the farmers wanted just barnyard wisdom, and the students preferred the Select Course of general studies with a Yale degree in three years and no Latin. Alternatively we might think about what happened to Eliot's "Elective System," which was *supposed* to favor the modern foreign languages and the new sciences. . . .

So we college people painfully learned (and I think we should remember today) that need may be one thing, public demand quite another, and public support still a country mile away. What is really needed is too often *not* what is demanded, and what is loudly demanded will often *not* get any public support.

This brings me to my second and third historical perspectives, which can be stated quite simply: our universities were very slow in coming, and they have never been popularly understood or appreciated. Let me offer a few words of explanation on each of these two historical interpretations.

III

Our universities were imagined almost from the beginning, but they were not achieved until after 1865, or almost 250 years after the hopeful plantations. The first true professional schools date only from the Revolution, or long after. The first true faculties or professoriat came only after 1815 and had to be trained in Europe. The first American Ph.D.'s came only in 1861, and it was not until the 1880s that we had enough professional scholars to form most of our professional associations, and not until after 1900 that we were able to perpetuate our own kind. Indeed it was not until after World War I (or even World War II) that research became a full partner in higher education, and the Remembering and Teaching College could grow into the Exploring University.

And who made all this possible? Not the public, nor the states, nor even the Congress, but the professors themselves and a few public-spirited millionaires. Thus in 1865, Cornell was begun on Andrew D. White's vision and Ezra Cornell's business acumen. Thus Johns Hopkins University was founded in 1876 on an unprecedented bequest of five million dollars put to unprecedented uses by D. C. Gilman. Thus it took Harper energy and Rockefeller oil to make the University of Chicago spurt up out of the prairies of Illinois. On the West Coast there were David Starr Jordan and a magnate named Leland Stanford; on the East Coast, Charles W. Eliot and the rich brahmins of Boston.

No, universities were not what the public demanded or supported or understood. I will even go so far as to suggest that, contrary to recent impressions, our universities have *never* enjoyed full public approval or understanding. It is true that no other nation has so many universities, in fact as well as in name. It is true that our colleges have been promoted to be the certificating agencies for general employment, and our professional schools for the giant corporations as well as for the professions. It is true that university science has become a factor in the world balance of power, and that university scientists, economists, and sociologists now staff federal bureaus. It is further true that our politicians no longer just scorn but now also fear our intellectuals. And it is true that the students, whatever they say about the university or do to it, pay it the compliment of trying to use it as their best weapon against society.

But it is also true, I believe, that Americans, by and large, have been and remain anti-intellectual; that the students would rather do than know; that professors, if no longer contemptibly ineffectual, are still not quite acceptable as 100 percent Americans; that excellence and elite standards are not quite American either; that academic pay scales, however impressive internationally, are still almost pitiable by comparison with business rewards—and one could go on and on. We academics pretend pretty hard, but in our hearts we know we're not quite accepted; so we put on symposia and public conferences to improve our acceptability.

People do tell us (and we even echo their misapprehension) that there was one period, 1946–1966, and especially the mid-1960s, when the universities were riding the crest of the wave. Indeed they were—by comparison with 350 years of neglect. But what kind of a wave was it? What made the wave? The GI Bill of Rights. *Sputnik* and space science. Health "breakthroughs" and biological warfare. In effect, the government was buying special services from the universities. And the endowed foundations did hardly any better. They did a little for the social sciences, and a little for the fine arts, providing they were performing arts. But for the humanities, they did almost exactly nothing.

Another way of estimating the balance might be to ask what percentage of the gross national product goes to Harvard or to Columbia or to Yale or to all of our fifty major universities? What share of the American dollar goes to the highest learning? The answers to these questions will make no academic feel secure in the public esteem.

Yet I hardly mean to imply that Americans have no use for universities. On the contrary: they have great uses for us—but on their own terms. In fact, they put us or try to put us to an astonishing variety of uses. Let me name just a few of the uses and abuses of the American university.

First of all, there are the parents and the students who see higher education primarily in terms of economic advantage and social mobility, and they'll do quite a lot to get to Elevator College or Forklift U.

Then there are the industries and businesses, great and small, who go to our universities for their engineering problems and technical personnel. They often think quite well of Short-Order U.

Again there are those who want to make the right social connections, either with or against "the Establishment"; these insist on either Cosmetic College or Camelot U.

And then there are the many, both inside and outside, who want the higher learning to serve all the major needs or desires of our society. And so we get the multiversity or Conglomerate U.

But what are or what should be our own priorities? Since I have already telegraphed some of my messages, I shall only outline my recommendations, and those rather categorically.

IV

1. The first clear priority is that we should accept and indeed insist upon diversity and pluralism of universities. Uniformity in the short run is impossible and in the long run unwise; conformity, whether imposed by Executive dictate, or Congress, or tyranny of majority opinion, or dogma of equality, must be resisted. A national university or a national educational pyramid, as European experience shows, is an invitation to disaster since it creates a national voice which then must be controlled. Diversity of higher education, however, creates varieties of opportunity, makes possible free choice, comparisons, and freedom to criticize or improve.

But to insist on the pluralism of universities requires this corollary:

we must multiply and diversify our sources of support. Financial centralization is extremely dangerous. The present tendency to siphon all our taxes and spare resources to Washington for redistribution to the states and cities and universities has to be countered. Somehow we must multiply and diversify our sources of support so that no single power holds an absolute veto—whether it be HEW or NIH or the city or the alumni or the students themselves. And because experimental science has become so ruinously expensive, this also raises the question whether we can any longer afford the proliferation of experimental sciences in our independent universities. Perhaps we will have to bite that bullet pretty soon. At least we must face the fact that one of the single greatest needs and most urgent current issues for our universities is how to get a safely diversified portfolio of financial supporters so that we may be allowed to continue to exist and do what we think we should do.

2. In the second place, while allowing others to shape institutions of all kinds, as their interests and constituencies suggest, we in the academy must do all we can to *preserve the liberal arts college in the liberal university.* For the liberal arts and humanities are concerned with the greatest issues of all: what is man, and what are his goals in life?

At the same time the liberal arts and humanities are extremely practical: they teach us how to observe and to count, how to reason and think, how to communicate in all the languages, ancient and modern, visual and oral, intellectual and physical. They teach us how to see things not in splinters but in wholes. They teach us how to balance feeling with reason and reason with human sympathy. Thus the liberal arts and the humanities free us from the twin tyrannies of blind passion and excessive rationalism. They also free us from three other great tyrannies: from the tyrannies of the past, or the present, or the future. Especially they free us from the false past, from unthinking habits, from inherited or unexamined prejudices. They free us from the even greater tyranny of the present, for they show us how transient each present is and how superficial or mistaken its concerns often are. Again they help free us from the fanatic prophets and futurists, from absolute threats of all kinds, even perhaps a little from the compulsion to "innovation," for sometimes renovation or recovery of old values is better.

On top of all this, the liberal arts and humanities, through history and literature and philosophy and the fine arts, give men their sense of values, their identity and goals, their experience, and their powers of imagination, above all their moral standards and their faith in man. Where else do we so well nourish our indispensable indignation against the Hitlers? Against Hiroshima? Against Vietnam or Watergate? It's we who are the enemies of the unscrupulous statesmen and the ruthless party bosses—and if we don't know it, they do. Or to put the matter quite differently: there can be no good society without responsible men, and no man is responsible who does not wish to feel and to know and to think about both.

And yet we must not confine the university college within the traditional languages and disciplines and subject matter. We need to give our ablest youth the newer tools, the newer languages, the newer perceptions. We need to teach them the newer power games with numbers so that they can live with the quantifiers and digital medicine men. We need to teach them the languages and philosophy of computers so that they won't lose their way in the deepening labyrinths of technology. We need to instruct them in the twentieth-century eye-languages of film and TV. We need to give them access to, and critical judgment of, the swelling knowledge industry so that they can cope with it intelligently—for most of their learning (after college as well as before) will be "sideways" learning, from the media and the daily news, and they will need to know how to handle all that. Meanwhile we need to expose them to society's failures, and help them to deal with the many kinds of injustice just as we need to organize for them what we have been learning about the body and soul of man.

New tools, fresh languages, keener perceptions: these must be built into our liberal education. But whatever the new devices, the purpose is not the device but the man, the individual man or woman. In the university college our first great duty is to find youngsters of character and intelligence and give them command of themselves. Such command, it seems to me, is our only true protection against "future shock." There is, we know, no complete or perfect education and no way to prevent or to neutralize change. But by the study of man through the liberal arts and humanities we gain our best preparation for the unexpected; indeed, one useful definition of the liberal arts and humanities might be just that: *Preparation for the Unexpected.*

3. Finally, I come to consider our crisis culture and the relevance

of the humanities. I think that urgent issues and contemporary problems are engaging to students, instructive to both students and faculty, and they may yield dividends also to the larger society, but . . . I would not tie the humanities *primarily* to *today's* problems. In this I'm afraid I have to differ a little with Wm. Theodore de Bary. I quote from his "Program of General and Continuing Education in the Humanities," as given on page 3 of the April 1973 issue of *Columbia Reports:* "For many students the relevance of the 'humanities' may be best appreciated in relation to the study of contemporary social, scientific, or technological problems in concrete form." Agreed. But how many of these "many students" do we have to have? Can we only preserve our colleges by giving in to the People of Plenty? Do we have to disguise the humanities or transplant into the liberal body a shop-keeping heart? Or, on the other hand, are we so tied to the innocent freedom-riders that we can teach only what they think important?

I have also a second objection. Contemporary is contemporary. There is a built-in obsolescence in relevance. The new revelation is soon inoperative. Instant salvation is gone already.

Today's problems may not last; they may not even have been important this morning. Yesterday's problems, and basic moral issues, may make better subjects for study. For example, "university governance" was once the great issue and activist study, but somehow it wasn't such good preparation, was it? And now, in 1976, shall we discuss the missile race or the old atom bomb? I submit that to study the dropping of the bomb in historical perspective will for all sorts of good reasons make a far more instructive study and a better preparation for tomorrow's terrors. I could go on to talk about Afro-American studies or of overkill by the ecologists or of the naiveté of trying to solve so many ancient problems by putting them into the new basket called Urban Studies—but, to illustrate my point, I would rather choose one cautionary tale out of my own experience.

In the late 1920s when I began teaching at Yale, I was assigned to the general U.S. history course (History 20), which we taught in sections, with a series of readings, from the colonial era to World War I. It was a survey course, required of those who had not had U.S. history. So it was hardly the course that John Dewey would have recommended. And as for the Progressive Education Association, which was then in full voice, it violated all their most sacred commandments. It was required, it was irrelevant, it did not focus on the important problems, or on issues that interested the students.

Ten years later I was still teaching that course and it occurred to me one day to reflect: if I had followed the counsel of the Progressive School men, what should I have taught? Well, in the late twenties I should have focused on the great boom periods of American development, and especially on the industrialization of the country after the Civil War, the age of the Robber Barons. And my sophomores would have graduated into the Great Depression. Being young and nimble, I would quickly have changed the focus to the panic of 1837, the railroad depression of the 1850s, the panic of 1873, and the hard times of the 1890s. And by the time I had got those materials into some sort of control, my students would have graduated into World War II. At that point I would have switched to Valley Forge, Lafayette's "lend-lease" of French noblemen, or Vicksburg and the Peninsula campaign. And my students, thus intelligently forewarned, would have opened the door from college to see a mushroom in the sky. . . .

And what had I been doing instead? Instead I had been teaching them a little about all those issues, even including the uses of new weapons, the *Merrimack,* or "damn the torpedoes."

So I would argue that present-mindedness, however tempting, has its weaknesses, and it may often, even generally, be better to try to understand the present by the past or by indirection.

In short, I would have liked to change the premise of the entire conference series, of which this book is the fruit, by one word and one shift of accent. The premise was "that the humanities give light when used as aids to the understanding of current urgencies." I would say, they *also* give light. I would italicize *understanding.* I might even put quotes around "current urgencies." So it would have read: "The premise of the program is that the humanities also give light when used as aids to the *understanding* of 'current urgencies.' " (And I would remind people that the light they also give may not always be welcome to Middle America or their Congressmen.)

Obviously I'm not being of much help. I don't think most Americans want a liberal education or trust the highest learning, and I wouldn't want to disguise the humanities to make them taste good to the materialists. So I wouldn't try, in our best universities, to be everything to everybody. Instead I would try simply to set some examples of excellence, in thinking and in the making of men, a few examples for others to imitate. Does this make me a total pessimist and un-American?

I hope not. For, looking back at the work of the liberal arts colleges and universities in America, I am impressed by

> how few have been the believers
> how hard their road
> but how magnificent has been their best work!

Let us not stop.

20

The University, Society, and the Critical Temper

In Response to George W. Pierson

WM. THEODORE de BARY

George Pierson's lucidly written argument makes his standpoint quite clear so I need not belabor it. Yet I have a few comments to make and some issues to join.

First, I wish to clarify an apparent misunderstanding. The passage he quotes from my paper on general education and the humanities does not, I think, serve well the point he wants to make. In that paper, I meant to emphasize that for some students—indeed, for many students—an approach to the humanities can usefully be made through a consideration of contemporary social and political issues. In discussing this, I was concerned with a pedagogic problem: I wanted to stress the importance of dealing with the contemporary in the perspective of history and in the context of the humanities so as to help the student understand more fully the issues and problems of both past and present. So I don't think there is disagreement between us on the essential point. The paper as a whole makes my position clear: I do not believe that education should simply be bent into the service of political activism or whatever happens to be the current educational fad.

It is important for us as educators to identify perennial concerns that have been expressed through the ages in many forms and in dif-

Wm. Theodore de Bary is Executive Vice-President for Academic Affairs and Provost and Horace Walpole Carpentier Professor of Oriental Studies at Columbia University.

ferent cultures. But we must also stress the importance of the seemingly irrelevant. There are certain subjects which are essential to the humanities and to a liberal education *precisely because* they are not immediately relevant. They open the mind to possibilities that we would be unaware of if we were simply concerned with what is insistently before us. And *this,* to my mind, is the only way we can prepare the student—or anyone—to deal with the unexpected.

We cannot, however, make a virtue of irrelevance either. We still have a problem with student activism—a problem which may not be with us as much today as it was a few years ago, but one which continues to pose serious questions. A concern for contemporary problems can be expressed in ways other than social or political activism; and in a university this concern should be expressed in terms of its distinctive functions: to know and to learn rather than to do or to act (in the political sense). Nevertheless part of that learning function is to help people understand what it is they are acting upon or what it is they are trying to do. There is a very close relationship between our learning or knowing function and what people around us are doing. We can ignore the implications of this, at our own peril. If the university looks the other way and fails to help society understand its problems, society may become unable or unwilling to help the university with its problems. And the understanding I speak of is not the solution to technical, research problems but the kind needed for citizens to meet their responsibilities in maintaining our key institutions. I do not think we are meeting this need educationally, and the reaction against political activism has become a further obstacle to efforts at reform.

Another point I would raise has special meaning for me, perhaps, because of my particular scholarly training. Pierson's topic is the university in American society, and he discusses it in the perspective of American history. As an Orientalist who must think in terms of millennia rather than centuries and who must confront the problems of education in rather different cultures, I am struck by the fact that the kind of university we are familiar with in the West has appeared relatively recently and under very special circumstances. Its academic autonomy reflects in part the extraordinary immunities which the church in the West enjoyed from political control and, more recently, the support that colleges and universities gained from the rise of the middle class. Pierson himself points to the significant middle-class origins of the American university, and I suspect that we will not fully understand the historic role of the university as a nurse of the liberal arts and of the humanities unless we have some better understanding of the middle-class culture and the pluralistic society which has produced it.

The absence of such conditions in other civilizations is very noticeable. They have had no independent church, no strong middle class, no universities. Even today, there are no real universities in China because they are closed—closed, that is, for the kinds of study I have been talking about, although they are open for other, largely technical and political, purposes. The contrast indicates how difficult it has been to create or sustain, in quite different cultural and social environments, the kind of institutions we take almost for granted in our society, and it becomes a large question how the values of the university can be sustained in the very different circumstances we are likely to face in the future.

It is a question which brings one back to Pierson's analysis of the various public attitudes towards the university. Our society no longer holds any institution in very great respect. Why this is so is not clear. It may have to do with our middle-class character, but it is certainly bound up with something Lionel Trilling was concerned with before his death, and which Steven Marcus and others are continuing to study, *viz.*, the searching critique of tradition and authority, or accepted values and models, which seems to be so fundamental an element in our own recent culture. Whatever else the university may be, it represents, perhaps primarily, the critical attitude: an attitude which engenders a disrespect for almost all institutions and accepted values, and now has turned on the university itself. The problem, then, is: how can the academic enterprise be sustained at all if it cannot be brought into some reasonably stable relation to the sources of its past and present support, if it cannot develop into a more constructive process, a process which does not eat itself alive?

If that process is not to be self-destructive, it must sooner or later come to terms with those traditional values and institutions which have provided the basis for the self-critical attitude itself. These can no longer be taken for granted but rather demand a positive commitment from us. In other words, the university cannot allow itself to celebrate or enshrine the "critical temper" alone, as if this could exist apart from tradition. To congratulate oneself on being emancipated from the past can be just as stultifying as intellectual smugness or ideological dogmatism. In such an atmosphere it is not surprising that history and culture become devalued.

If this kind of liberation is not to uproot us from the values which sustain institutions or to isolate the scholar from the society upon which the university depends for its support, the academic enterprise must make a comparable and complementary effort to identify the roots of a deeper humanism that is self-sustaining as well as self-critical. It seems to me this need is pointed to in Steven Marcus' paper for this series, which I take to imply that, amidst the collapse of the nineteenth- and twentieth-century middle-class culture identified with modernism, we must look for and reaffirm the underlying humanistic values which engendered that modernism and critical temper but go beyond them.

Young people today, I believe, sense that need. They feel a powerful urge to affirm and not just to criticize. From this standpoint it will not be enough simply to find fault with the aberrant tendencies rampant on our campuses in the sixties. Lacking the securities—spiritual and intellectual as well as economic—of the established scholar, young people could not be content with a situation in which the dominant "critical temper" had cut us off equally from past, present, and future. Academic life had become divorced from its supporting traditions and institutions and removed also from the crucial issues of the times. In this sense, student protest had a positive significance—reminding us that every exercise of the critical intellect must at the same time be seen as an affirmation of both past experience and future possibilities.

21

Some Questions
in General Education Today

STEVEN MARCUS

When I was asked to write a paper for the conferences on Education and the Good Society, the questions originally proposed for me to address were these: "What is a good society, and what can established educational structures (whether real or possible) do to achieve or maintain it?" Now, if I knew the answers to these questions, I would certainly not be here. I would be in heaven along with the sages and patriarchs who have debated over these matters in our civilization for the last two and a half millennia. Indeed one reading of the statement might suggest that it represents among other things a familiar if amiable kind of obfuscation; that such questions are being put in such a way largely because at the present moment our ability to approximate for ourselves formulations that even remotely resemble satisfactory—if temporary—answers is about as low as it has ever been. And so I think it is.

The statement therefore solicited in me an impulse to try to find and focus attention upon what I take to be one source of the particular feebleness and absence of self-confidence that I sense everywhere about us nowadays. Although research and graduate study have their due share of problems and vexations, I do not think the primary troubles are to be located there. They are to be located, I suggest, in undergraduate education. And as with all things academic, these trou-

Steven Marcus is George Delacorte Professor in the Humanities at Columbia University. From 1974 to 1976, he was Director of Planning for the National Humanities Center and is now chairman of the executive committee of the Center's trustees.

bles are both real enough in themselves and symptomatic of other and probably wider developments. What seems to have happened is that the idea of liberal or general education has expired.* It died without anybody killing it, and so it may be said to have died a natural death. Due exception may be taken to this general statement—which I offer of course only as a hypothesis. Certain undergraduate institutions may be pointed to in which liberal or general education continues to be pursued in a state of rude and vigorous health. That may be true, but it doesn't seem to me to be generally true; moreover, it doesn't seem to me to be generally true of the better or elite undergraduate institutions either, so that we cannot limit or define the situation by such conventional means as stratification. Another objection to my assertion may be voiced in the counterstatement that just as there are no new ideas in education, so too no educational idea ever perishes as well—it merely leaves the scene for a while and sooner or later reappears in a new guise or costume. This is received wisdom, the lore of our craft or mystery, and we should pay it the regard that its venerable condition merits. It may be that the eclipse of liberal or general education—and what is equally important, the belief in it—is only temporary, and that if we are patient and bide our time and attend to our duties it will make an appropriate reappearance among us. That may be true. But it is also true that real changes occur, that we live in real time, that cultures and societies and institutions actually alter and develop, and that these alterations and developments have consequences. I will therefore adhere to my hypothesis in a modified form, and say that liberal or general educa-

---

* When I use the term *general* or *liberal education* in this text, I am referring, I realize, to a wide array of phenomena. I am aware of the heterogeneity of kinds of undergraduate programs, how they differ at different levels of the American system of higher education, and how each of them has changed internally over periods of time. Nevertheless, I believe it warranted and useful to discuss the matter as a whole. What I am referring to is *not* any particular program, such as undergraduate education at Columbia College, or at the University of Chicago, or at Harvard or the University of Michigan. What I am referring to is a series of ideas that have had to do with undergraduate education, and in particular with the first two years of that education. The ideas in question have had to do with certain kinds of study that were not directly connected with preprofessional or pretechnical training, although they were not incompatible with such training. They were also not directly connected with a system of majors, although again they were not incompatible with such a system or systems. Although such programs have taken, as we all know, many forms, they were bound together by a number of ideas and assumptions. One of the purposes of these comments is to discuss some of those ideas and assumptions.

tion in America today, if it is not extinct, is in a state of decline, that its vital signs are not encouraging, and that it isn't certain that intensive care is going to assist it to survive—or that we know what such survival might entail.

I

If we mount an inquiry to try to begin to determine how this state of affairs has come about, there are any number of circumstances, moments of time, occasions and places that may be chosen as points of departure. Unless one goes back at least as far as the *Meno,* there will be something arbitrary in any one choice—and indeed there is something arbitrary in going back as far as the *Meno* as well. The point of departure that I shall choose is also a point of both professional and topical convenience. The occasion that seems to me diagnostic is the first great debate in the modern era that set the tone and framed the terms for much future discussion. I am referring to the spirited exchange that took place in the early 1880s between T. H. Huxley and Matthew Arnold.

In 1880 Huxley delivered an address at the opening of Sir Josiah Mason's Science College at Birmingham. It was printed in 1881 under the title "Science and Culture." In this essay, Huxley attacked with good reason the standard classical education of the period and argued for the intellectual and cultural relevance of the study of science. (He did not of course neglect to mention its practical usefulness as well.) In the spirit of laissez faire, he went out of his way to deflate "the pretensions of our modern Humanists to the possession of the monopoly of culture," and in a related context in the same essay stigmatized such humanists as "monopolists of liberal education." In addition to the study of science, he argued for the cultural value of the study of modern languages and literature, and ended with the firm recommendation that the study of sociology should also be a central part of any first-rate program of education—thereby foreseeing and predicting in one short essay what has become the conventional tripartite disciplinary structure of the modern university.

Although the dead heads of classical education were the chief (and easy) objects of Huxley's attack, he did not stop with them. And at a number of moments he went out of his way to take aim at the most eminent and high-minded of the apostles of culture of the period. Huxley observed that, although it would be mistaken to identify Arnold's opinions with those of the hidebound classical dons and schoolmasters, "yet one may cull from one or another of those epistles to the Philistines, which so much delight all who do not answer to the name, sentences which lend them [i.e., the opinions of the hidebound] some support." And he proceeds to cull.

Arnold was prompt to reply. His famous essay, "Literature and Science," was first delivered as the Rede Lecture at Cambridge University and was then published in the *Nineteenth Century* in 1882. (He subsequently revised it considerably and delivered it as a lecture during his American tour of 1883–84, publishing it in this form in 1885 in the volume, *Discourses in America.*) Arnold proceeds according to his custom to take the very highest line. As for the aims of education and "the motives that should govern us in the choice of studies," he does in fact go right back to Plato—significantly, however, not to the *Meno* but to the *Republic,* from which he takes the following sentence. "An intelligent man will prize those studies which result in his soul getting soberness, righteousness, and wisdom, and will less value others." He assumes that there is a consensus on this definition and that it applies universally, "whether we are preparing ourselves for a hereditary seat in the English House of Lords or for the pork trade in Chicago." The line of argument he is going to follow, he characteristically remarks, is "extremely simple." His defense of the primacy of humanistic education is based on something so simple as "the constitution of human nature," and "the instinct of self-preservation in humanity."

Arnold's argument in brief is as follows. The largest part of our ordinary or secular education consists in the building up in us of our capacities of intellect and knowledge. Yet we have other fundamental capacities and needs as well, and these have to do with conduct and with pleasure in the form of beauty. Out of these Arnold constructs a paradigm. "Following our instinct for intellect and knowledge," he writes, "we acquire pieces of knowledge; and presently, in the generality of men, there arises the desire to *relate* these pieces of knowledge to our *sense* for conduct, to our *sense* for beauty—and there is weariness and dissatisfaction if the desire is baulked" [italics mine]. And he goes on to observe how "everyone knows how we seek naturally to combine the pieces of our knowledge together, to bring them under general rules, to relate them to principles; and how unsatisfactory and tiresome it would be to go on for ever learning bits of exceptions, or accumulating items of fact which must stand isolated." Science, for Arnold—however systematic it might be; and I think that Arnold had only a rudimentary awareness of the systematic character of science—remains in the realm of fact and knowledge. But literature, humanistic learning, takes such knowledge and puts it in relation to "our sense for conduct, [and] our sense for beauty." And it does so by reason of its "undeniable power of engaging the emotions," although Arnold is unable to give a precise account of how this comes about. Literature and humanistic study, Arnold continues, have "a fortifying, and elevating, and quickening, and suggestive power." They represent "the criticism of life by gifted men, alive and active with extraordinary power at an unusual number of points"— and they solicit an equal life and power in those who devote themselves to such studies. The effect of humane letters upon its students will be to "call out their being at more points"; indeed such studies "will make them live more." Hence he cheerfully concludes, "so long as human nature is what it is," the attractions of humanistic studies "will remain irresistible. . . . And a poor humanist may possess his soul in patience . . . and still have a happy faith that the nature of things works silently on behalf of the studies which he loves, and that, while we shall all have to acquaint ourselves with the great results reached by modern science, and to give ourselves as much training in its disciplines as we can conveniently carry, yet the majority of men will always require humane letters; and so much the more, as they have the more and the greater results of science to relate to the need in man for conduct, and to the need in him for beauty." On these grounds, he contends, humanistic studies in general and literature in particular will continue to occupy the center of education.

I do not think it to be too much of an exaggeration to remark that there are those of us who were, so to speak, brought up on this essay—or on some modified version of it. To my mind it represents in nuclear form the idea or ideas that were or have been behind most schemes of general or liberal education, although these schemes developed themselves historically along a diversity of related lines. If we ask ourselves what it is that Arnold is saying in this essay, or how we can translate his utterances into the terminology of contemporary critical discourse, we can derive a number of connected formulations. First, in insisting upon the powers to relate—or the relational powers—that he finds to be of central importance in humanistic education, Arnold is laying stress on what today would be called the integrative functions of such studies. These integrative functions are at work in a number of areas. They are at work in the different domains of knowledge itself, and Arnold regards the humanities as among other things functioning to integrate the various separate and vertically developed realms of intellect. Since all these domains and realms exist within living persons, Arnold also notes that humanistic education works to integrate us personally—that it answers our inherent needs for structure, wholeness, and coherent generality of view. In addition, since all this is taking place in a world, Arnold regards such studies as having an integrative function socially as well—indeed that complex notion is one of the principal explicit theses that he advances in *Culture and Anarchy*. I do not imagine that anyone will doubt that historically one of the chief and vital purposes of any established system of education has been the purpose of social integration. The ground of contestation in this matter has to do with such questions as who is being integrated, by whom, by what means, and to what ends. Education is, after all, one of the instruments or institutions through which societies and cultures reproduce themselves, and in this context one cannot separate for long the integrative functions of education from the socially reproductive ones. And that reproduction means or includes the reproduction in time of determinate social structures—including classes, elite groups, differential statuses, elaborate arrays of role relationships, and the like.

Second, Arnold singles out one specific form of integration as bearing special weight. What literary or humanistic studies tend to do, he remarks repeatedly, is to relate or integrate all our knowledge "to our sense for conduct, [and] to our sense for beauty." Such studies, in other words, bring knowledge, or knowledges, into integrated relation with certain values or groups of values—or, as we would say today, with certain systems of values. *Values* is a word that is much in currency these days. Terms of this sort always tend to be when people or societies begin to become aware of the circumstance that they are living in the midst of a large historical crisis, and the coin has been accordingly both debased and inflated. It remains nonetheless true that Arnold has a point, if not the only point or the whole point or all the points. And it is equally true that some such idea is embedded somewhere in every scheme of liberal or general education.

Third, Arnold specifies further the means by which humanistic education will effect this integration. It will do so, he asserts, by reason of the "undeniable power" possessed by humane letters "of engaging the emotions." The logic of this is that humanistic studies will function as religion once did but no longer can. Arnold is quite explicit on this score. "The Middle Age[s]," he replies to one of Huxley's charges, "could do without humane letters, as it could do without the study of nature, because its supposed knowledge was made to engage its emotions so powerfully." Modern science has exploded that supposed knowledge and all that it entailed—revealed religion, special creation, God's in his heaven have all become inaccessible and impossible. Only humane letters, only humanistic pursuits, have anything like the comparable power of engaging the emotions, and therefore, according to Arnold, the importance of such studies "in a man's training becomes not less, but greater, in proportion to the success of modern science in extirpating what it calls 'medieval thinking.' "

Arnold has been much chided for putting forth this view. And the chiding has come in from a variety of quarters. He has been accused of trying to sacralize what cannot be sacralized—hence of having committed heresy. (He was anticipated in this forward-looking deviation by Coleridge, who advocated his idea of a clerisy along similar lines.) He has also been charged with encouraging one kind of institution to do work for which it was not designed (at least in its modern forms), and to permit a displacement into it of functions that it cannot perform. In Arnold's scheme liberal or humanistic education comes to suffer from what today we rather quaintly call functional overload. And our modern secular forms of liberal or general education have as a rule made a studious show of trying to avoid most of Arnold's terminology and of skirting around it. Liberal education, for example, aimed to develop "the whole man." Or it was centrally concerned to develop our cognitive but nonquantitative competences in a range of disciplines. Or its primary purpose was to develop in undergraduates a variety of skills in different kinds of inquiry. Or, as it was touchingly put in a document that issued from my own institution, it was meant to "include all studies that contribute to the art of living, as distinct from the channeled preparation for making a living." Or its goal was to expand our cultural horizons, increase our sense of the options in life that are open to us, lead us toward some kind of autonomy. Or its purpose was to develop in us the capacity for appreciating different kinds of experiences and to recognize empathetically in others who seem alien to us "centers of consciousness" as authentic and as human as our own. Or it sought to acquaint us with the canon, the great ideas and the great books, to help us to begin to know—to use a transportable phrase of Arnold's—the best which has been thought and said in the world. There was and there remains something admirable in all of these conceptual formulations and in the efforts and programs that accompanied them, however unevenly they may have each eventually fared. Nevertheless, I am convinced that as soon as we begin to press with moderate force on any one of these formulations, a residuum of Arnold's view pops up. To be sure that view is more secularized than Arnold's; it tends to be held with greater tentativeness; it is diffuse and sometimes attenuated; when it is articulated it is sotto voce; and it is indeed a residuum. Still, it is there, and importantly there.* As long as we are talking about that part of organized higher education that is not professional or preprofessional or vocational training, I don't see how it can be avoided. Nor do I think it should be. How it is to be confronted is another matter again.

Finally, there is a fourth connected consideration. This consideration takes the form of a conviction that is both a precondition and a consequence of everything else that Arnold has said, and he puts it out front. Today that conviction, when it is held at all, is held covertly and in silence. That conviction is to be found in the sentence that Arnold annexes from the *Republic*. " 'An intelligent man,' says Plato, 'will prize those studies which result in his soul getting soberness, righteousness, and wisdom, and will less value the others.' " In other words, there is a moral idea behind it all, a belief in the moral efficacy

* Consider, for example, this passage from the famous Harvard report, *General Education in a Free Society:* "There is doubtless a sense in which religious education, education in the great books, and education in modern democracy may be mutually exclusive. But there is a far more important sense in which they work together to the same end, which is belief in the idea of man and society that we cherish, adapt and pass on."

of education. This belief can be and has been articulated in a diversity of forms. A liberal education, it was said, would make better men or citizens of us. Or, it was once believed, that reading good books was good for one's character. Or, it was in part conceded, that if virtue could not be taught it could at least be learned. Or, to put it in the idiom of the rationalist-utilitarian mode, education will lead toward the promotion of a better life on earth; it will do so by leading us toward becoming more rational and therefore more wise, more just, more virtuous, and more happy than we were before. This idea, I have said, is a moral idea or belief. It is in fact a matter of faith and an article of faith. And it is this faith I have been implying all along that is passing if it has not already been extinguished.

If we ask why this faith is failing any number of explanations spring quickly to hand. (I am assuming, of course, that there can be no single explanation that is satisfactory for a cultural situation that is so complex and overdetermined). For example, it will be said that no one who has lived through the experiences of the twentieth century, and who has learned that an appreciation of Goethe's poetry is by no means incompatible with a predilection for lampshades made out of human skin, is likely henceforth to place much faith in the moral efficacy of humanistic education. A similar observation may be made about what is probably the most important movement in architecture and design in our century, the movement known as the *Bauhaus.* The ideal that directed the practice of this group was that social conduct could be improved and rectified by having it take place within an environmental system of design that was total, rational, pleasant, and humane. I have recently read an article characterizing this ideal as a heroic illusion, an *ignis fatuus* of avant-garde thought. And the commentator went on to remark that in any case "no one really becomes less wicked or more rational by living in an International Style building." There is certainly some truth in these observations, but they do not seem to me to account adequately or in themselves for what I have referred to as a failure of faith. After all history is replete with instances of people continuing to hold faith in ideas and ideals in the face of mountains of disconfirming evidence and disillusioning experience. That's what faith is for—ask any Marxist.

There is another way of coming at these problematical circumstances. At one point in "Science and Culture," Huxley singles out a passage from Arnold as demanding special attention. It comes from Arnold's great essay, "The Function of Criticism at the Present Time," and is one of those many passages in Arnold in which he tries momentarily to define his general position—many and momentarily because Arnold's general position was by intention and necessity insusceptible of being unitarily defined. The kind of culture and the kind of criticism that he is advocating, Arnold writes,

is a criticism which regards Europe as being, for intellectual and spiritual purposes, one great confederation, bound to a joint action and working to a common result; and whose numbers have, for their proper outfit, a knowledge of Greek, Roman, and Eastern antiquity, and of one another. Special, local, and temporary advantages being put out of account, that modern nation will in the intellectual and spiritual sphere make most progress, which most thoroughly carries out this program. And what is that but saying that we too, all of us, as individuals, the more thoroughly we carry it out, shall make the more progress?

Huxley reproves Arnold for his apparent neglect in this passage of the claims of the modern, and in "Literature and Science" Arnold handily defends himself and convincingly demonstrates that in actuality he meant no such thing.

"The Function of Criticism at the Present Time" was written in 1864 and is almost certainly the most important document of its kind in English written during the latter part of the nineteenth century. Some fifty years later in "Tradition and the Individual Talent," an essay of comparable importance for the modern period, T. S. Eliot undertook to restate Arnold's notion from a historically altered perspective. His notion of the modern, Eliot argued, was compounded in considerable measure out of an awareness of and a grasp upon tradition. Tradition, however, is not according to Eliot something that can be inherited. Quite the opposite, he asserts; if you want tradition,

You must obtain it by great labor. It involves, in the first place, the historical sense, which we may call nearly indispensable to anyone who would continue to be a poet beyond his twenty-fifth year; and the historical sense involves a perception, not only of the pastness of the past, but of its presence; the historical sense compels a man to write not merely with his own generation in his bones, but with a feeling that the whole of literature from Homer and within it the whole of the literature of his own country has a simultaneous existence and composes a simultaneous order.

This complex historical sense, he goes on to argue, is what makes a writer—particularly a modernist writer—simultaneously traditional and acutely conscious of his own contemporaneity, difference, and uniqueness. When a new work of art or a new cultural creation occurs, Eliot continues, something happens simultaneously to *everything* that has gone before.

The existing monuments form an ideal order among themselves, which is modified by the introduction of the new (the really new) work of art among them. The existing order is complete before the new work arrives; for order to persist after the supervention of novelty, the *whole* existing order must be, if ever so slightly, altered; and so the relations, proportions, values of each work of art toward the whole are readjusted; and this is conformity between the old and the new. Whoever has approved this idea of order, of the form of European, of English literature will not find it preposterous that the past should be altered by the present as much as the present is directed by the past.

Prepared by his earlier academic pursuits with a knowledge of Hegel, in part via F. H. Bradley, Eliot is writing with greater theoretical certitude and penetration than Arnold; yet the major point that both are making remains remarkably the same. Eliot, the chief architect in English poetry and literary criticism of that radical movement of departure in culture known as modernism, is arguing precisely that part of the continuity of tradition is also to be found in the radical breaks and discontinuities within it. Moreover, Eliot is right; for when we look back from the vantage point of the present upon that great cultural movement that is widely and loosely known as modernism, we see two things. We see the modernist movement as a whole tending self-consciously to represent itself as radically breaking with the past, as indeed it often did. But we now can see as well—as Eliot warned us we should see—the deep continuities that also existed between the chief creative figures of modernism and the great creative figures who were their predecessors; these continuities were at first invisible to most—so powerful was the impression of newness that the modernists initially made.

Eliot therefore seems to me to hold firm theoretically, and this holding extends back to and includes Arnold. Yet something else has to be noted as well. Although Eliot, and the passages in question, may hold theoretically, they do not, in my judgment, continue to hold experientially or existentially. Those passages could not, I believe, be written today—nor could anything resembling them in tone. They breathe a self-assurance, a confidence of historical location, and a consciousness of intellectual centrality (even though they were written in the midst of World War I) that one could not possibly hope to expect much less find today. Once again more interpretations of this circumstance are at our disposal than can be possibly brought forward on the present occasion. For example, Eliot is writing in England and from the point of view of a society in which the literary culture had achieved an ascendancy that it never did in America. In addition, since the time in which Eliot wrote, vast changes of every kind have occurred. We all know what some of them are and what powerful influences they have exerted. They include changes in the structure of knowledge, changes in the structure of our institutions of higher education, and changes in the relations of those institutions to both society and culture, whose structures have themselves undergone significant alterations. Higher education at the great American institutions of learning, for instance, is no longer primarily concerned with its functions of integration and social reproduction—though those concerns do continue to persist, and those functions are attended to. The university and higher education as a whole have become auxiliary institutions of production. In this newly emerging context of the university as part of the system of production, the role of liberal or humanistic education becomes increasingly problematic. And we may well ask what is the essential role of such an education in the production system of a technobureaucratic order whose dominant values are characteristically expressed in forms of utilities and commodities. What is the essential or overriding role of such an education in the culture of such an order? The answers to such questions, I venture to suggest, appear on the part of humanists to be about as forthcoming as the answers to the questions which have provoked these comments.

Yet important and complex as all these circumstances undeniably are—each of them merits extended discussion—they do not for me represent by themselves the decisive influences that are involved in the failure of faith, the failure of belief in the moral efficacy of liberal education, that I have been referring to. Perhaps nothing in such an enormous context is decisive, but I should like in closing this section of comments to bring forward two further considerations. When Arnold speaks of "Europe as being, for intellectual and spiritual purposes, one great federation, bound to a joint action and working to a common result," and when Eliot speaks of the poet writing "not merely with his own generation in his bones, but with a feeling that the whole of the literature of Europe from Homer and within it the whole of the literature of his own country has a simultaneous existence and composes a simultaneous order," they are both expressing and referring to an assumption that they hold in common. This assumption has to do with an essentially unquestioned assurance about the hegemony of European culture. In the interval between Eliot's essay and the present, that hegemony has been lost. The cultural world that we live in today is presided over by hegemonic powers that are not European in the sense referred to by Arnold or Eliot.

The last form in which the hegemony of European culture realized itself is what is known in Europe as bourgeois culture. In America we call and called its equivalent high culture, though they aren't quite the same thing. And, as I have recently written elsewhere, what we have begun to be aware of is that the modernist movements in art and thought—now that they are defunct—seem to have been the final phases of bourgeois or high culture. (It is worth noting in this context that the modernist movements were also essentially European in inspiration and composition, although Americans did make significant contributions to them.) And they were never more so than when they were most adversary, critical, apparently subversive, and elitist—obscure, hieratic, mystagogical, outrageous. They were the positive fruits of that culture, even when and especially when they were negating it. Indeed one can say that at its greatest moments middle-class—or bourgeois—culture was able to conceive of its own transcendence or self-transcendence. I am referring to the development and elaboration on the one hand of the idea of socialism and on the other the great and multifarious projects of modernist thought and art.

What has occurred since is something else again. When a phase of culture comes to an end or to a kind of end, it does not simply disappear and leave not a wrack behind. What it usually does is to go into a state of decomposition. It is that state that we are passing through now. We are, I believe, in the midst of the beginning of the decomposition of bourgeois culture, or of high culture, or whatever it is that we want to call that phase of culture that accompanied the development of capitalism—industrial and then advanced capitalism—in the West until 1950 or thereabouts. Since then, as the modernist phase of culture began to decompose, what we have been witness to is the increasingly rapid diffusion of the decomposed elements of modernism through the larger culture and society. This diffusion was made possible by the technologies of mass production and mass consumption characteristic of advanced industrial society. The agent of this process was what is known as the culture industry, those large institutions in modern society which are devoted to the industrial production of culture for mass consumption. The culture industry was and remains the chief conduit for the further and continuing diffusion of debased fragments and decomposed remnants of what was once a great if corrupt bourgeois-modern social and cultural tradition.

In a number of surprising ways, the idea of liberal or general undergraduate education was in part an analogue (or even at some moments an expression) within the university system of this last phase of bourgeois culture. Orginally, of course, it functioned to reproduce a social-cultural elite. And like modernism, it had a frankly elitist conception of knowledge. Yet it was at its best a liberal elitism and proposed the possibility that higher education had other purposes than turning out functionaries who would serve the institutional needs of the social world as it exists. Located at the very center of society, it still provided enough space for young men and women to turn about in and question the dominant social and cultural values of the world they were being prepared to enter. Within this space it was possible to cultivate a spirit of critical inquiry and to begin to imagine a different future. The ambiguities and contradictions of both bourgeois culture and its modernist phase were reproduced in the ambiguities and contradictions of liberal undergraduate education. On the one hand, it could train and turn out an intellectual and social elite that could be as arrogant and unfeeling as any of the elites that were its predecessors—and at times it did so. On the other, it could stimulate and promote a spirit of critical and intellectual inquiry and of critical doubt of a kind that did not in America previously exist—and at times it did this too. At its best it supported a notion of autonomy, of rigorous intellectual effort which was not in the service of inequities, however stringent and elitist its standards of judgment might be.

I have been suggesting that the failure of faith in the moral efficacy of liberal education, the decline of liberal undergraduate education, and the beginnings of the decomposition of bourgeois culture are entailed in one another. The entailment is complex, and I do not want to propose simple causal sequences or priorities. Nor do I want to rule out influences and developments of other orders and of equal importance—such as those that have to do with economics, social structure, and the structure of knowledge. What I want to do is to point to and begin to identify a particular historical nexus. This activity should not be construed as a lament for what seems at the moment irretrievably gone; it is not intended to be a swan song for a culture that almost no one of high intelligence was really happy with when it was around; and it will not, I hope, be heard as the disconsolate hooting of the owl of Minerva, as the world outstrips the thought of the world and passes it by. It is an effort of self-clarification and illumination, an effort that is peculiarly appropriate to the humanistic tradition, whose efforts are always in some part propaedeutic. They are, I would hope, propaedeutic to the answers that others will be able to propose to the questions originally asked.

But I do not want to close this part of my comments on so flat or solemn a note. Let me conclude it therefore by quoting a favorite passage. It comes from the greatest of bourgeois social philosophers, a sometime humanist, and the prophet of doom upon the society and culture out of which he arose and whose nature found idiosyncratic expression in him.

The ancient conception, in which man always appears (in however narrowly national, religious or political a definition) as the aim of production, seems very much more exalted when contrasted with the modern world, in which production is the aim of man and wealth the aim of production. In fact, however, when the narrow bourgeois form has been peeled away, what is wealth if not the universality of individual needs, capacities, enjoyments, productive powers, etc., created through universal exchange? What, if not the full development of human mastery over the forces of nature—those of humanity's own nature as well as those of so-called "nature." What, if not the absolute working-out of his creative dispositions, without any preconditions other than antecedent historical evolution—i.e., the evolution of all human powers as such, unmeasured by any *previously established* [or pre-determined] yardsticks—an end in itself? What is this, if not a situation when man does not reproduce himself in any one determined form, but produces his totality? When he does not strive to remain something formed by the past, and that he has become, but is in the absolute movement of becoming? In bourgeois political economy—and in the epoch of production to which it corresponds—the complete working-out of what lies within man appears as a complete emptying-out, this universal objectification as total alienation, and the destruction of all fixed one-sided purposes as the sacrifice of the human end-in-itself to a wholly external compulsion. Hence on one side the childlike world of the ancients appears to be superior; and this is so in all matters where closed shapes, forms and given limits are sought for. The ancients provide satisfaction from a narrow standpoint; whereas the modern world gives no satisfaction; or, where it appears to be satisfied with itself, it is *vulgar* and *mean.*

It was Marx's faith that even in this darkest hour of dehumanized existence, and even in this abyss of contradictions, mankind was nonetheless closer than it had ever been before to achieving the classical humanist ideal of free individual self-realization. He thought he knew what a good society was. I am not certain about how much thought he devoted to established educational structures. Had a good society, according to his lights, been really in the offing, I suspect he would have had something interesting to say about such structures.

II

If we turn for the moment from these theoretical considerations to matters of practical interest, to questions that involve possibilities of implementation—of implementing at least *something;* and as Americans I don't see how we can in good conscience fail to do so—I find myself drawn back once again to trying to compose what I have to say from a historical perspective. What I want to suggest has to do in the first instance with secondary education and in the second with dispelling a myth that currently seems to attach to the idea of it, or to whatever it is supposed to fail to be. If we look back again to the nineteenth century, one of the more striking things about the institutions of education in America was the weakness of secondary education. For example, as Jencks and Riesman have pointed out in *The Academic Revolution* (1968), in 1870 there were only 7,064 male high school graduates in America, yet in 1874, 9,593 male students obtained the B.A. Similarly "in 1890 the high schools gave diplomas to 18,549 men, while in 1894 the colleges reported awarding 17,917 B.A.'s to men." This situation was to continue for some time; and indeed "in 1914 male college graduates outnumbered 1910 high school graduates two-to-one." This remarkable structural imbalance was not eliminated until after World War I.

What these figures suggest, among many other things, is that, from very early on, the United States had failed to develop a system of secondary education which was in any way comparable to the American system of higher education—a system by the way which, however inadequate we may judge it to be, is surely the best that has yet existed. This weakness becomes even clearer when we compare secondary education in America during this period with the great systems of secondary education that had been built up in France and Germany, and, in a different but related context, Great Britain. The strength of the European systems of secondary education was an expression of the strength of the bourgeoisie or middle and upper middle classes *as classes*. This education aimed to achieve a number of ends. First, it ensured that the children of the bourgeoisie were safely isolated from the children of common or working-class people. Second, it aimed to appropriate for itself the cultural rights, privileges, and prerogatives of the former ruling classes. Third, it prepared the sons of the middle classes for roles as members of the new ruling order and for roles that were closely related to the state. One of the results of this arrangement was the normally strict separation of secondary and higher education on the one hand from technical training on the other. It was this separation that Huxley was deploring in his essay. And it was this separation that did not generally exist in America. In the American system there has never been a firmly drawn boundary between higher education and vocational, technical, and professional training. And in America as well, the colleges and universities have—for at least the last one hundred years—been the center of the system of education, while secondary education has not. This has created a much more "open" system of higher education in America—just as American society has been more open than

European society; and it has served both the cause and the ideology of social mobility through education as well. What it has not done, and what it can never do so long as our historical circumstances essentially persist, is to create a system of secondary education that is in any way commensurate in quality, force, range, diversity, and organization with the American system of higher education. Hence the nearly universal and quite understandable feelings of status deprivation that high school teachers in America notably suffer.

What I am suggesting is that there may be a certain element of cant—or of hot air—in the perennial and current lament that the high schools in America are not doing their job. When did they ever? Due exception must always be made historically for certain special schools both public and private, but we are speaking here about a national cultural problem that exists in similar structural configurations on a variety of levels within the system of higher education. And since a considerable part of the volume of this chorus of lamentations comes precisely from members of the professoriat and licensed practitioners of the higher pedagogical arts and crafts, I suggest that we stop bellyaching for a bit and see if we can determine what it is that we might do.

One of the first things we might do is to admit that the work of higher education in America—particularly in the first two years—has always to some extent been reparative. The degree of reparation has varied both historically and according to what level of the system of higher education one is examining; but it has always been there. What has not always been there is the willingness on our part to admit that this is what we are doing, and the even further unwillingness to admit that this is proper and legitimate and rewarding work for specialists in medieval literature or French history or the logical theories of signs. If such an admission is made, then a step toward demystification will have been taken. And if this step can be followed by the acknowledgment in all good cheer and with appropriate academic irony and skepticism—that much of the work that we have to do in the first two years is the work that is not done in secondary education and will not be done in secondary education, and that it is our proper work to do, then a step away from demoralization will have been made as well.

A half-step in this direction has in recent years been taken by the

University of California. Nearly all freshman students at the University of California campuses come from the top 12.5 percent of their graduating high school classes. This year, for example, testing revealed that 48 percent of the entering class at Berkeley and 45 percent of the entering freshmen at UCLA needed remedial instruction in English. As a result of such testing the University of California began a few years ago to institute a course to remedy these deficiencies, a course which they delicately named "Subject A." The students cottoned on at once to this euphemism, and in reality the course is widely known as "bonehead English." That the students are correct in their ascription is supported by the fact that the University assesses an extra fee of $45 upon those students who are relegated to taking this course. (I understand that this fee will soon be dropped.) I regard this as a half-step forward for several reasons. The University of California—and there may be others as well—has recognized openly and in public that the problem exists and has begun to make systematic provisions for dealing with it. At the same time, however, it remains stuck with the idea of remedial English or remedial work and what is implied by such a conception. It is this idea that I should like to see banished if not abolished.

The idea of remedial work throws the substantive onus of failure upon the individual student. It implies that he is personally falling below a norm or standard that his fellows and contemporaries regularly attain. It is in action a punitive notion—witness that extra fee of $45—and in fact a false one. For what we are dealing with in America today is an entire culture that needs work of this kind, and the sooner we bring ourselves to face this circumstance the less harassed and bedeviled we will feel and the less guilty and bewildered will our students perhaps be. As a result, what I would first propose would be a course—how and at what level it would be taught would to be sure vary with the institutional context in which it was set—in which the reading and analysis of selected important and primarily nonliterary texts would be accompanied by a great deal of writing: in my mind the two processes are inseparable. And I do not think that this course need be staffed entirely by English departments; what would be done in such a course does not have an exclusively special relation to the study of literature. It has to do with the study of anything that requires the use of natural language.

I have three further courses that I should like to see tried out in an experimental way on undergraduate students. I should like to have freshmen take a course in history—almost any kind of history will do, as long as it is history. I say this for two reasons: first, because of all the major humanistic disciplines, history seems to me today to be in the least problematical shape, though any historian will, at the drop of a hat, tell you that his professional calling has had its recent share of troubles. My second reason has to do with the circumstance that, after English, history, in my opinion, is the most important foreign language for students—especially American students—to acquire, and that they can make this acquisition most conveniently in a setting in which they have access to historians. I should like to see two further courses experimentally tried out in the second year as provisional requirements. The first would be an introductory course of philosophy. Once again, I would not specify what kind of philosophy, but simply make sure that it is philosophy. The fourth course probably comes out of utopia or the utopian future. I would like to see a course instituted in which students are asked to read, discuss, analyze, and write about a number of the most important works or documents of social theory. I would say of this course, as I did of its counterpart in the first year, that it ought not to be staffed exclusively by sociologists or social scientists. Indeed, I would make it a requirement that the staffing not be so exclusive.

Such courses, or anything like them, will not save the world, nor will they restore the idea of general or liberal education. They would not be easy to teach or to take; they would be "noncreative," in the current sense of the meaning that is applied to the term *creativity*, and they would be meant to be. They might even on occasion be dull. They would not, however, visit upon students the indignities of taking noncourses in nonsubjects in an institution that has on occasion appeared to be a supermarket for the mass consumption of poorly packaged intellectual goods. And they would not treat the individual student as an object that requires personal or idiosyncratic remedial intervention and attention. The remedy would be directed at the culture in which we all perforce live.

And there, alas, is the sticking point. At the risk of hyperbole, I will now undertake to say directly what I have been implying all along, which brings us back to where we started. It seems to me that there exists today a widespread if quasi-conscious doubt among humanists in particular about whether in fact they have a culture to transmit. This doubt may be only partly articulated, but it is not to be confused with the poised and critical skepticism that characterizes the humanist tradition at its moments of strength. This kind of doubt may express itself differently at different levels of the system of higher education and in the society and culture at large, but it is distinctly perceptible among the elite groups at the great universities and the larger cultural enclaves with which they are symbiotically connected. If in fact humanists cannot sustain their belief in the value of the culture they have inherited—and which is professionally within their care—or in the value of its values both immanent and overt, then it seems to me to follow that they will be severely hindered in arriving at judgments and decisions about priorities within their own disciplines and judgments and decisions about allocations of resources in both research and teaching—about curriculum, the structure of advanced degrees, and other internal matters—not to speak about judgments and decisions that pertain to the substance of their own continuing work and vocation.

The belief that seems to be passing, as I have said, is the belief in the intrinsic or transcendent value of a historical culture and in the intrinsic and transcendent values that attach to the activities by which it is sustained, transmitted, reproduced, internalized, and modified. The erosion of such beliefs undermines the humanities in particular if for no other reason than that in them the teaching function is extremely intimate and peculiarly vital. And that function is most vital and most demanding in the first two years of higher education. The problematic of those first two years, then, remains inseparable from the larger problematics I have touched upon in these comments. That theoretical inseparability does not, however, mean that nothing can be done. It does mean, I think, that we should remain aware of the character and magnitude of what it is that we are facing.

Let me finally conclude by referring to how another of our great humanist ancestors tried at one moment to realize his awareness of the character and magnitude of what it was that he in his time thought had to be faced. The passage I have in mind comes from the text, "On Fortune and Misfortune in History." It was the last in a series of lectures that Jacob Burckhardt gave at the University of

Basle between 1869 and 1871, and reappears in revised form as the last chapter of his *Reflections on World History*. Burckhardt looks out on the European scene before him and finds himself driven to these remarks:

Everything depends on how our generation stands the test. It may well be that frightful times are ahead, and an age of deepest misery. We should like to know on which wave we are driving forward—only we form part of it. But mankind is not destined to perish yet, and nature creates as liberally as before. But if happiness is to be found in the midst of our misfortunes, it can only be a spiritual one: to be turned facing the past so as to save the culture of former times, and facing the future so as to give an example of the spiritual in an age which might otherwise succumb entirely to the material.

Well, the frightful times were indeed ahead, more frightful than even Burckhardt could have imagined. And we learned through those frightful times that everything does *not* depend on how one generation stands the test, although each generation must stand the test on its own and do what it can for those who come after it. And if the succumbing to the "material" that Burckhardt refers to has also taken place to a degree and in a depth that even he was quite unable to foresee, yet the idea of the "example" that he sets forward, and that his own being and career in life embodied, is still in large measure the idea by which humanists justify their vocation. For although it is probably not true that being a professor at Basle is preferable to being God, yet it remains true that being a humanist means that one has a vocation. And that it is a vocation is what constitutes its justification, or such justification as it may, even in these advanced times, continue to claim.

22

Some Inconsistent Educational Aims

ONORA O'NEILL

It is a common ambition of parents to want their children to be happy—and successful. Educational institutions concur. Their students are to be prepared for life—and for careers. They will, therefore, receive general instruction—and also vocational training. The college-educated will have a liberal arts curriculum—but also a major or preprofessional training; their education will be broad—but also (in some ways) deep. Everybody is encouraged to seek fulfillment—and success.

I am going to argue that no educational system can coherently aim at happiness and success for all its students. To do so in a few pages demands a regrettable peremptoriness. I shall not vindicate each premise of my argument; my hope is mainly to get through the argument. If it turns out that there are no educational systems which are committed to the happiness and the success of their students in the sense in which I shall define those terms, then I shall not have shown that there are any educational systems whose aims are inconsistent. So much the better. But I fear that there are indeed systems which are aiming for happiness and success.

To aim for a person's happiness is to aim to fulfill (a reasonable proportion of) his or her desires. In an educational institution (family or school) the happiness which is aimed at is not only happiness in the present but also, and sometimes exclusively, future happiness. Children or students may even, on some theories, have to be less

Onora O'Neill is Lecturer in Philosophy at the University of Essex.

happy in the short run so that they may be happier in the long run. (The ultimate vision of gratification postponed sees the school of life as a painful prelude to eternal rewards). Because of their relatively long temporal perspective, educators can try to secure students' happiness in two ways. Either they can try to develop and encourage those desires in students which are likely to be fulfilled, or they can try to ensure that the lives their students will be able to lead and the capacities they have are likely to fulfill the desires they have. Most likely, they will try to do a bit of both. Students will be encouraged to desire those things which are relatively accessible and will be equipped to acquire or achieve some of what they desire. This does not add up to any guarantee of happiness (even assuming the education works), but it is likely to rule out some major discrepancies between desires and achievements which would lead to deep unhappiness. A guarantee of happiness would require either a complete restructuring of desires so that desires never exceeded capacities (a Stoic approach, revealingly dubbed "sublime sour grapes"), or a complete extension of capacities in the direction of omnipotence, or some combination of these two.

In making these points about happiness, I advance no substantive theory about what makes people happy; I merely offer an account of the relationship between the concepts of desire, capacity, and happiness. I believe that different people have different sets of desires (though there may be a number of desires which all or almost all people have), and so different people are made happy by different things; I also believe that there are certain limits to the degree to which it is possible to redirect desires or inculcate capacities. But even these very general empirical beliefs are not required by my argument.

To aim for a person's success is to try to make him or her more capable at some task or career than others are. An educational institution or family aiming at a person's success seeks to inculcate capacities which will give that person the best chance of excelling at some task. For success, capacities must be adjusted not only to the requirements of some task or tasks but to, or rather above, the level of capacities of other persons doing that task. The successful gymnast not only can do gymnastics but is better at it than most other gymnasts; the successful doctor is not merely able to practice medicine but excels other doctors in some way or ways (e.g., skill or earnings or reputation). Success is necessarily selective because it requires not competence nor even excellence at some task but preeminence.

In making these points about success, I have tried to avoid any empirical claims about the capacities needed for success of various sorts or the frequency with which successes of various sorts are attained. My argument needs only an analysis of the concept of success.

Since success requires preeminence rather than excellence, only some persons in any line of endeavor can be successful. Hence it is inconsistent for the educational policy of any society to aim at success for all students. There is nothing inconsistent in aiming at universal competence at some task or even universal excellence at some task; but universal preeminence at some task is not a coherent aim, unless no two people have the same task. A particular educational institution can consistently aim at success for all its graduates; a particular family can aim at success for all its children. The graduates of the one, the children of the other, could be preeminent when ranked against the graduates of other institutions or the children of other families. But nobody could aim for the preeminence of all children or of all graduates. The success of some requires the failure of others; if we seek to eliminate failure, we would also eliminate success.

It might be said, however, that schools and families do not aim at, though they may desire, the success of those whom they educate. Educators can aim only to give those whom they educate the possibility of success by ensuring that their capacities are equal to the opportunities that may arise. But educators cannot control whether or not those whom they educate will choose to grasp opportunities. Hence the aims of educators must be restricted to equipping people for success. I believe that this objection fails because it is not the case that schools or families affect only their students' capacities. They also try to structure desires, to foster ambition, to present desirable "role models," to convince students that certain sorts of jobs and certain types of preeminence are desirable. Though they cannot guarantee students' success, they try to confer not only the capacities but also the desire for success.

While particular schools and particular families can give all students or children both the capacities and the desire for success, no policy could aim at inculcating such desires and capacities in all students unless it aimed at disappointing some students. Since success is necessarily selective and requires that some must fail, an educational system which fosters a desire for success in all students must disappoint some students.

Such disappointment cannot be avoided by aiming to give different students the desires and capacities for different sorts of success. Specialization of training cannot eliminate disappointment for two reasons. First, there is an order of success within each line of activity: even in small pools there are larger and smaller fish. Second, comparisons of the success achieved by people in different lines of activity are possible.

If these arguments are correct, the most ambitious educational policies aim at success for some and either disappointment or absence of success for others. Such policies may seem reasonable if they aim also at something desirable for the unsuccessful. Even if success is necessarily selective, most of the other more general goals of educational policy are not. Happiness and self-fulfillment, in particular, are not necessarily selective. There is no impossibility in the idea of a society of happy people. So it should be possible to aim at happiness for all and success for some. There is no general incompatibility between happiness and success. Indeed they are often linked in a person whose desires have been partly directed to success. Such people if they are successful are also (to some extent) happy; notoriously, if they are unsuccessful they are to some extent unhappy.

Yet it is not a coherent aim for an educational policy to seek universal happiness and selective success if the successful are to be selected by the standard processes of testing, competitive examinations and selective access to desirable sorts of training and jobs. Once an educational system takes upon itself the task of selecting for success, the format of education is altered. Students are taught not just to be competent or excellent at some things but to try to excel; they are graded and ranked not simply for their competence at tasks but for their preeminence over other students. In an educational system within which selection for success goes on, all students are encouraged to desire certain things which are known not to be universally available. The hidden curriculum of an institution which runs on a competitive basis alters the desires of all students—not just the desires of those who do best. Even those who fail may have been taught to desire success, to fear failure, and to perceive themselves as failures in situations where they are less than preeminent. The latter predicament hits the successful also. The ultimate preeminence is to be first in the field, and most of those who have learned to rank themselves against others and to value preeminence will not rank first in a field even when they are successful.

So aiming at success for some detracts from the aim of happiness for all. For the redirection of desires produced by competitive selection procedures and educational systems which are geared to these produces desires which can (necessarily) be satisfied only in a few cases. Happiness for all and success for some are inconsistent goals, unless the successful could be selected without exposing all persons to the changes in self-perception and desires which are produced by competitive procedures. If, for example, success were based largely on heredity or on a lottery, then the successful could be selected without awakening unfulfillable desires in the unsuccessful. Nobody's happiness would be jeopardized by the necessary selectiveness of success.

There are no doubt strong objections to selection procedures for the successful which do not involve all persons at some stage of their lives in institutions which encourage people to strive for success. To confer success on any basis other than demonstrated preeminence of competence seems to most people both unfair and inefficient. For if success is earned or merited, then the preeminence must have been achieved by a person's own capacity or hard work and not by discriminatory procedures for preventing others from entering the field. And if success is conferred without demonstrations of preeminent competence, then those who succeed may not be those who are (or would be with equivalent training) the most competent within each field. Hence if considerations of fairness and efficiency lead to an educational policy which exposes all to competitive selection and its pressures, we must expect to pay the price in disappointment. Desires will be awakened which cannot be fulfilled; not only success but happiness must be selective.

23

The Disestablished Humanities

ROSEMARY PARK

At a recent symposium in Washington on the form of science counsel to be offered the chief executive, Jerome Wiesner of M.I.T. asked whether any optimism was justified today about our capacity to manage contemporary society appropriately.[1] The questions raised by my topic relate to the same theme: what confidence can we feel that sources of knowledge exist, which, if applied, would enable us to operate our complex society appropriately? The symposium in Washington focused on natural science capabilities; I will be concerned with the possible contributions of the humanities, particularly as they are cultivated outside established educational institutions. In addition I shall suggest that some relationship may exist between these now disestablished areas and the concept of "a good society." That phrase can be usefully equated with Wiesner's hope for appropriate management since both ideas involve a decision about values to be realized in the society of the future. The scientists lament the fact that they possess today no such blueprint for development as Vannevar Bush provided almost a generation ago in his report to the president entitled *Science—The Endless Frontier*. Humanists, of course, have never been called on for a similar assessment either because their concerns do not appear to affect society so directly, or, should they

Rosemary Park is Professor Emeritus of Higher Education at the University of California at Los Angeles. From 1969 to 1975, she was a member of the Advisory Council of the National Endowment for the Humanities.

be seen as relevant to society, because there is no drama in that relationship.

For science, frontiers exist which can be enlarged and at which confrontations may take place. However quiet and peaceful the process in the laboratory, the final result can have literally earth-shaking consequences. The government therefore treats with the scientist, seeks his advice on some occasions, and is in general respectful. The humanist does not work in terms of frontiers and confrontations. His efforts are less sharply defined and concern matters about which a variety of opinions is possible. Often, though not always, he seeks for wholeness rather than for precision in a limited area and the wholeness he aims at may be attained by nonrational as well as rational procedures. Imagination serves not only to form hypotheses as in science, but the stimulation of the imagination may be the principal end sought. The interrelations of these different styles of learning and forms of knowledge present a fascinating history of human intellectual endeavor which can be traced through educational theory and practice with considerable clarity.

I

Many discussions of this sort would now refer to the etymology of the word *education* as deriving from *educo, educere,* "to lead out," a derivation familiar to us from countless commencement orations. *Leading out* is no doubt an interesting conceit, but a more authentic source for the word is from the first conjugation verb, *educo, educāre,* which meant, according to the distinguished Harvard philologist George Lyman Kittridge, not to lead out but "to bring up," a quite different exercise. It is well to keep this correction in mind as we try to identify the particular contributions of the humanities to "the good society."

In simpler cultures, education or bringing up entailed exposure of the younger generation to the elementary facts of their community and of literacy. Once this was completed at home, it was expected—in ancient Athens, for instance—that the young person would continue his education by participating in the life of the city itself. It is therefore no idle talk when Pericles calls Athens the school of Hellas or when Isocrates speaks of Athens as the *didaskolos* of all Greece.

Education was not then synonymous with the school. In medieval times and into the Renaissance the royal courts performed somewhat the same function. There selected youth, after minimal instruction at home, were received for further development in manners, in sports, and in cultural activities. As in ancient times so in these later epochs, association or *sunousia* of older men with younger in adult activities was the educational form.

In the New World, in America, however, there were no cities, no courts, no landed estates to begin with and the responsibility for bringing up or educating the leaders of a new generation rested with the colleges, which in the Puritan dispensation and later outside it, sought to prepare young men for service, as Yale's founders put it, "in Church and Civil State." If one examines the early college curricula, it is apparent that the aim of instruction was to insure a student's understanding of the theological commitments of his society. He was trained to be articulate about his belief in strictly logical form, and since the Church, the College, and the Society agreed on the nature of man and the universe, explication rather than creativity was cultivated. Latin as the *lingua franca* of the professions and Greek and Hebrew as keys to the scripture were tools each student was expected to master. These languages and a minimal reference to literary texts, combined with the theological and logical preoccupation of the curriculum, produced an education almost totally confined to the humanities, which were then synonymous with scholarship itself.

As we all know, the material problems of a developing country and an increase in toleration soon tended to disestablish a particular theology as the heart of the curriculum. Educational innovation in those later days came to mean planning engineering programs and eventually introducing agricultural and other experimental sciences into higher education as provided in the Morrill Act of 1862. Thus after years of dominance in the curriculum and in intellectual life, humanistic studies very gradually began to yield some place to scientifically oriented activities. The humanities continued to be sure as a traditional offering, appearing occasionally, in the Darwinian controversy, for example, to oppose scientific advance but in reality seeking to appropriate aspects of the scientist's working methods for humanistic studies, especially in the pursuit of linguistic and historical evidence. In the midyears of the nineteenth century, the tradition of scholarship, once coterminous with the humanities, became associated with certain scientific studies as well. The subsequent efflorescence of science in the succeeding century can be compared only with the magnificence of the humanistic Renaissance five hundred years before.

In our day the resulting encroachment of the sciences on the educational hegemony of the humanities has led to the displacement in public esteem of the man of letters by the man of science and to the increased importance of novelty as a characteristic of knowledge. Scholarship has come to concern itself with the new, with the highly specialized and in consequence has encouraged proliferation of curricular offerings in the universities and the rise of that modern phenomenon, intellectual obsolescence. Though the humanities experienced the same proliferation of specializations as the sciences, they were less subject to obsolescence or at least less likely to be condemned on that charge alone. These changes in the nature of knowledge have had very profound effects upon educational institutions and upon the humanities, once the core of all higher education. It is comparison with that former status which justifies the use of the term *disestablished* in characterizing their present condition, both inside and outside the educational institutions. Both areas of learning, however, have been called on in dealing with the obsolescence which has resulted from the quantitative increase in modern knowledge.

Originally, and late into the nineteenth century, the colleges could provide each new generation with most of the intellectual capital required for a lifetime and in the early days, in Massachusetts Bay and New Haven, with the moral standards and norms of behavior as well. Today, however, we no longer recognize a core of common learning and even our specialized competencies cannot sustain us through a lifetime. I was trained as a philologist; the very term is no longer valid. I knew Gothic grammar but never suspected the existence of generative grammar. When I first learned about Mendel, there was, of course, no one to alert me to DNA. Similar experiences were shared by most of my contemporaries, who thus learned that education today is not a complete package for youth but a continuing process for a lifetime.

For the scholar or professionally trained person, the inadequacy of early training is particularly obvious. As our society increases its dependence on technology, however, all occupations will be affected by obsolescence, which in its pervasiveness may appear a peculiarly modern phenomenon. Actually, history can prove it to be a more ancient occurrence. I would even suggest that the earliest universities were attempts to cope with obsolescence. What, after all, were the lawyers about, who flocked to Bologna in the twelfth century, if not retooling and compensating for their ignorance of the Justinian Pandects by listening to Irnerius? Or what brought the crowds to Paris at the same time but to hear Abelard on the new logic? These were, however, isolated attempts at professional improvement, and the university, once it was established, tended to concern itself with the young man who was beginning a career rather than with the mature professional who was renewing or updating his intellectual capital.

Similar problems of obsolescence faced the working man, particularly after the industrial revolution. By the nineteenth century, programs to remedy skill deficiencies were emerging. The People's Colleges in New York and elsewhere, the later lyceum programs, and most famous of all, the Chautauqua offerings, all represent early attempts to retool, improve, and modernize personal competence. Considerable initiative was required to participate in these scattered offerings even when the universities eventually institutionalized them as University Extension.

From the early days of such programs at Chicago and Wisconsin, university extension has come to be recognized as a way of bringing a variety of educational experiences to broad sections of the population who do not attend the regular courses of the university. At UCLA, for instance, 100,000 people in a year take advantage of classes in the humanities, sciences, and social sciences, at their own expense. A recent survey in the county of Los Angeles showed that 17 percent of the population would be interested in a program of part-time degree studies. When offered, it was apparent that the degree programs attracting the largest groups were those relating to professional improvement, a clear indication that the expansion of knowledge has affected the occupational life of the country at many levels.

A question still to be answered is whether university extension is seen as a cultural experience or primarily as a vocational opportunity. A difficult job market and the recent emphasis in the schools and in the country at large on career education may tend to reinforce the idea that increased professional training is the prime use for extension. A final decision may be determined by the capacity of extension to reach outside the middle class and into quite different groups in the population whose interest in the university is in the first instance neither for professional training per se, nor for cultural enlargement, but merely for proof of a legitimate right, documented by the degree, to be recognized part of society.

This demand for participation in the society and its constituent organizations may be, as Irving Kristol has said, a veiled need for authority arising from the disharmony between public and private values.[2] As the demand relates to the university in the form of universal access policies, it envisages a revolution within the university itself and a capacity in the institution to remake society—assumptions which must be questioned. Thomas Jefferson, as we know, was confident that the citizens of a republic, by means of education received in the lower schools, would be enabled to recognize ambition when they saw it and would therefore avoid entrusting the state to unworthy men. This hypothesis has hardly been sustained by history, especially by recent history. Though almost one-half the college-age group is in college, we nevertheless observe that more than half the qualified voters did not cast a ballot in the 1972 national election. Some cynicism about the effect of formal education on political judgment, not to say participation, is surely justified.

On rather different grounds a well-known critic argues that the schools have failed and that we must now extricate ourselves from this repressive institution and take education of all sorts back into our own hands. Ivan Illich, the proponent of this doctrine, urges a new amendment to the Constitution which would provide that Congress should make no law with regard to the establishment of education.[3] It is perhaps fair, at this point, to remember ancient Athens and the educative functions then ascribed to the great city itself. While conceding the capability of the university to produce well-trained specialists in all professional and advanced fields, is there not perhaps a sense in which Illich and Athens may be right or at least in which they may represent an alternative path to be explored? Before we open the university in a gesture of universal welcome, we need to be more certain that greater frustration may not ensue. The university has no *allein-selig-machende Kraft* and can work only in concert with other agencies to remake society in a juster form. If so, then we need to know what other institutions could serve today as the city did in other times to bring up and to enlighten on a continuing basis. And particularly, for the purposes of this paper, we should consider what contribution the humanities outside the schools can make to the process of appropriately managing society.

II

Our modern cities have no Parthenons, except for Nashville, but they have magnificent museums of art, of technology, of natural history. In our cities too are historical societies and libraries, theaters and cinemas, concerts and symphony orchestras, churches and temples— in short, cultural opportunities are open to us today about which Athens, rich as it was in these matters, knew nothing. We are not disposed to think of these institutions as primarily serving the young as does the school; they appeal to some in all age groups and in all socioeconomic and ethnic enclaves who are willing to learn the language or symbolism in which they speak. They offer opportunities to participate with others in elevating and enjoyable functions which enhance our own well-being by the peculiar kind of pleasure or inspiration they afford. Kant spoke of this pleasure as *interesseloses Wohlgefallen,* a pleasure which could be experienced freed from any lust of possession. The activities of these museums and societies are not quantifiable in terms of cost-benefit analysis nor is any certificate of attendance granted or expected. Each of us develops his individual relationship with these public programs, thus in effect controlling his own education. Today we have a vast network of these cultural, humanistic institutions in America which provide, as Kant's definition indicates, a countervailing force to the materialism of an affluent society.

That the state grants most of these institutions tax-exemption implies that legally they are seen as promoting the general welfare. In addition, some government funding is assigned to them in recognition of this mission. The discrepancy between the public monies assigned to science through the National Science Foundation and those granted the humanities is of course substantial. For the fiscal year

1974, $65.5 million was appropriated to the National Endowment for the Humanities and $570 million to the National Science Foundation. Such monolithic assistance to science by government can be justified in a variety of ways. One should not overlook, however, an advantage to the humanities in this low level of government funding: it means that many more persons must be called upon to help maintain them, must therefore understand the need and mission of these humanistic institutions. They are then closer to the community and freer to program in response to local interests and needs. However they may be supported, the humanities represent a different set of values from those of affluence and thus fulfill a significant role in the total society.

Open to similar widespread participation is that prime tool of humanistic study, the printed book. Reading could be said to be institutionalized in libraries and in schools, but those aspects of the activity still leave room for reading pursued for pleasure or for personal development quite independent of institutionalized programs. Indeed those who have had the most of school or college programs often seem the least interested in continuing serious reading as a humanistic attempt to enlarge personal horizons. A recent study of college graduates published in the Carnegie series reports that 39 percent of a fair sampling of the class of 1961 frequently read (but do not necessarily finish) a nonfiction book; 26 percent of the same group frequently read (but do not necessarily finish) a work of serious fiction. Of the same sample, 8 percent go to concerts and 11 percent visit museums or art galleries with some frequency. The authors of the study observe that it appears to make little difference what type of college one attended. "Sex," they report, "is a far stronger predictor of serious reading and interest in serious music than is college quality"—which means that college women more often engage in these activities than do the men.[4] Such a report, like that on voter participation, seems to point to genuine deficiencies in our college programs and accents the need for other agencies in the community to share in this educational responsibility.

When the low figure for serious reading is compared with the fact that printing and bookbinding have become major heavy industries, as the *Washington Post* reported recently, [5] then one must conclude that serious reading is by no means confined to the college-educated, but is much more widespread than might be supposed. In the twelve years between 1958 and 1970, Fritz Machlup tells us, book publishing grew by 8.9 percent, the GNP by 6.5 percent.[6] Though recent growth of the industry has not been as spectacular as in the 1960s, nevertheless 1970−71 showed a 5 percent increase over the previous year.[7] The 3 percent increase of 1972 is called the smallest growth in twenty-odd years although that figure stands for sales receipts of close to $3.2 billion.[8] The money spent for books compares favorably with the $4.75 billion spent for television and radio and strikingly with the $1.43 billion spent for movie tickets during the same period.[9] Of special interest is the phenomenal development of the trade paperbacks which grew by 22 percent between 1957 and 1970;[10] taken with the other figures, this seems to indicate that books as purveyors of pleasure and information have not been superseded.

A large portion of the expenditure for books is no doubt attributable to purchases by libraries through which many readers procure their books. The financial problems of libraries are particularly great in this time of inflation and economic crisis. In local areas, public funds have been supplied to support library facilities—and usually more directly than to museums and churches. The large grant recently made to the New York Public Library by the National Endowment for the Humanities and the Endowment's studies of the problems facing other great libraries document a genuine concern for the survival of these institutions at the present time. As the heart of the scholarly enterprise in the humanities, the great collections in the United States must, of course, be sustained and developed. They have now become a matter for genuine national pride and are convincing testimony that a popular democracy can understand and will support library development as the kings and aristocracies did in earlier times. American libraries, like the museums, are further evidence of the existence in our society of values which are not materialistic and which are recognized and supported as a matter of public policy.

A similar source of enlightenment, though more expensive and occasional, can be found in the media which depend, not on the written, but the spoken word. The fact that a special association has come into being to express concern over the effect of media on children (Action for Children's Television) reveals the lack of consistent quality in television and radio programming and the predominance of the profit motive over any intention to enlighten or improve public taste. Nevertheless, the ubiquity and power of these media are unsurpassed, both for good and for ill. They can present certain kinds of educational material with great effectiveness and their contribution to political instruction has been enormous, especially in recent years. Also with great success television has presented theater and drama, opera and symphony of the highest quality to a nationwide audience—supported in this effort by both public and private funds. The nine-part television version of *War and Peace* was seen by at least one million households in New York City alone. That such great works of humanistic achievement can succeed with the self-selected audience and also with the mass audience is merely testimony to the genius of the artist. That both industry and government find it compatible with their interests to support such productions speaks for the continuing improvement of public taste, a development one would like to attribute to education—except that the evidence cited above seems to indicate how relatively small a role the college-educated play in this field.

Unlike the schools, the media offer few structured programs. The listener is free of restriction or prerequisite and enjoys a genuinely universal access. Unless he can provide a meaningful framework himself, he is obliged to accept enlightenment or obfuscation at the pleasure of the sponsor, whose prime, though not only, concern has to be profit. Where personal motivation is high, the TV station, like university extension, the museum and the book, can be a source of information and pleasure. A recent review of the new season for commercial television, however, shows no regular programming in the humanities so that one is still obliged to see TV as an occasional not a regular humanistic source.

Even more commercial than the TV offerings is the whole area of advertising. With some exaggeration, one can nevertheless almost say that good scientific education could be secured by reading the advertisements of the science-based companies. Humanistic material is also offered to promote a product's sales but considerably less often and less effectively. In a restaurant recently, I was surprised to find a small packet of sugar with a reproduction of a modern painting on the outside and a note telling where one could find a print. A glance at the magazines of a generation ago reveals in startling clarity how far advertising has progressed since the personal testimonies for

Lydia Pinkham filled those same periodicals, and how well advertising reflects the tastes of our democracy which have so steadily improved. Though one grants that the aim is still to sell sugar and not to explicate paintings, nevertheless the fact that this and not some other form of advertising was used shows that the company believes the buying public will respond to this picture rather than to mere prettiness. There are, of course, other less praiseworthy uses of great works of art to enhance profits. Uneasy as the practice makes one— these masterpieces are defenseless objects in the public domain—one can also acclaim the increasing attempts by industry to raise and not lower public taste in the process of turning a profit.

In addition to these various activities which on occasion at least have a humanistic focus, one could mention newspapers and periodicals as well as specific forms of theater and cinema. No one expects the theater to be what Schiller envisaged, a moral institution, *eine moralische Anstalt.* Instead we dissect our times and our souls with a skill which is sometimes considerable but which results in analysis not insight. Courses run by newspapers represent a new aspect of the old extension idea of bringing learning to the people where they are. In 1973, 264 newspapers carried the first newspaper course for credit on "America and the Future of Man." That the response to the project was so immediate and of such national dimension makes one wonder whether in the initial enthusiasm over television as an instructional medium, the effectiveness of the written word was neglected. In this respect Great Britain's "Open University" has made a most successful combination of media and books for instructional purposes. The word is the chief tool of the humanities, and though the humanist may use the pictorial image, the musical sound, and the choreographic form on occasion, it is finally to the word that he returns.

One cannot know whether this verbal culture of ours is about to disappear and be replaced by expressive grunts and mathematical formulae, statistics, and a few pictures. Sometimes we seem close to such a development as the schools find it increasingly difficult to teach English expression, either oral or written. Then one could fear for the great tradition of humanistic learning which however displaced from its former dominance still represents a vital element in a good society.

III

In addition to the variety of their forms and the essential nonmaterial values they embody, the humanities share with science a capacity to surmount international and intercultural boundaries. The delight which may be derived from art and literature or from philosophy and history is not barred to those who lack prior training. Not all readers are scholars, nor all visitors of museums art critics; nevertheless, the humanities are accessible at varying levels of sophistication and knowledge. They are, as it were, more catholic in their appeal than science. Another advantage they possess is the capacity to make of cultural boundaries an enhancement to enjoyment rather than a barrier to be cleared and forgotten. The great Russian, Chinese, and French exhibitions of the past several years have attracted millions of Americans who in some form responded to these cultural treasures of another people by thinking, "If these foreign peoples delight in these works as we do, perhaps there are other areas in which we can also find commonality of aspiration and experience." As the interdependence of the world increases, the common language of the arts and humanities may prove of inestimable value in building widespread respect, if not understanding. For this reason alone we might conclude that a good society will not neglect these activities in the future, particularly since their products outlast all changes of taste and even the great miracles of science.

More certainly than we have yet discovered, a good society will have to develop what is called the multidisciplinary approach. Most scientific analyses of the energy problem conclude by declaring that its ramifications are so numerous that no existing branch of learning can expect to fashion a solution by itself. And if further proof were needed, we have only to consider the tragic consequences of Egypt's Aswan dam which, though an engineering feat, nevertheless destroyed the ecology of the region for lack of medical, botanical, and zoological contributions. That our problems are complex and increasingly susceptible only of multidisciplinary solutions may well give the humanities an increasingly important role since the traditional approach of the humanities has always been toward wholeness. The famous phrase of Terence, *Homo sum: humani nil a me alienum puto* ("I am a man and deem nothing human alien to me") is an ancient formulation of the humanist creed. To Shakespeare the idea was equally familiar: "What a piece of work is man! how noble in reason! how infinite in faculties! in form and moving how express and admirable! in action how like an angel! in apprehension how like a god!" One might conclude that the humanities were *multidisciplinary* before that term was invented. Perhaps knowledge had to develop through the extraordinary richness of modern specialization before the need to reinstate the merit of wholeness and breadth became apparent again. We still tend to believe that the analytical and precise are somehow truer than the general and comprehensive. The task of knowledge in the future may be to unite these two styles of learning. Indeed the mandate of the humanities in history may well have been to preserve this concept of wholeness until we developed by specialized knowledge the proper substance for the concept. Taking man as the representative of multifaceted experience, the humanities serve not only to fashion a form of multidisciplinary learning but also to judge the options a mature science will offer to society.

And finally, the pleasures and knowledge of the humanities are private cultivations. As no one can learn for you, so no one can appreciate or be inspired for you. In the humanistic experience the individual is not eliminated in order to reach truth but is always conscious of the uniqueness of his experience. Much of modern life must be for efficiency and profit, must concern itself with the mass, with what Walt Whitman called "the mean flat average." The humanities are accessible to the "mean flat average" but also to the extremes in sensitivity and wit, and to both they bring a strengthening of the person by virtue of the multifaceted experience of life they reflect.

To return to the initial question, to the possible relation of the humanities to a good society or to the appropriate management of our modern complexities, the response must be affirmative but with special understandings. The humanities are not tempted to claim that only through them will our problems approach solution. Rather they may claim that modern society will not evolve toward the good society without them, both as a component of formal education and as an enlightening and joyous activity diffused through many agencies of the community emphasizing the wholeness of man but at the same time sustaining the uniqueness of each individual person.

24

A View from the Ivory Tower

In Response to Rosemary Park

ROBERT W. HANNING

Rosemary Park's stimulating survey of the disestablished (or non-university-based) humanities in our culture and her confidence that the humanities "may claim that modern society will not evolve toward the good society without them" provide an establishment (i.e., academic) humanist like myself with plentiful food for thought. In this brief response to Park's essay, I wish to reverse her rhetorical emphases while entirely agreeing with her that "the university . . . can work only in concert with other agencies to remake society in a juster form." Where she has ably exposed the historical and continuing limitations of the American university as an inculcator of lifelong commitment to the humanities (or to other civic values, such as political participation), I wish to point out what I shall call the vulnerabilities of the disestablished humanities at this moment in our history. On this basis of mutual incompleteness, the universities with their corps of trained specialists and the cultural institutions that comprise the disestablished humanities can explore, realistically and creatively, ways to carry forward together the banner of "the Good Society." In the latter part of this essay, I will review some of the ways in which the universities can contribute to this ongoing, crucial cooperation—can serve, that is, as brokers between the disestablished humanities and the largest possible number of beneficiaries.

Robert W. Hanning is Professor of English and Co-chairman of the Program of General Education in the Humanities at Columbia University.

I

The first, most evident, most pressing vulnerability of the institutional sector of the disestablished humanities is its *urban-centeredness*. The cities are and must be the strongholds of the "magnificent museums of art, of technology, of natural history" and of the other societies, institutions, and performing groups whose labors and exhibitions help tell us who we are by showing us where we have come from and what we accomplished there. Yet as the cities suffer financially, caught between inflation and a shrinking tax base, between the continued influx of rural poor and the flight of corporations and the middle class, their cultural institutions suffer with them—all the more so as they are invariably placed below police, fire, and sanitation services on the scale of municipal priorities. Thus, the fiscal plight of urban America forces us to face the phenomenon of the great repositories of our humanistic legacy being located in precisely those areas of our society that can least afford to support them.

The disestablished humanities in their noninstitutional, mass-media forms—books, movies, television—are vulnerable in a second way, namely, to *manipulation*. Park is aware of the possibilities for commercial exploitation of the "defenseless masterpieces," but I am thinking more of the tremendous power for taste formation and manipulation placed in the hands of those who produce mass entertainment by the market analysts of the advertising industry. For example, to ensure the success of enormously expensive but artistically worthless films like *Mandingo, The Exorcist,* or *Jaws,* moviemakers join forces with admen to create a climate of expectation for their product (especially through liberal use of television advertising); if it seems advisable, the theaters of an area can be saturated with the movie so that its very ubiquity gives it a specious air of importance or excellence. There is a dangerous interplay here between prediction and propaganda—two "arts" borrowed from the realm of politics: a movie is made to accommodate the lowest common denominator of its prospective audience while an audience with that low level of expectation is then created by the manipulative strategies of packaging and advertising.

Moreover, the very success of the "hype" in selling a movie or a book (for the same techniques are at work in the field of mass publishing) seems to be decreasing further our chance for a salutary variety of mass-culture entertainment to choose from. Thus, fewer pictures are now being made than in recent years because the great cost of a *Jaws* and of the advertising package in which it has been gaudily wrapped makes its producers unwilling to have other of their films compete against it. Publishers, meanwhile, are taking books with a small but steady volume of sales out of print, pulping them, and producing in their stead (and with their paper) big sellers that do not take up expensive warehouse space. Such a policy is understandable, and more unfortunate than iniquitous—publishers must make money to survive—but it testifies to yet another way in which the mass-culture humanities are vulnerable in our society.

The last vulnerability of the disestablished humanities I shall mention differs from its predecessors in being more abstract and removed from crass financial considerations. Park observes that in other times and places the task of inculcating an interest in evolving a good society—i.e., of training citizens for a lifelong involvement in their culture at a level above that of merely seeking to earn a living—was not borne by the universities; that in this country the university has not in fact discharged this responsibility well, in part because it has played another role, as an avenue of vocational advancement. In particular, Park notes that Periclean Athens and the "royal courts" of the Middle Ages and Renaissance "brought up" the young to be contributing members of the polity without requiring them to undergo the academy's routines and rigors. Linking the Athenian experience and Ivan Illich's radical critique of our contemporary educational systems, she asks, "Is there not perhaps a sense in which Athens and Illich may be right" in suggesting that education for creative citizenship should be carried on outside the universities, in part under the aegis of the disestablished humanities?

This is an attractive idea, made more so by the seductive appeal of Athenian democracy and its artistic greatness to democrats like ourselves. (I leave aside the example of the medieval and Renaissance courts, for in fact they were, from the twelfth century onward, full of learned men—clerics, later humanists—who were the products of university or proto-university training. Even courtiers who were not university-trained themselves were taught and influenced by those who were.) The problem is that key components of the Athenian experience are not available to us, and their absence in effect places the disestablished humanities in a much weaker position as trainers of enlightened citizens. That is, they are *vulnerable to* (or, more precisely, the victims of) *historical change and evolution.*

Park indirectly states one aspect of this vulnerability when she says that although our cities possess no Parthenons, they have cultural institutions and opportunities about which even Athens knew nothing. The Parthenon, of course, was a temple to the city's tutelary deity Athena, and its place within the Acropolis, the hilltop citadel of Athens which contained the city's treasury and several other temples, reminds us that at the focal point of Athenian civic life, there was a shared sense of divine sanction for the city's political order and institutions. No museum, however magnificent its collections, profits from or prompts such civic piety.

Consider the intermingling of civic and religious impulses glorified at the end of Aeschylus' *Oresteia* trilogy. Athena herself has presided over the first civic jury trial and cast the tie-breaking vote to acquit Orestes of the charge of matricide brought against him by the Eumenides, goddesses of the family who take revenge on those who kill their kin regardless of the circumstances. The establishment of the jury system assures Athens of a new kind of rational and nuanced justice to replace the heroic-age bloodfeud. But Athena also placates the defeated and angry Eumenides and offers them a key role in the administration of Athenian justice henceforth; they will live in a shrine beneath the hill on which trials will be held (symbolically, the relocation of the Furies within Athens expresses how some visceral fear of violating taboos is a necessary underpinning for a rational system of justice). As the trilogy ends, the whole new arrangement is celebrated with a civic procession that guides the Eumenides to their new home. The procession, of a kind in which most of the trilogy's original audience would frequently have participated, links drama and ritual, politics and religion, in a final all-embracing image. As Richard Kuhns puts it in his study of the *Oresteia,* "myth is endowed with the cares of communal life, and the human city reaffirms its dependence upon a cosmic order reflected in myth." [1]

Needless to say, the humanities, in or out of the academy, have no such imagery and mythology available to them today for celebrating or preaching the possibility of a good society. Our sense of our society is national rather than civic and, except in certain fundamentalist and right-wing quarters that tend to distrust or ignore the humanities, pragmatic and analytic rather than religious and celebratory. Consequently, we lack an appreciation of civic rituals that might bring us together in search of a good society and provide opportunities for both new and old works of art to inspire us in our quest by the profundity with which they address the issues. The only public events that bring large groups of people together and fashion some sense of communal identity are sports events and ethnic parades. The former offer a fantasy model of intercivic strife that sometimes explodes into real physical violence on the field or off; the latter are subcivic events, exclusive in essence, and evocative of a divided cultural allegiance.

I would argue, then, that the humanities must suffer, as agents of a good society still to come, from the lack of a context of civic piety in which to challenge and inspire us, communally as well as personally. Park suggests that the great museum shows of foreign collections and achievements prompt in us a sense of solidarity with other cultures who use "the common language of the arts and humanities." This is true at a certain level of abstraction; in our experience of the cultural events themselves, however, the presence of thousands of our fellow citizens more often than not serves as a handicap rather than a stimulus to our enjoyment. Having to wait in line six hours only to move fairly rapidly through a rich collection so others can see it too does not usually promote our apprehension of the civic pleasures of the humanities. We like our ball parks full, for the excitement of the crowd adds to the event, but we like our museums empty so we can indulge to the fullest our personal contemplation of great art.

Obviously, I am not arguing that we should invent rituals or cultivate religious fervor about the arts and impose them on the institutions of the disestablished humanities. Neither holy German art nor socialist realism holds any attraction for me. I am simply pointing out the lack in our culture of a crucial element that was present and effective as part of the Athenian experience of civic involvement. Instead, we have the opposite ideal with regard to the humanities: the attractiveness, exploited by snobbish mail-order solicitations, of enjoying the great books, music, and paintings of the world "in the privacy of your own home," away from the discommoding masses of one's fellow citizens and perhaps of the attentions of the pickpocket as well. This is hardly the attitude upon which to base one's faith that individuals, as a result of humanistic pleasures, will actively work to fashion those pleasures into a better, more humane social model for all Americans.

II

It has not been my intention in the foregoing, somewhat polemical remarks to sound a death knell for the great cultural institutions of our cities or for the possibility of finding and enjoying a good book or movie in America. Even less have I wished to glorify the experience of other cultures (specifically the Athenian) at the expense of our own. My catalogue of the vulnerabilities hindering the disestablished humanities from being a self-sufficient midwife of a good society hopefully clears the ground for happier considerations, *viz.*, how the university, with its traditions—however faulty—of teaching and training in the humanities, can work with those who are outside its walls to strengthen the institutions and media that bring the testimony of humanistic values and achievements to all who wish them, and thus to strengthen the humanities themselves in our society. The following points seem to me important; the first two concern internal procedures, the other three involve the university working with other groups and institutions.

1. First of all, the universities can teach the humanities well and centrally and help young scholars formulate a pedogogy for doing the same. Colleges and universities would do well to reinstate humanities-related degree requirements, in art, in music, in foreign languages, where these were abandoned, as a palliative to restless students, in the late 1960s—or to create them where they have not existed. Although Park documents a decline among college graduates in the use of cultural opportunities, it has been my experience over the years that many of my contemporaries, having been introduced for the first time to the joys of Bach and Mozart, Michelangelo and Bernini, in required college courses, have remained remarkably involved with such material even if their location (an important fact not taken into consideration, perhaps, in her statistics) makes it difficult for them to visit a major museum or hear a concert.

2. The universities must overcome departmental suspicions about loss of autonomy and quarrels over budget allocations and press ahead with increased interdisciplinary training for undergraduate and graduate students alike—though not, I hasten to add, at the cost of destroying competence in specific, traditional disciplines. Nearly ten years ago, Daniel Bell, in *The Reforming of General Education,* proposed "third-tier courses" for the undergraduate senior "that would 'brake' the drive toward specialization by trying to generalize his experiences in his discipline." [2] Today the rhetoric has changed, but the need for multidisciplinary approaches remains and has been too little satisfied. Columbia's Program of General Education is now bringing together scholars from various arts and sciences disciplines, and putting them in touch with professional school personnel as well, in a laudable attempt to foster a cooperative attack on large problems confronting our society. Park is correct in seeing our need, society-wide, for "the multidisciplinary approach"; her comment that "the traditional approach of the humanities has always been toward wholeness" is somewhat belied by our higher educational experience in this century, but I would argue that the university is one of the main (if not one of the only) places where the multidisciplinary approach can be inculcated thoughtfully and responsibly.

3. At a purely political level, the universities and the great institutions of the disestablished humanities must learn to work together in making a claim at all levels of government for support of their endeavors. Both groups of institutions are equally threatened by the urban financial crisis and by a (hopefully temporary) lessening of our willingness as a nation to support the services and institutions that contribute to our cultural life and thus remind us of aspects of existence beside economic and political competition. Joint appeals and joint programs to give exemplary rationale for them may well be a crucial development in the relationship between the academic and nonacedemic humanities during the next decade.

4. The university must make its students conscious in new ways of both the advantages and the problems of the disestablished humanities. Recently, the Columbia College Alumni Association offered its members a course of lectures on great paintings in the Metropolitan Museum of Art, to be delivered each week at the museum by a different member of the art history department of the university. This

excellent initiative became an unwitting exercise in consciousness raising when New York City's financial crisis caused the museum to shut down on the evening scheduled for the lectures. Colleges and universities might well increase their programmed use of local cultural resources in courses of instruction and might also make the growth, function, and problems of cultural institutions the subject of multidisciplinary study involving historians, sociologists, and humanists, perhaps helping thereby to foster more prolonged and frequent use of these institutions by their students in later life.

5. Potentially the most exciting function for the university as a broker for the humanities involves work in what might be called "boundary areas"—areas in which all or part of society finds its interests served by working together with the university in pursuit of some version of the good society. An example will clarify my meaning. At Columbia, the university's Casa Italiana has broadened its mandate from simply that of presenting the Italian cultural heritage to the university to include the establishing of a creative, ongoing relationship with New York's Italian-American community. The Casa now includes in its programs lectures and artistic programs devoted to the Italian-American experience and thus provides a cultural dimension of self-definition for an ethnic group that is working through political channels to upgrade its traditional neighborhoods and improve social services for its members. In addition, through their connection with the Casa, Italian-American civic groups can also be introduced to the great achievements of Italian civilization in the arts and can thus be encouraged to partake of the pleasure and inspiration offered by the humanities to all citizens.

More initiatives of this kind, with full participation by the institutions of the disestablished humanities, not only may result in heightened sensitivity to and support of the humanities; they may even contribute to the development of a communal dimension of our experience of the humanities, that crucial ingredient, now largely missing, that will keep humanistic values at or near the center of our attempts "to operate our complex society appropriately." I am thinking here specifically of our bicentennial celebrations, which had great potential as an occasion for stimulating civic or national rituals enacted in a climate not simply of self-congratulation but of earnest self-analysis. In helping us look backward in order the better to look forward, the humanities should have a key role, and the universities could be of great use in clarifying and presenting that role. Working both intra-murally and in concert with alumni, civic, and governmental groups, and cultural institutions, the universities could lay out a map of American history that links celebration of our genuine achievements with an anatomy of problems faced but not resolved. In the process, the place of the arts and of humanistic values in our culture could come under scrutiny, and some of the problems I have already mentioned, such as the plight of the urban humanities and the manipulation of mass culture, could be placed before audiences that had not thought of such things before. I am not imagining town meetings in urban sports arenas sponsored by local colleges, museums, and historical societies, but more modest initiatives that could, however, reach relatively large numbers of people through educational television and local government sponsorship. Perhaps it is now too late to do more than ponder opportunities lost, but the potential for cooperation between universities and nonacademic institutions in the humanities implicit in the bicentennial makes it a good example of the "boundary area" concept.

III

I will close these comments by offering a comparison between two possible modes of our involvement with past and present achievements in the humanities. The comparison takes the form of juxtaposed descriptions of two paintings I have recently encountered. The first hangs in the Metropolitan Museum of Art in New York; it is one of several paintings by Titian (and his workshop) with the shared subject, *Venus and Musician*. A young courtier, fully clothed, plays a lute while gazing intently at a reclining, nude Venus, who is being crowned with a floral wreath by an attendant Cupid. Venus holds in her hand a recorder; in the right hand corner of the painting, a viola da gamba stands propped against Venus' couch. As I understand this lovely painting, one of its aims is to suggest the performance of a trio, only two members of which are depicted in the image: The viola da gamba player is outside the frame; his instrument is there for him to pick up and begin playing at any moment. We, the educated spectator, are the third member of that trio, and our instrument—our contribution to the painting's meaning—is the knowledge of Renaissance thought about love that we bring to our experience of Titian's beautiful images. For love in all its complexity—sensual desire, Platonic idealism, and every stage in between—is the subject of the painting, but its precise meaning will vary—and did vary, we can be sure, with the learning and predilections about love that each of the sophisticated courtier-humanists who were its first audience brought to it. For the fullest aesthetic enjoyment of *Venus and Musician,* the spectator must bring to the work a cultural preparation that enables him or her to play in the trio.

The second painting hangs in a private home in Florida. On a large, rectangular canvas, the artist has painted, in apparently random distribution, an array of labels from detergents and other commerical products. Superimposed on the canvas is a raised piece of plastic in the form of an hourglass, sealed at both ends and containing a supply of what seem to be detergent crystals. An arrow above the painting directs the viewer to turn the picture upside down (it is attached to the wall on a swivel), whereupon the detergent crystals run through the hourglass to the other end, and the process can be repeated, ad infinitum. In this case the viewer's participation is prescribed, sharply limited, and absolutely fixed, from viewer to viewer. The movement of the crystals suggests the emptying of a box of detergent and thereby links response to the work of art and a mindless consumerism. The whole strategy of the artist and painting makes it unclear whether we are manipulating the painting or it is manipulating us, by calling forth a specific, limited response that gives gratification without in any way challenging our mental or moral faculties. Finally, the content of the painting—commercial product labels—and the action that makes it work—a rotating movement rather like turning a very large knob—remind me, designedly, I believe, of our relationship to our most accessible medium of mass culture, television, where the turning of a knob produces (all too often) mindless entertainment accompanied by manipulative advertisements for commercial products. This second painting, I would argue, is a parable of our conditioned, unindividualized consumption of popular culture pressed into the service of the profit motive.

I do not wish to argue which of these paintings is better; rather I offer them as symbolic of the choices we have in responding to the arts and to cultural manifestations generally in our daily life. Our participation in the disestablished humanities will have to come much closer to the first model—the trio in which we play a vital, personal role—if we are to appreciate the place of humanistic traditions and values in our search for a good society. And, with all its faults, the university still leads the way in our society in providing us with the knowledge and the self-awareness that will make our relationship with the disestablished humanities at once harmonious and lively and save us from the more nearly passive role of merely turning the picture over and watching the crystals run through.

The Humanities and Public Policy

In the school of political projectors I was but ill entertained; the professors appearing in my judgment wholly out of their senses; which is a scene that never fails to make me melancholy. These unhappy people were proposing schemes for persuading monarchs to choose favorites upon the score of their wisdom, capacity and virtue; of teaching ministers to consult the public good; of rewarding merit, great abilities, and eminent services; of instructing princes to know their true interest, by placing it on the same foundation with that of the people; of choosing for employments persons qualified to exercise them; with many other wild impossible chimeras, that never entered before into the heart of man to conceive; and confirmed in me the old observation, that there is nothing so extravagant and irrational which some philosophers have not maintained for truth.

Jonathan Swift
Gulliver's Travels, 1726

25

A Philosophic Perspective

ABRAHAM EDEL

The easy way to enter our topic is through the analytic gate: to reflect on the meaning of "public policy" and of "humanities" and hope that the types of relations encased in the "and" will stand out in bold relief. We must not, however, expect the results to be more than suggestive.

Policies formulate directions for conduct at a moderate level of generality. Too general, they turn into aims; not general enough, they are rules. But these are formal distinctions and do not constrain the content. Any content significant enough to deliberate about can under some conditions turn into a policy.

Three general components in policy decision may be distinguished. We have to have aims, purposes or *values*. We need *knowledge*. And we require some account of the situation to which value and knowledge are being applied; let us call this third component the *practical context*. Now the first temptation is to correlate value with humanities, knowledge with science, and practical context with the field that calls upon us for decision. A kind of synthesis of the three would then yield the policy. But this is too neat. For one thing, values are riddled with factual presuppositions and assumptions. For a second, knowledge even in its scientific form has a purposive base and is permeated with selective elements. For a third, the kind of structure to be ascribed to any particular context for action itself involves both knowledge claims and particular purposes. And fourth, decision is only in special cases an intuitive synthesis, just as it is only in special

Abraham Edel, Distinguished Professor Emeritus of Philosophy at the Graduate Center of the City University of New York, is Research Professor of Philosophy at the University of Pennsylvania.

cases a technical computation. Most of the time it is at least a dramatic rehearsal of alternatives and a rough estimate of their consequences.

Now for the "humanities." Fortunately, Congress has given us a good and broad-minded start. The humanities, we are told, include "the study of philosophy, history, literature, language, linguistics, jurisprudence, comparative religion, ethics and archaeology; also the history, theory and criticism of the arts and those aspects of the social sciences concerned with values." In short, not the sciences, not the arts themselves, but a great deal of the rest, and most of it on a reflective level. "Humanities" is thus taken to designate a body of subjects, a way of regarding it often found in academic life.

A second approach to the humanities would be to think in terms of a set of human faculties or functions to be stimulated by such subjects. Just as knowing and reasoning are linked to science, so imagination and feeling and creativity are linked to the humanities. (Insight and understanding hover on the border.) A third approach would characterize humanities by the humanistic, no longer a body but a spirit; so freed, there is no field to which it might not be applied. We could envisage a school of humanistic engineering, where physical science was taught in the context of the history of science as an imaginative enterprise (much as Jacob Bronowski saw it); where ideals of technique and practical control were infused with the spirit of Prometheus and the sense of human well-being and the quality of life; and where organizational ideals breathed the air of democracy and community of effort rather than the authoritarian confines of either corporatism or technocracy. And throughout would run the philosophy of history of human beings making themselves, with a full sense of their responsibility for the products. Indeed, it is this sense of "humanities" which often triggers the complaint that much of what passes for the subject matter in humanities can be as narrowly technical as any applied science.

One more preliminary reflection: how practical are the humanities? At once we think of Plato's warning that whoever controls the music of the young controls their character and their fate more surely than they are determined by the laws. And by "music," although he refers to the heroic Dorian mode and the effeminate Lydian mode, he really means the whole range of the humanities, the domain of the

Muses. In our time, similar questions are raised, not about the Homeric gods as risky models for youth, but about nursery stories and comic books, riotous music, violence and aggression on TV. All this is too familiar to need discussion here. Yet it is worth noting on a broader social scale that departures in the arts often signal (if they do not control) a restlessness that moves to social change. One thinks of futurism in Italy playing into the hands of fascism and its glorification of strength and antirational spontaneity. Or one thinks of writers like T. E. Hulme in England before World War I and Oswald Spengler in Germany during that war, who traced such connections with a kind of grim joy or pungent gloom. Hulme's criticism in art and the humanities was directed to ferreting out *every* possible source for glorifying violence and revolt against rationality that could be seen as a forerunner of an antihumanism: the force of Jacob Epstein's sculpture, the retreat from reason to the intuitionism of process in Bergson (Hulme translated the latter's *Introduction to Metaphysics*), even the philosophy of G. E. Moore and Husserl as betokening a break with the subjectivism and relativism of humanist ethics. Hulme was a self-styled reactionary and finally found his violence (and met his death) in World War I. And Spengler too, though he uses a method of riotous analogy, has a sense for the practical relations of arts and modes of thought and feeling. I take from all this no more than the minimal warning that the values to be seen in the humanities at a given period may be foreshadowing if not actually governing the rising practice of the emerging world.

There is another, quite different sense in which the humanities have shown themselves to be practical: the fancies of one day may be the practical problems of the next. Aristotle, in defending slavery as necessary for the work of the world, says it would come to an end only if spinning took place by itself and the lute played of itself, and perhaps not even then. Today, with automation, the end of exploitation is thus overdue. John Stuart Mill, in arguing for qualitative differences in pleasure, asks the speculative question whether one would choose to be a dissatisfied Socrates or a satisfied pig. Today, lobotomies and psychosurgery, as well as new modes of operant conditioning, make such choices a highly practical matter; less dramatically, in ordinary life, there is always the choice of a tranquilizer to lull the anxieties of facing problems socratically. It is worth noting that the science of tranquilizers—*ataractics*—derives its name from the Greek for lack of disturbance or peace of mind which was the great goal of the ancient Epicureans, much like the "apathy" of the Stoics. Yet there is a great difference—one might almost say a world of value—between tranquility and tranquilizing. I forego what we might do with Schopenhauer's speculative urging that men refrain from procreation to bring to an end the blind restless striving of the Will on earth, now that we can really poison the atmosphere with nuclear energy. . . .

So much for preliminaries. The general subjects of the preceding essays preeminently represent areas of burgeoning policy issues that tap a diversity of central problems, human, global, and contemporary. It is well to keep them before us: Justice and Human Equality, Private Rights and the Public Good, Technology and the Ideal of Human Progress, War and the Social Order, Education and the Good Society. I want to tackle our central problem—what the humanities can do for public policy issues—not by taking each of these in turn but by following a number of theoretical themes that keep weaving their way through these papers, no matter which of the five fields their authors focus upon. Here are the topics I shall deal with:

1 Sibling Rivalry: The Relations of the Humanities
   with the Psychological and Social Sciences
2 Rationality and Beyond
3 Intimations of a Natural Order
4 The Hold of the Past and the Hope of the Future
5 Symbols, Interpretation and Reality
6 The Study of Values
7 Mine and Thine or I and Thou?
8 How to Be Practical though Humanistic

1. Sibling Rivalry: The Relation of the Humanities with the Psychological and Social Sciences

So basic a theme calls for a myth—for surely the intellectual life is entitled to its own myths. Zeus, it is said, seeing the philosophers worried only about questions of physical science, sent Socrates to instill a wider set of questions. But Socrates got so entranced with the phenomenon of asking questions that he never even got to asking all the questions on the list that Zeus had given him. Zeus therefore had Socrates put to death and gave the list to Plato. His instructions were peremptory: Plato was to sow the seeds of the humanities and the social sciences, and he was to write a report—no oral account this time. Plato's report is the *Republic,* which asks all the questions in all their magnificent interrelations. But when Zeus came to read it—or had one of his Danymede-like scribes do it for him—he found that Plato had become so interested in the questions that he also gave all the answers, and of course such *hubris* brought its own *nemesis:* most of the answers were wrong, though mankind would take millennia to find it out. (In fact, Thrasymachus had tried to steer Plato into social science at the outset, but Plato had violently snubbed him.) Zeus therefore, making the punishment fit the crime, condemned Plato to write another report on the answers, which no one would ever bother reading! This of course was the *Laws.* And ever since then—for every explanatory myth must end with an "and so"—three things have happened: the humanities ask the questions and offer only intimations of the answers; the social sciences give the answers but never bother much about the questions; and the foundations insist on written reports but never bother much to read them!

I shall make no attempt here to defend the literal truths built into much of this myth. The *Republic* does ask most of the questions relevant for social policy and does it beautifully, especially in the disorder with which it goes from one field to another, which is the natural way to see all the presuppositions of one's questions. And it does give cursory answers because it begs numerous issues that require a psychological and social science to clarify and furnish evidence. And the sibling rivalry (humanities is the older sibling) has been continuous. Witness the way in which social science long shied away from value problems on the ground that values were unscientific; or the way in which social sciences narrow their path to questions that their tools are capable of dealing with, as if the aim of science were merely to exercise tools; or the way in which departments of psychology and of political science and of sociology have split within recent decades in the United States because one side wanted to hold on to value issues and policy issues while the other was entranced with behavioral operationism or with simulating human behavior in computerizeable models. Or look at the same phenomenon from the side of the humanities: so anxious are the humanists not to be scientists that they set up rival methods of symbolic analysis and interpretation and grasp every weapon of phenomenology that comes sauntering by as if to say, "Look, we have our own science which is not science, and it's better than my sibling's, and it gives you a real truth, not a spurious one." And if more evidence is needed, look at the way scientists rush to answer humanistic questions in their off moments, or after retirement, or occasionally in a collateral utopia—for example, *Walden II*—as if these really important issues did not require their full scientific concentration.

I say all this not to complain but to suggest a thesis about the relations of the humanities and the psychological and social sciences. They are both cut from the same cloth, and the designs run through them often without a break; we must not mistake historical divisions of labor for metaphysical dichotomies of subject matter or for more than differences of degree in method. Of course, we can manufacture sharply contrasting pictures if *every* time a scientist uses his imagination we call him a humanist, and *every* time a humanist looks for evidence we say he is turning scientist. But then *our* dichotomies, not nature's or reality's, will be showing.

I suggest therefore the unity of the human endeavor in these domains, without at the moment presenting it as a structured and analyzed hypothesis. Some considerations along these lines will emerge from the treatment of themes later on. But perhaps a brief case study will clarify what I am driving at.

One of the massive works in psychology and social science at the midpoint of our century was *The Authoritarian Personality* (1950) by Adorno, Frenkel-Brunswik, Levinson, and Sanford. It made broad and extensive studies of the antidemocratic personality, identified initially in the ideology of antisemitism. It sought to establish that there was such a configuration of personality, to identify its marks and degree of generality and typical ideological manifestations, and to correlate its occurrence with familial and institutional patterns. It had all the expertise of scales and tests and statistical correlations. There was, of course, criticism of its foundational ideas and methods, but such criticism is a common phenomenon in controversy over a scientific project that is large-scale in scope, especially where it has broad policy implications for institutional reconstruction; it is not pertinent here to ask how far the criticism was justified.

To the same era belongs Sartre's "Portrait of the Antisemite," which appears in his *Reflections on the Jewish Question* (1947). It asks the same kind of question as the scientific study; that is, it wants to know what is going on in the antisemite. It is anecdotal: for example, it tells about the man who was anti-British and came to life only when someone referred to an Englishman. It has literary references, for example, to the picture of the Jewess in literature. It frames no tables of evidence but gives conclusions in an impressionistic way: that antisemitism is not an opinion but a passion involving the entire personality; that the antisemite is a man afraid of himself, his freedom, his responsibilities, indeed of everything except the Jews. Ultimately, Sartre interprets antisemitism as a fear of man's fate, as a desire to be a thing—anything but a man. The essay is eloquent, passionate, thoroughly humanistic. Its conclusions are not far from those of the scientific treatise.

I make these comparisons not to argue that a humanist can reach the same results without being scientific. He might, but how would we know he was correct? Nor do I intend to deny a difference in method—whether this be in kind or in degree. Sartre's thesis is not formulated with technical precision, it is not sufficiently refined, and it is intended to lead us in a different direction, to his philosophy of freedom and man's fate. But why should we be so *either-or?* The humanist and the scientist are here doing the same job but pushing in different directions.

It is interesting to note as a historical matter that the scientific study of the authoritarian personality had its origins in value and policy issues. Obviously it was stimulated by the phenomenon of Nazi Germany and the desire to understand what happened to a whole people. Moreover, the roots of its major participants lay in German culture and the experience of the authoritarian family. The Frankfurt school from which Adorno came was fascinated by the phenomena of authority and very early did studies on authority in the family. This in turn had repercussions in the humanities as well as the sciences. For example, it is this background which probably explains why Erich Fromm, coming from the same school, focuses on an aspect of the Oedipus myth entirely different from the usual Freudian one: he emphasizes the relation of Oedipus to his sons (as evident in *Oedipus at Colonus* and *Seven Against Thebes*) rather than simply his relation to his mother. The dominating fact is paternal authority. I should not be surprised if in turn the roots of ego psychology, as in Erickson's work, are fed at the same spring.

It may be thought that illustrations from the soft sciences would naturally find the humanities and the sciences close together since depth psychology and personality psychology are sometimes more humanistic than scientific. But the same lessons could be derived from economics, particularly when it is looked at in relation to policy. In his *Proposed Roads to Freedom* (1919), Bertrand Russell suggested that society set up a "vagabond wage," that is, a subsistence minimum which anyone could get if he preferred not to work—for example, if he wished to spend his time on philosophy and poetry. Russell added that enough people would want more than the minimum wage to ensure the production the world required. Half a century later, Milton Friedman proposed his "negative income tax" which would, by income tax rebate, bring every family to a subsistence minimum. Russell's view was romantic and humanistic and philosophic; Friedman's is based on hard science. Yet both embody values: the common one of acknowledging some social responsibility for diminishing poverty; the further one on Russell's side of encouraging creativity; the further one on Friedman's side of enhancing liberty by getting rid of state controls in welfare systems and the like. Russell guessed that there would be enough people wanting more to satisfy production needs and that society could afford it. Friedman could give hard figures for manpower in production and for costs. Both were realistic. Russell's impressions of the way his plan would enhance freedom were quite matched by his estimate shortly after of the authoritarian potential of the Russian revolution. Friedman's realism was evident in his blunt remark before a congressional committee (reported in the press) when he was asked whether beneficiaries of a negative income tax should not lose their vote; he replied that if putting one's fingers into the public till meant losing one's vote, businessmen would be the first to lose it.

Actually, the difference between Russell's proposal and Friedman's is accounted for, not by a contrast between humanist and scientist, but by the experience of half a century, the tremendous growth of production, the lessons of both economic waste and human indignity in the workings of the welfare system, and so on. Both show humanistic imagination in making a form of property ("wage" and "income tax rebate") out of the routing of tax funds, and neither loads it with "charity" or even "distributive justice." There is no hard line here between humanist and scientist in their dealings with public policy.

Since these remarks have stressed the community between the sibling disciplines, it is all the more important to note how they diverge when science becomes *technical*—a term, after all, that betokens only the disciplined precision of *technē* or craft. Such divergences can best be seen by dealing with evidence. Imaginative impressionism will no longer do; the responsibility to the real world is as severe as the responsibility to consequences in action.

Take, for example, the broad question that arose in several papers about whether men at bottom really value conflict and war. In "War and the Clash of Ideas," Adda Bozeman culls materials on war and violence from many different cultures to illustrate her thesis of the natural acceptance of these phenomena throughout history, as compared to the recent emphasis in peace research on conflict control.* (I was reminded in this of Santayana's remark that philosophical manipulation of history is like a man looking over a crowd to find his friends.) Now clearly, if the issue called for evidence, we would have to ask whether one could also cull materials on the opposite side; whether all Bozeman's materials concerned men, not women; whether hers came predominantly from preindustrial societies; and so on. Too often, in the humanities, one finds the impressionistically striking rather than the evidential cautions which the sciences have stabilized over a long experience of inquiry.

A similar question arises in a somewhat different way when Elisabeth Hansot suggests, in her "Reflections on War, Utopias, and Temporary Systems," that peace involves "the absence of certain human capacities or passions which have come to be partially associated with combat." At the very least, we would have to inquire whether this absence held only for certain types of societies such as ours—individualistic societies which set only competitive challenges and not hard communal challenges. The

* Originally presented at one of the conferences on War and the Social Order, Adda Bozeman's "War and the Clash of Ideas" has appeared in *Orbis* 20 (April 1976): 61–102.

range of evidence is not enough; it involves the analysis of the
conditions of the evidential.

Similarly, William McNeill's comparison of historical empires, "On National Frontiers," from which he seeks to
show how great has been the role of polyethnic empires as
against the limited role of the principle of national self-deter-
minism, would give only a start to inquiry. We would have to
know the conditions under which each occurred and to evaluate
the successes and failures (as well as the costs and benefits) with
respect to the maintenance of peace; and if interest focused on
the future, whether the conditions of the one were likely to pose
the problems for action or the conditions of the other.

On this question of evidence the sibling disciplines would cooperate most successfully if rather than jockeying for priority they stressed
mutually integrative results. They have been seduced too often by
oversharp contrasts between global-synthetic and analytic-statistical,
or the various contrasts between portrait-painting and generalization
that have been argued over in the philosophy of science under such
captious captions as "ideographic" and "nomothetic." But to pursue
this here would take us too far afield. In any case, they are born
together in the problematic field, and they come together in the determination of policy, though they move apart in the intervening
processes of methodological differences and technical requirements.
Yet even in these intervening processes the similarities may be not
without significance.

2. Rationality and Beyond

There is a class of humanists who take their stand (or rather their
comfortable chairs) on sentiment (not sentimentality) as distinct and
above rationality. In vain, for them, did William James try to bridge
the gap by seeing rationality as itself a sentiment. For them, rationality is on the other side of a great gulf. They would, if they could,
revise Emerson's oft-quoted remark about a foolish consistency by
calling consistency "rationality" and leaving out the "foolish." It
would then read: "*Rationality* is the hobgoblin of little minds." Rationality is practical or rigidly logical; it is useful enough, but utility is
the slave of value, and the humanities are the value bosses. I am talk-

ing, of course, of the ideal type; for some of our contributors the shoe may pinch a little, but it surely will not fit.

There is always a touch of ideology—usually conservative ideology—about this view. Perhaps we should say of it what Oakeshott says of a conservative philosophy, that if there is a felt need to develop a conservative philosophy, conservatism must be already slipping. So too, a lofty elitist humanism has a certain charm as an expression of truly cultivated spirits, but when it begins to defend itself it is surprising how often we find quotations from the French-monarchist intellectuals still revolting against the revolution, or the picture of culture in T. S. Eliot and a distaste for the masses, or a Bergsonian berating of a static intellect, or a Nietzschean abuse of John Stuart Mill as a blockhead. Still, humanist sentiment must not be condemned by association simply because it attacks rationality. The question rather is: what is wrong with the picture of rationality that cultivated sentiment should shudder at it? It may be simply, as Santayana remarks in a chapter on mechanism, that poets shudder at skeletons and scientists give us the skeletal structure of the universe. Or more likely, there is as much of ideology in concepts of rationality as there is in the attacks upon it. For after all, the history of Western thought resonates loudly with the battles of reason against faith, of science against religion, and all the familiar rest. If sentiment is tied to the aristocracy, reason and science are tied to the bourgeoisie, but our problem is not—or is it?—a question of evaluating the style and quality of life in different social classes.

There is an occasional attempt to treat the issue of reason and beyond as itself a scientific question. Lévy-Bruhl in his *How Natives Think* (1925; French original, 1910) tried to fathom the prelogical mentality that makes its inferences by some feeling of participation rather than by the law of noncontradiction. But Franz Boas in his *Mind of Primitive Man* (1911) early argued for the logical powers of preliterate peoples; we tend to be misled by differences that are in the assumed premises, not in the processes of thought. Again, in the social sciences, there is the effort to develop interpretation as a distinct category of inquiry, fit for the humanistic; this we shall examine below.

But what of rationality itself and its many portraits? Its minimal sense is of course consistency. A man says, "War is terrible," and the next moment he rhapsodizes, "War is glorious." We confront him with himself and make him say, "War is terrible *and* war is glorious." Then he must look for an out: sometimes terrible, sometimes glorious; terrible in one aspect, glorious in another; glory itself has terror in it; anything but to be led to the brink of the utterance "p and not p." Now take Homer's scene in which Hector says what is to be his farewell to Andromache while holding his young son in his arms. He has just finished saying that he is fighting an unjust war and will continue fighting only to defend his family; the sense of war as senseless is obvious. As he holds up his son, we might have expected him to say, "Let there be peace when this boy grows to manhood." But instead, he prays that the boy's valor may be greater than his father's and that his mother's heart may rejoice when he brings in the bloody spoils of his slain foe. A contradiction, yes. But contradictories cannot both exist, and here we have a people for whom violence is the source of prestige and status and yet whose violence brings destruction to them in the search for glory. Obviously, the meaning of contradiction has been enlarged. Here it is contradictory objectives, in the sense that there is inner conflict and unavoidable frustration. When Dostoyevski's underground man says that something will yield the greatest happiness but that precisely for that reason he spurns it, is this self-contradiction, or a critique of happiness as the human end, or a rejection of calculation in the good life, or just a clinical symptom?

It would be a long story to trace the path from minima to maxima in the idea of rationality. We could move from conflicting aims to an inner conflict in a system. After all, was not the whole of Marx's *Capital* a theoretical structure to show the underlying conflict in a system of apparent harmony and expression of agreement between buyer and seller (of commodities, including labor), much as two innocent neighbors drilling for oil might not realize that they were tapping the same reservoir? We could follow a different path in which not so much the demand for unity is built into the idea of rationality as the bare notion of utility. Rationality now starts from mere efficiency, the least wasteful use of means for an end. It goes on to include concepts of bureaucratic rationalization (as Max Weber analyzed them) and even often in a disguised way to incorporate the abstract end (power, money) to which the utility is directed. Or, letting people choose their own ends, it develops notions of maximization, scrupulously avoiding, with Pareto's blessing, any interpersonal comparisons. In a more aseptic analysis, limited to the nature of scientific processes, rationality will grow from logical consistency to inductive rationality which incorporates the trend of evidence, and a rational man will be one who follows the evidence.

Clearly then, to reject rationality is to reject the special form of thought and institutional orientation that is elaborated under its broad umbrella. But this does not mean that there is a pure rationality captured by a special pattern. Perhaps ultimately there never has been, and cannot be, a concept of rationality which does not, once it goes beyond logical consistency, embody some view of man, his world, his faculties, his aims, and at least their general requirements. The concept of rationality is therefore a growing one, demanding evaluation at every stage. As we learn the lessons of method in the growth of knowledge, our concept of rationality itself expands. For example, as we come to make fairly reliable predictions over longer stretches of time, a rational person is one who takes the long view, not the short view. As we learn the inner mechanisms of self-deceit, a rational person is one who discounts for his inner distortions. As we come to understand the influence of class and its ideologies, a rational person is one who tries persistently to see things from the other point of view as well, instead of only through any one entrenched perspective. (Is this not the lesson of contemporary liberation movements?) As a global perspective begins to show the confusions and frustrations into which anything less than a global view will get us (at least on some questions such as energy and agriculture and pollution of the ocean and the air), does not rationality begin to embrace a notion of taking a global perspective where a global perspective is essential to answering the problem—even if nationalism has all the strength some of us have attributed to it? Finally, as philosophic wisdom grows and shows us that the kinds of answers we give depend in measure on the kinds of questions we ask, does not rationality begin to embrace a theory of the appropriateness of questions?

Theoretically, there is no reason why the direction of an expanded rationality should not be sensed by humanists with their eyes on values and the imagination sometimes long before scientists who are immersed in present data under a complex establishment or else extrapolating their curves from present dilemmas. Often, however, the scientist and the humanist may be preaching the same lesson in different ways. Take, for example, the great contemporary lesson of the human need for an open system as against a closed total world view. The scientist has long incorporated it in his methods and his probabilism in the face of growing human knowledge. In religion, there has been the shift from a dogmatic assurance about God's will to an existentialist recognition that we act in our own responsibility without a dependable knowledge of what is demanded of us. The emphasis on transcendence in Jaspers, or on self-transcendence in Neibuhr, is on openness. For that matter, when Sartre takes Kierkegaard's analysis of the episode of Abraham and Isaac, he does not see it as illustrating Abraham's depth of faith; rather he wonders why Abraham should not have questioned the command to kill his son as unlike his God, and possibly coming from an imposter. Even Marxism, it should be noted, though often applied in a closed way, contains in its metaphysics the principle of the transformation of quantity into quality, which means the recognition of critical points in historical processes at which new levels of phenomena come into being. (Marx always opposed the structured plans of "utopian socialists.") Although there may be different social stresses in these various views, there seems little difference in the message of contemporary science and contemporary religion and contemporary humanism about the rationality of keeping the system—whatever system—open. The desire for a closed system is, in Dewey's phrase, a quest for certainty. We can sympathize with Santayana's preference for Dante over Shakespeare, because Dante has a total world view whereas Shakespeare only traces the ramifications of fragmentary themes as they generate from his magnificent display of human passions. But we know now that a closed total world view is not a mark of rationality but its opposite.

Today the concept of rationality needs reassessment in every area of human endeavor. Rationality in law, rationality in ethics, rationality in politics—all call for analysis and evaulation. Was what Watergate revealed the culmination of rationality in politics? When a student says to a professor, "Thank you for your time" instead of "Thank you for your criticism and ideas" (which started, I think, somewhere in the 1950s), is that the culmination of economic rationality in intellectual relations? Is cost-benefit analysis based on the postulate that there is nothing that cannot be overbalanced by some aggregation of other values? With such tasks of inquiry on our collective hands, there is full scope for the humanities and the sciences and practical life to recognize their kinship in the ideal of rationality, rather than to equate it with its historical shortcomings.

3. *Intimations of a Natural Order*

In none of these essays is there any loud proclamation of a natural order. This is remarkable, considering how long philosophy and theology and law and economics centered on the concept of nature and the natural and how persistent the tradition of natural law has been. Nevertheless, we are not without some residual intimations that call for exploration. There is a marked propensity for perennialism in the humanities.

Perhaps the notion of a natural order is covered by constant talk of future shock these days. At any rate, the idea of change has grown to tremendous proportions since the nineteenth century. It is still possible, however, to ignore that idea by asking questions about the nature of man and not analyzing the questions. We then get debates between opposing answers. In this fashion we find debates about whether man is naturally aggressive or naturally affiliative, with occasional gimmicks thrown in like territorial imperatives derived from our kinship with animals. Neither side questions the basic concept of a nature or an essence, but we do not notice this because they argue so heatedly as opposites. Even the view of man as plastic has not been cast so much as a denial of a nature as an affirmation of a culturally versatile nature. Again, there are desperate efforts in philosophical anthropology to provide a definition of man. Is he a rational animal, a tool-making animal, a being capable of ideals, a purely historical being, or what else? Perhaps a good answer would be: Man is a being who asks the question "What is man?" and declines to answer it. Why should man not refuse to be defined? "What is man that Thou art mindful of him?" was not a request for a definition but for a ground of divine concern.

In the next section, I shall discuss how the nineteenth-century discovery of history and change gradually crept in. Objectively, it was involved in the biological concept of evolution and the different explicit philosophies of history: man came to be regarded as a historical being at the very core—far beyond the eighteenth-century notions of indefinite progress toward a fixed ideal. Subjectively, the shift is seen in the emergence of insecurity in men's feelings toward the world. An eighteenth-century Hutcheson could rest morality on sentiment because he felt secure that God would not go changing the basic instincts and passions of men. He probably would not have understood a Nietzchean demand that man surpass himself. But already Kierkegaard rejects the socratic maxim of "Know thyself" because presumably there is nothing secure to know. He substitutes "Choose thyself," and we can be sure that the choosing is done in fear and trembling. In the twentieth century one finds at last talk of man's nature *changing,* and even—in existentialist views—the denial that man has a nature on the ground that a nature is incompatible with man's absolute freedom. Perhaps this is less revolutionary than it sounds since man is said to have a condition, if not a nature, which imposes invariable problems on him (he lives in the world and chooses, works, dies). In the scientifically minded history we find an occasional archaeological vista which prompts a scientist such as V. Gordon Childe to title a book *Man Makes Himself* (1936).

There is one very definite appeal to nature and the natural order in these papers, apart from such residual intimations as allege perennial verities about man's lust for war or suggestions of proper functions for education, and it lies in Roderick Nash's conception of an original and pure Nature.

In "Do Rocks Have Rights?" Nash proposes an environmental ethics that will go beyond human beings to embrace animals, plants, rocks—in short the whole of our natural world. (The whole view will concern us later.) On the crucial question of the needs and wants of existents, he notes that slaves can speak for themselves, but "what, after all, do rocks want? Are their rights [being] violated by quarrying them for a building or crushing them into pavement or shaping them into statues?" For the time being he compromises on the question, agreeing with Aldo Leopold that "a thing is right when it tends to preserve the integrity, stability, and beauty of the biotic community. It is wrong when it does otherwise."

Now what might justify such a view of rocks? Why should not the statue be a fulfillment of the rock's nature?—Aristotle is often quoted for the view that any stone longs to be a doorstep. Nash's attribution of an essence to rocks is less a Newtonian theory of inertia than a conservative assumption that nothing is to be changed. In his case—that of an environmentalist's battle against corrupting interests in land development and pollution—we must not confuse conservation and conservatism, but neither should we allow one to glide into the other. His mediating idea can only be the concept of a pure unspoiled nature whose restoration is the goal rather than a critical evaluation of competing goods among which the wisest decision is to be made.

If today we have to talk of the natural and the essential, we have to talk of growing essences or changing essences. But the question is not so simple. Why is it the case that though we have the full vista of change in our scientific speculations and give lip service to change when pressed, still in the humanities (and in a great part of the social sciences too) we frame our questions and look for answers in terms of the older appeal to a natural order? Perhaps it is because change in an essence seems spread over a long period of mankind, so that a changing essence is still an essence for a while. And we live in that while. What does it matter that further back our "essence" came into being in an evolutionary process and some day in the future that essence will be different? We know of history but we live in a present stretch. And so the changes that concern the humanities are the cycles of life and death of our own daily and annual rounds. Is that why Aristotle gave poetry a deeper philosophical status than history; why we respond readily to T. S. Eliot telling us that April is the cruellest month in its contrasts of life and death, its mixing of memory and desire, but feel no palpitation if the historian tells us that the battle of Tours in A.D. 732 ensured a Christian rather than Mohammedan West?

On the other hand, it may be that this luxury of ignoring the matrix of change is running out. Whitehead noted that hitherto great changes were cumulative over many generations, but in the twentieth century they were beginning to come within a single lifetime. Our generation, Margaret Mead has pointed out in her *Culture and Commitment* (1970), is the first to have a generation gap in time, a gap comparable to the culture gap between an immigrant group and its first native-born generation in the new land. The strength of her educational ideal—to breed a generation that can bear to be free and make its own decisions about what its life will be like—comes from the fact that it is unavoidable when the rate of change passes beyond a critical point.

If ordinary experience is at last catching up with change and development because the latter are now permeating daily experience, what does this mean for the humanities in their reckoning with the Heraclitean flux?

4. *The Hold of the Past and the Hope of the Future*

To include history among the humanities was a stroke of congressional genius. Too often colleges put history in the social sciences. In any case history is just taken for granted. As a discipline it has been around for a very long time; everything has a history, and if you are interested, get a historian to work at it.

If the search for a natural order has indeed abated, there is an explanatory vacuum for filling which history is an obvious candidate. Even more, as a humanistic discipline it may be called on to carry the sense of history into all of human life. I suspect that the first who would have to be convinced of this are the historians themselves. They have been too accustomed to an ancillary role, satisfied by the fact that their assistance was everywhere applicable. Perhaps a philosophical glimpse of the rise of history in the past two centuries would serve some inspirational purpose.

Consider the ahistorical scene under the reign of the Newtonian model. The tendency was to ignore developments and concentrate on universal laws; the physical world had them, the psychological world was getting them, and society would quickly catch up. But about the middle of the eighteenth century there are stirrings, even for the history of the physical world. Kant's work on the theory of the heavens (1755) launched the Kant-Laplace account of the development of the solar system and the formation of the earth. Kant was interested not only in origins but in the continuation of change; for example, he has a little paper in which he tried to figure out how much the earth is slowing down under the friction of the tides. Now an earth that came into existence will need a historical account of the origin of life and then of civilization. Once set on this historical track there is no going back. In this vein it seems almost a pity that Kant's tremendous vigor was diverted by the three *Critiques*! He did come back late in his life to the human side in the topic of universal history and the development of freedom and peace.

Hegel's contribution to the sense of history is analytically very rich, and this despite attempts to see him as obscurantist. He adds several explicit features: the search for unity in the life and culture of a people at any historical cross-section (this, already found in Vico earlier, was to beget the idea of cultural patterns in later social science inquiry); the search for *stages* of development in a historical process; the formulation of the historical inquiry in terms of a *world-picture;* and the delineation of a pattern and its elements in each society in terms of what the developing world-pattern requires. All this was then, and is perhaps even now, seen as philosophy of history. But it is much more: it is a set of guides for the sense of history in the description of what is going on.

Darwin's theory of evolution has, of course, been recognized as vital to the growth of a historical sense at least for over the longer time-span, and its impact has reverberated in all fields of thought. In our own day, genetics and the study of mutations produced by radiation (and those engineered in genetic transplant) bring it into the present moment, not merely the long-range future. Hegel might have thought it the cunning of history to have the study of the causes of cancer tied to that of genetic change, as if the gods were using our immediate concern with health to give us an immediate consciousness of long-range unfolding. In the late nineteenth and twentieth centuries the growing consciousness of change and development moved beyond that of the species and into the sense of the individual; this is usually remarked upon in tracing the intellectual antecedents of Freudian psychoanalysis. The minutest scrutiny of the individual's early history (as reflected in his consciousness) became a significant factor in understanding the psyche, in striking contrast to earlier disinterest. Similarly, there is the growth of interest in biography, which presents in its most focused context all the problems of the theory of history.

I do not underestimate the tremendous strength of the antihistorical Establishment in the humanities. Its slogans are familiar enough. It brands any interest in the historical context of ideas as a

genetic fallacy when one should be busy analyzing the ideas themselves. In literary criticism or in biography it accuses the historically
minded of substituting the life for the works. It is ready to tolerate historical study as a side interest so long as it does not pretend to affect
the *meaning* of the work; history must limit itself to interesting stories
of externals.

We cannot here engage in a battle that is being waged along the
whole front of our intellectual life today. Let me simply take an illustration to suggest what I mean by the way in which history affects
meaning, whatever be the theory of meaning—and I take the illustration far away from the front of the interpretation of literature and art.
In *Sex and Temperament in Three Primitive Societies* (1935),
Margaret Mead describes how the Tchambouli males primp and dec-
orate themselves and takes it as evidence of the capacity of males to
have the standardized female roles of other societies. I have somewhere seen a critical comment to the effect that decorating had been
part of the ritual in preparing for war, but the colonial power had
abolished the war, so the preparation went on without the performance! I am not concerned here with the correctness of the criticism,
but with the illustration of what plunging an item into its historical
context can do for its meaning, how it gives us an altered sense of
what is actually going on in the men who are decorating themselves.
It is the altered sense of the present that is at stake in historical interpretation, not merely past accumulation or causal suggestion.

Suppose the humanities today were seriously to devote their energies to working out the implications of the shift from essence to history in the various fields of human endeavor. What reorientations in
work and attitude would be involved?

> Arno Mayer's "The Lower Middle Class as Historical Prob-
> lem" is an excellent example of the historical sense at work in a
> present problem. He uses history at full strength and with full
> complexity, tracing the step-by-step development of the form he
> is studying: the rise of a first lower middle class, then of a second
> kind in an altered economic context; the different composition
> over time; the type of consciousness that accompanied each
> change; the uses made of this class in different theories; the role
> played by this class in the onset of war; its central alliances and
> probable future. (This brief summary scarcely outlines the

richness of the paper.) In short, Mayer is not simply using historical data for general social or cultural conclusions; he is attempting to clarify historical process, including of course any general lessons. In this mode, history as humanistic discipline can help the cultivation of the historical sense, giving us the ideational equipment for studying the movement of peoples and the movement of ourselves. (Would a small shopkeeper who read Mayer's paper actually *see himself* as trying to roll back the wheels of history, and pause in his action?) For example, if we pay attention only to the spread of the behavioral phenomena, we are reduced to studying the voting behavior of different groups in different districts, and at best the surface changes. If we approach the subject with the depth of a historical sense, we are directed to the relation of class and subclass aims, behavior, and consciousness. Understanding of action becomes clearer, and predictability acquires a deeper base.

It is not surprising that contemporary thinking about the future finds itself almost immediately involved with methodological questions about the philosophy of history, with the feeling that how one goes about viewing the past and the present will be vital in formulating questions about the future. A sense of history in its fuller sense begins to focus attention on human beings in the whole developmental process, to see the emergence of specific human problems under specific historical conditions, and to view actions and policies as experiments in the achievement of aims and the solution of problems. This is a fairly new way of looking at man-in-history. Matured in the historical-dialectical outlook, it shares the activism which the pragmatic philosophers—particularly James and Dewey—brought to the understanding of human action. It fits precisely the relation of history and policy, as distinguished from simply history and prediction or history as fact and value as ideal norm.

Louis Henkin's paper, "Privacy and Autonomy under the Constitution," suggests what could happen if a historical sense were given full scope in the theory of law.* Henkin distinguishes the concept of private rights from the specific right of privacy. He traces the birth of this right in judicial decision, its ambiguities,

* Originally presented at one of the conferences on Private Rights and the Public Good, Louis Henkin's "Privacy and Autonomy under the Constitution" has appeared in the *Columbia Law Review* 74 (1974): 1410–33.

the diversity of things swept into the same basket, the finer shades expressed or overlooked. But instead of seeing it as a challenge to the historical sense, he sees the outcome as rather a logical mess. Suppose instead he had traced it as a juristic experiment: how this new concept might help solve in a systematic way problems in our society of electronic eavesdropping, control over one's own body in sex and marriage, safeguarding the individual against intrusive pressures of commercialism, and so on. Could it do things that older concepts of liberty and trespass were failing to accomplish? On such a view, the Supreme Court's action on abortion would be a bold experiment, comparable to its decision on integration in education. The concept of contract once had a similar origin in unifying a variety of legal ideas and institutional procedures, apt for the growth of commerce. Is the concept of privacy likewise on the historic rise, apt for new men and women in a new world? Whatever the correct answer, such a way of asking the question would be quite different. And it would express the changed intellectual attitude in the conjunction of a humanistic history and a humanistic pragmatic epistemology.

Such a shift in orientation does not predetermine the kind of activism that need ensue. There are intervening premises about what is possible, what men need and what they are like, which would have to enter into the demonstration or conjecture, as well as premises about the degree of conscious intervention that is desirable. Activism at its minimum is an interest in the possibilities of control even where modes of actual control may not be available. An activist reads the course of events as if it were an experiment from which he could learn for future planning. Yet strikingly different attitudes are found. On the one hand, for example, there is the outlook of social engineering with its mechanisms set to go and its projects planned to the last detail. On the other, there is the fear of power and manipulation and a faith that people can be trusted to devise and improvise, as in the anarchist reliance on mutual aid and cooperation. At the present time these are broad humanistic attitudes that are far apart—neither has decisively established itself. We work with something in between, or oscillate. Meanwhile, the clarification of both attitudes to people and the desirable patterns of human interrelations constitutes one of the great issues on which the humanities can make their contribution.

5. Symbols, Interpretation and Reality

The history of the social sciences is replete with attempts to deny the continuity of man with the rest of nature. After speculative philosophical arguments based on freedom of the will were thrust aside, the battle shifted within the sciences themselves, taking methodological forms. First a line was drawn between physical-natural science and social science. Then it retreated to a contrast of all science and history. Then, particularly as the social sciences for a while turned rigidly behaviorist, the whole realm of subjectivity was coopted in the fray. Appeals to empathetic understanding as against objective description, to phenomenological method as against causal analysis, are familiar enough, as are the arguments to show that a more liberal version of science than the behaviorist can accommodate the analysis of experience from the agent's point of view. These phenomena in the history of method represent, of course, extreme swings and slow returns. Think how hard R. C. Tolman had to work to reintroduce a respectable notion of purpose in dealing with animals. But the resistance is understandable when we recall the long history of loose teleological explanation in biology.

The most recent attempts to deny the continuity of man with the rest of nature would have the humanities take over a large share of the field, in the name of symbolism and interpretation. They are no longer the old attempts to appeal to *Verstehen* as a distinctive human mode of awareness, though they may still invoke it or draw on its capital.

Perhaps a glance at the problem in anthropology, which has been one of the central battlefields, is helpful. In the older days, when Tylor's all-encompassing definition of culture prevailed, culture consisted of all the habits acquired by man as a member of society. But gradually all sorts of different formulations arose, in both anthropology and sociology. R. M. MacIver's distinction between *civilization* and *culture* put into the former all cumulative techniques, social as well as physical, and left the symbolic for culture. This had, built into the concepts, the contrast of the external and the internal, as well as that of means and ends, natural causation and spiritual value—all pairs correlated. MacIver, in thinking of culture as values, even went on later to think of its symbols as myths. Talcott Parsons distinguished *culture* as symbols, *social structure* as institutions functionally viewed, and *psychology* as impact on individual development. Meanwhile, Ruth Benedict and Margaret Mead continued the integrated and comprehensive descriptive concept of culture which concentrated on the totality of a people's ways—pattern the discoverable unity. Their interest in personality psychology and education tied in with questions of how cultural pattern became embedded in individuals and later with problems of cultural change. The integrated character of their anthropological method is seen in Benedict's view, as against sharp academic boundaries, that sociology is simply the anthropology of more complex cultures.

When Kroeber and Kluckhohn came to review the concept of culture in their *Culture, A Critical Review of Concepts and Definitions* (1952), they found it too multiple to do more than list and describe. They did not probe far into underlying theoretical presuppositions about inner and outer, or about values as ends and means, and how the picture of culture varied correspondingly.

The situation was even more complicated by the rise of phenomenology in the psychological and social sciences. A bold counterattack against the neglect of direct experience and meaning in behavioristic science, it made good its claim that one should look at the field of awareness and describe it fully and analytically without interrupting with physical conditions and psychological reactions. This is the operative meaning of the famous Husserlian *epochē,* the bracketing of the natural world. But after elaborating some useful techniques and scoring many a useful hit, phenomenology fell into the temptation to become a school. It stalled on removing the brackets and seeking phenomenological-physical and phenomenological-psychological relations, and it turned every question, particularly about humans, into a phenomenological question.

We cannot go further here into the controversies of schools. I am merely suggesting that when a social scientist attempts to isolate the cultural side of anthropology and set it along the structural-functional, not just as a rough division of labor but as a separate field with its own method—contrasting the interpretation of symbols for the one with the scientific for the other (as Clifford Geertz does in his *Interpretation of Cultures* [1973], developing a semiotic theory of culture)—he is taking on the whole burden of a tangled century's conflicts on method. And when in the humanities one talks generally of symbols and their interpretation as a distinctive humanistic task, one assumes also this dubious inheritance. For the crux of the matter is how symbols themselves are to be dealt with in interpretation, and this simply reopens the old issues. Can symbols really be explored as self-enclosed, ending in a self-sufficient state of conscious feeling and awareness, capturing an inner essence removed from the outer natural or social domain? Or does the very exploration of a symbolic web rest on all sorts of presuppositions about the context of its employment? These, if spelled out, would subvert the distinction between inner and outer and underscore the integrated character of knowledge; and this would open the door to the cooperation rather than the partition of methods.

Man is born and bred in symbols. But language is a system of symbols, and religion is a system of symbols, and science is a system of symbols. A semiotic discipline has great importance, but it cannot be captured for any one distinctive method. To ask how differently nouns and prepositions function in language, or the cross and the flag in religion and nationalism, or different scientific concepts in different sciences, and to make transdiscipline comparisons is as fascinating as it is important. But can it bear the weight of a diremption between scientific objectivity and humanistic subjectivity?

There is considerable proclivity to this subjectivism in literary approaches. For example, in "On Privacy and Community," Emile Capouya looks at the right of privacy not in terms of judicial decision but with a sensitive appreciation of what goes on in the human being whose privacy is being violated. Offenses of searches and seizures are "insults to the soul." Privacy itself is regarded as "communion with our fellow beings when we are not physically in their presence." In a similar vein, Benjamin DeMott in his "Equality and Fraternity" asserts that a fabric of feeling is basic to institutions and ideas. Indeed, he goes so far as to say that equality and justice bemuse us because they are concerned with arrangements whereas fraternity and charity involve subjectivity. Since any system rests really on a grain of feeling, he argues that equality properly is an attitude of feeling—that of respect—or else nothing.

To locate the reality in this fashion in inner consciousness as against social arrangements is strongly reminiscent of the way in which Clement of Alexandria reinterpreted Jesus' call on the rich man to give up his worldly goods if he would be saved. Says Clement: it does not mean actually to give up the wealth but only the spiritual attachment to wealth. That will suffice for salvation.

The alternative view for the analysis of symbols—as talking of the same world as science does but in a different way—can be illustrated from George Santayana's naturalistic treatment of the relation of myth, science, and religion, in his *Life of Reason* (1905). Myth is a kind of poetry. When it intervenes in human affairs to order life, it is religion; when it supervenes on human life, giving expression to our fantasy, it is poetry. When it is stripped to its fighting weight and valued only for what it points to, it is science. To think of Apollo driving his chariot daily through the sky is a myth, an ingredient of poetry. To pray to the sun at stated intervals in organizing our life is religion. To see the sun as rising and moving through the sky is common or everyday belief, still holding a residue of metaphor. To plot the curve of the relative positions of sun and earth is science. Again, to describe heaven and hell is not a celestial geographic science but a myth symbolizing the moral truth that present action has fundamental importance and should be viewed under the guise of eternity.

> In many respects Roderick Nash's paper on environmental ethics can be seen as a glorious symbolic myth in Santayana's sense. It condemns the ethical cut-offs which limit membership in the moral community to humans and projects our kinship with all of being—including land, air, water, rocks. Its map of the rise through time from environment to life to mankind and on finally to family and self (narrowing at every step) and then upward in reverse on the march of ethical expansion from self all the way to environment, fits easily into an old tradition of the alienation of the individual self from the world and its long return. It is of the same order as Plotinus's story of the history of emergence from the primal One and the endeavor of the lone soul to be reunited to its source. As a myth it would not have to argue whether rocks have life or whether they can be as eloquent as babies. The story simply tells a moral truth of our neglect of our natural environment. It projects a set of human attitudes to our cosmos and its contents. Comparative studies of such world views are by now familiar in the anthropological field.

What of the analysis of a specific symbol? Does the symbol operate to yield an affective insight, or does it break open paths that are amenable to reflection and science and policy as well as to clarified consciousness?

In his "Structures and Machines: The Two Sides of Technology," the article summarized in the first section of his essay on "The Structuring of Cities," David Billington offers us a symbolic contrast for understanding technology: the dyke and the "fast-moving, smoke-belching, harshly shrieking" locomotive machine. It is a striking comparison and Billington spells it out as signifying two aspects of technology: the static permanent structure and the dynamic machine. This becomes for him the basic dichotomy of the engineering field: he outlines different laws for each side, sharpens the contrast (static-dynamic, permanent-changing, individually produced as against mass-produced, and so on) and tries to correct the American overemphasis on the machine. By thinking of structures as works of art, he says, we could conscript in principle the humanistic tradition for awakening the appreciation of structures.

Note then what the symbolic contrast does: it reinforces an opposition of categories by the striking difference of the image. And yet, there are other ways of looking at the dichotomy of structure and machine. Why should not a structure be seen as a machine doing slower or steadier jobs? (The dyke is holding back the constantly beating waves.) Mario Salvadori, in his "Response to Billington," is even ready to use a machine cooling system to maintain the proper temperature of the structural components, a crossing of categories which would horrify a strict structural engineer! Perhaps the differences are not categorical, but only matters of degree of motion or stability. It is clear that the insistence on the separate categories rests in part on different practical principles in construction. But other common features could equally be selected; pragmatically different features are selected in the light of different purposes. Further, Salvadori sees all engineering as essentially defying nature, while others may see it as artistically using nature. One might even suggest that we should work toward the older tradition that did not distinguish art from technology.

The lesson of this inquiry is that the symbol selects features of the existential context and accentuates them. It elaborates them and gives them an identity. A network of similarities and metaphorical transfers carries a set of implicit inferences steeped in feeling. But the content is in effect an hypothesis about the way of analyzing, categorizing, determining relevance, assigning values for the material under scrutiny. The symbol is poised to move in many directions. It opens, it does not close, inquiry. And interpreting a symbol consists in following some of the many lines laid open.

> Perhaps the most direct theoretical confrontation about the character of consciousness is found in the papers of Bozeman and Mayer, in dealing with war. Bozeman thinks consciousness gives the reality in itself so that to have people's thoughts about war gives the realities that perpetuate the phenomenon of war. Mayer, however, sees consciousness as a phenomenal surface to be explained. It is not without effect, but it reflects more basic processes in the real world of which it is only a part. His study of the ideas and values of the lower middle class and of the conditions under which their rigidity swings the balance of forces toward war (referred to above) illustrates both the effects of consciousness and its relation to fundamental processes.

Symbols move humans deeply and set much going. But unless we are ready to dissolve all reality into acts of consciousness, it is better to recall the age-old distinction between the dreams that divide men and the waking life that brings them into a common life. We are today in a period of awakening in which we realize how far we have been locked in one-sided perspectives. A humanistic sense of the plurality of outlooks and the multiplicity of perspectives can help prepare us for constructing a common reality.

6. The Study of Values

Value has a wide reach and voracious appetite. It is used on both descriptive and normative levels as well as in explanatory contexts. It has served many functions in theoretical controversy and found its way into humanities and social science alike. It is grammatically dexterous: a verb in one context, it is a noun in another; as a noun it may indicate a process of evaluation (just as *price* sums up a market relation) or sink back into substantival inertia, almost as if one were to say: "These are my jewels, those are my values." So versatile a conceptual fellow runs close second to *being* or *is* in ambiguity, systematic or otherwise. My impression is that historically, it was a favorite in

Germany and Scotland and America but long ostracized in England, where a fine sense of linguistic differentiation wanted every fine shade to be expressed by different terms in different contexts.

In ethics, *value* may denote a person's likings and dislikings, his criteria for evaluating these (what is good and worthwhile), an objective property of events, a person's obligations (what he ought to do), his generalized standards (ideals and norms). In aesthetics it bundles beauty with all the other aesthetic attitudes or qualities or sentiments (quite literally spanning the sublime to the ridiculous). Religious values may cover both attitudes to the divine (such as awe or love) and virtues (charity, forgiveness). In economics where it early made its home, it first covered use-value and exchange-value and then became almost synonymous with the latter. In sociology and psychology it fluctuated between a descriptive notion of a person's interests and preferences and a normative notion of his grounds or criteria in judging these. (Whichever way it went, the term *norm* took up the other way.) More recently, in social science, my impression is that the critical component in the use of the term has been winning over the preferential component. In metaphysical conflicts, the notion of value on the one hand became the standard-bearer for what distinguished spirit from nature (all consciousness is selective-valuational); on the other hand, it became the common feature of all life (animals as well as humans) out of which by differentiation and functional specialization the higher values could be seen to develop. It thus served both antinaturalistic purposes in some and naturalistic purposes in others.

For a long time values belonged to the humanities and philosophy, not to the sciences. Science avoided value judgments. The psychological and social sciences studied human behavior, not human values which were taken to be inner and subjective. But by the midtwentieth century it became clear that man's selective-valuational behavior was a fertile field for scientific study. One of the most fascinating large-scale efforts in this direction was shepherded by Clyde Kluckhohn at Harvard, whose special genius it was to remain humanistic when he was being scientific and not to lose his scientific bent when he was being a humanist. The Harvard Values Studies concerned five cultures in the Southwest of the United States (Navaho, Zuni, Spanish-American, Texan-American, Mormon) and engaged in comparative exploration of values in different areas of life. For example, they included analysis of value themes and value trends in religion, law, attitudes to property and acquisition, music, family, readjustment of veterans to native life, self-orientations, physiological reactions, and of general value-orientations to nature, time, and fellow-men. Similar studies were meanwhile going on under Robert Redfield and Milton Singer at Chicago, into the comparative values of large cultural traditions (Chinese, Indian, Mohammedan, et cetera), while at the University of Pennsylvania Irving Hallowell investigated more particularly self-orientations in the building up of a viable personality in different cultures. All these studies were at the frontier of what had been a dispersed approach to values from many different perspectives—for example, historical studies of changing social ethics in R. H. Tawney and Max Weber on the rise of the Protestant ethic. What the large-scale social science plunge into the field accomplished was to focus the need for comprehensive value study by all available methods and to stimulate greater systematization.

It is perhaps against this systematic treatment that considerable sentiment in the humanities is gathered. Values are regarded as inner, individual, personal. They are the core of subjective reality, what a man stands for, lives by, and dies for. That such a perspective of the individual's phenomenological exploration (though only one of different possible patterns) need not be in conflict with social roots, historical development and social role, is one of the first lessons of wide value study. The comparative projects had no methodological bias. They sought light in every way possible. Observational reports were only one path, linguistic analyses and symbolic analyses were welcomed, experiential reports (for example, in comparative Rorschach studies) were included as a matter of course, aesthetic reactions had their place. It is surprising how dogmas of method tend to evaporate when there is a generous yield of results by all, which turn out to fit well together.

To illustrate the breadth of approach needed in dealing with important cultural values, let us consider *equality,* which was the subject of several of our papers.

> The papers by Robert Nisbet, Herbert Gans, Benjamin De-
> Mott, and William Vickrey deal specifically with the costs and
> humanistic impact of the ideals of equality and justice. (The way
> DeMott pits an inner attitude of respect against outer institutional

arrangements for equality has been touched above.) In "Justice, Equality, and the Economic System," Vickrey seriously canvasses the economic techniques for reducing inequality and finds that some are available and ingenuity would provide others which could be used without seriously impairing economic processes and initiative as they operate in our system. He does not consider basic revisions in the system itself if it fails to yield a greater equality. He finds the ideal of justice fairly useless and likely to lead to perverse results; but this came after examining Rawls's theory, and Vickrey's turning to utilitarianism may be construed as a demand for a utilitarian theory of justice instead. An interesting critique of the concept of justice is his remark that since it is so often defined in terms of existent expectations, it tends to be an essentially conservative concept.

The full-blast attack on contemporary concepts of equality come from Nisbet in his "Costs of Equality." Here he attacks the New Equalitarians as wanting *equality of results* and therefore revolutionary redistribution of income, property, power, status. He is particularly vehement against Rawls, whom he interprets as demanding equality of results, apparently because Rawls assumes differences have to be justified, and that justification has to show how the lot of the most disadvantaged would be improved. Nisbet canvasses especially the cultural costs of the demand for equality in terms of the untrammeled opening for envy, the attacks on status and prestige, the weakening of the family, and so on. (DeMott, in one of his well-turned phrases, also says that refusal of deference diminishes respect for respect.) Surprisingly, while Nisbet's argument would seem to lead to a defense of meritocracy, he does not approve of the attack on inequality in the name of moral worth or merit, but follows Hayek in allowing reward to go to those who succeed simply as such.

Herbert Gans, branded by Nisbet as one of the New Equalitarian triumvirate, responds to Nisbet's paper in his "Costs of Inequality." He contends that to want greater equality is not to insist on complete equality of results; that the costs of such reform are less than Nisbet envisages; that inequality is breaking up families and greater equality would mean greater familial stability; that Nisbet's account of the costs of equality fails to include the gains; that the costs of continued inequality are vastly greater. He tends to ascribe the differences between his position

and Nisbet's to a disagreement on means rather than a broad
philosophical conflict.

Actually, what we are now witnessing is a broader and more revolutionary moral movement than this interchange suggests. Conceptions of the good life are being questioned for their overemphasis on
competitive success and conceptions of justice for their overemphasis
on meritocracy. The historical development—liberty, then political
equality, then demands first for a career open to talent and after that
for social equality in the removal of discriminations, then demands
for wider positive means (such as education and higher education) to
open doors of greater opportunity—would if extrapolated yield a
complete meritocracy. But this is precisely what is being questioned,
not in the sense of rolling back the whole development but in a new
ethical insight of the worth of persons. No doubt its basis includes the
fact that the means of production no longer require the exploitation
of a part of the population, and the fact that the spread of some competences under universal education and the automation of industry
remove large areas from the need for unusual skills. But the impact is
primarily ethical—the sense of a new community of people with an
all-human morality and a concept of individuality that stresses the
diversity of capacities of living beings rather than selecting a few for
elitist grandeur. We cannot here enter into the exploration of these
new moral trends—this is one of the areas in which the humanities
could be most fruitful—but nothing is to be gained by assimilating
them to the nineteenth-century battles between liberty and equality
or relying on the clichés of the older mistrust of the masses.

> In addition, arguments and proposals, even well-intentioned,
> will miss the point if they assume as presuppositions the very
> premises that are being challenged. For example, in "Some
> Questions in General Education Today," Steven Marcus takes
> for granted that competitive excellence is a basic aim. Similarly,
> in her "Some Inconsistent Educational Aims," Onora O'Neill
> seems to take education to aim in part at a success requiring pre-
> eminence rather than excellence. It is no longer possible to
> argue from such premises as unchallengeable features of con-
> temporary life.

What is more significant is that contemporary antiequalitarian arguments continue to be formulated in terms of the individualist tradi-

tion. But the new equalitarianism is not bound to such formulations nor even to the individualist character of the utilitarian concept of the greatest happiness. It does not have to reckon and sum individual benefits or even think in Rawlsian contractual terms. It can formulate directly common goals of a quality of life for the community. In such an approach the dichotomy between public and private itself requires reassessment. Such a vast reformulation cannot be reached by looking into inner feeling alone, for this consciousness is likely to reflect the developed individualist tradition. It has to be grasped by a full sociohistorical study of the development of ideals and conditions.

7. Mine and Thine or I and Thou?

If a society does not always get the categories it deserves, it at least gets the ones it cultivates. Our social tradition has worked hard for the last few centuries to reach the point where the only alternatives to stand out are private or public, self or society, me or the rest. This is, of course, the familiar intensified voluntaristic individualism that permeates our economics, our political theory, our ethics. Instead of norms of an economy that will support the good life of a community, we reckon how individual interests may be maximized without affecting entrenched holdings. Instead of systems of organization based on the common participation of a free people, we have the polarization of autonomous individual and threatening state. Instead of the virtues of a common good life, we have virtue bisected into self-regarding and other-regarding and so the task of ethics becomes one of balancing egoism and altruism.

All of this is usually attributed by the economic historians to the type of economy developed by the bourgeoisie in the rise of commerce and industry out of feudalism, on an individualist pattern. It is attributed by sociological historians to the breakdown of intermediate groupings between the state and the individual as well as to the specific phenomena of urbanization. It is attributed by metaphysicians to the exaggeration of extreme aspects: Bradley thunders against the vicious abstraction of the pure individual and the pure social and looks to status and its duties (now respectably scientific as role theory) for an expression of the operative totality; Maritain finds both individual and state to be heresies, yielding voluntaristic pursuit of the material on the one side and totalitarian Rousseauesque common wills on the other, and he calls for recognition of the person as a God-oriented being. It is attributed by Marxists to the dominant bourgeoisie who needed individualism to break through the older restrictive economic and social patterns and who perpetuate it as an ideology to ensure their own exploitation.

I have no doubt that most of these accounts are more or less correct, and I will not pursue here how the variety of conceptualization can be integrated with the variety of social processes. Certainly the family of dichotomies that goes under the contrast between individual and social or private and public has not always been with us. In Aristotle's ethics the private means simply the share in the good life, the fitting roles and tasks that the individual should have in virtue of what he is like and able to do. When Aristotle discusses whether a man should love himself, his answer is that it depends on what he is like: a good man has something worth loving, a bad man has not. Nietzsche, who had a keen sense of the ethical character of intellectual categories and their historical careers, records the remarkable success story of self-interest: it starts as an outcast from morality (selfishness, self-aggrandisement), becomes a respectable member—in fact, a good half—of the team of self-regarding and other-regarding virtues, and ends up as dominant in ethical theory on the Benthamite summit. Bentham's utilitarianism, in fact, shows us the full face of our problem. It is not even a question of private or public since the idea of community is a fiction; there are only individuals, only a reckoning of the sums of individual interests. The social ideal is the greatest happiness of the greatest number, not the structure of a common good life. Perhaps this is why we still find political philosophers, even of opposite persuasions, when faced with "liberty, equality, fraternity" gladly elaborating upon liberty, quarrelling about equality, but very suspicious of fraternity, as if it smacked of totalitarianism. . . .

The fact is that the real content of private vs. public has not been individual vs. group, but me vs. the rest. Only a specialist in the humanities—particularly in mythology—could do justice to the history of attempted reconciliations of mine and thine. There are unseen providential hands working to adjust the actions of individuals so that self-interest pursued reasonably will yield general welfare and conversely to guarantee that pursuing the general welfare is a good investment with high dividends; or again, there is Spencer's optimistic notion that evolution will take care of the integration of egoism and altruism. There are also, of course, realistic attempts to analyze how interests affecting us all generate self-conscious and organized publics—for example, Dewey's *The Public and Its Problems* (1927).

The basic issue of private and public takes two forms. One is the concept of individual rights *vis-à-vis* the public good. The second is a system of private individual property as against collective or social property.

Charles Frankel penetrates to the heart of the first. In his "Private Rights and the Public Good," Frankel notes our common belief that there are some rights which no government ought to infringe short of an emergency and some not even then. Then he distinguishes the line between public and private from that between what affects others and what affects oneself alone, and recognizes it as a construction. But a construction must be governed by ends, and so we are not surprised that after listing a set of values underlying a system of private rights he decides that these values in the system themselves "define some considerable portion of the public good as it should be understood in a liberal society." Hence the weighing is not of private against public, but "of different social values against one another and their constant readjustment"—in the light of a complex ideal of civilized life.

Hannah Arendt's "Response to Charles Frankel" similarly breaks through the limited dichotomy of private and public, but in a different way: she sees many of our rights as themselves public not private because, like service on a jury or peaceful assembly, their object is a common one. Arendt relentlessly categorizes in terms of the content, not the legal form or the subject whose right it is. Thus when she pits private against public it is quite literally in the sense of privacy or being left alone in an activity; and she rejects talk of reconciling individual interests and the common good because of the "urgency of individual interests." Actually, her reconciliation would lie in the same individual sharing in the common as well as having the private—being a citizen in the ancient sense as well as urgently individual.

Both Frankel and Arendt are thus dissatisfied with the intense individualism in which our tradition has culminated. Through one or another different construction they would turn back toward a public good.

The importance of such constructional tasks and the practical impact of our categorial selection is seen in every debate on public policy. Even in the papers on equality, there is the tendency to think of everybody having as much as anybody else rather than in terms of a common life with full participation.

> An exception to this narrowed vista is Paul Freund's essay, "Equality, Race, and Preferential Treatment," wherein he calls for the preferential consideration of blacks. He bases his argument firmly on the need for the whole community to bring into participation a minority that was so excluded as to have no stake in the common life, its processes, and its benefits. It contrasts strongly with the interpretation of compensatory justice that Herbert Deane criticizes in his "Justice—Compensatory and Distributive," for that interpretation limits its focus to the conflict of individual claims. Deane's analysis is particularly interesting in two philosophical respects. One is the recognition of the importance of the rubric selected; for example, he wants no-fault insurance also removed from the category of compensatory justice and treated under some different rubric. The other is the clear pragmatic appeal in the selection of categories to those that will help solve the problems at issue rather than engender greater dissent and conflict.

Once a system of private rights (or for that matter a system of public goods) is viewed as a conceptual instrument, an overall conception of a good life is implicit—or the quality of life in a community, or some other higher order, ends in terms of which a choice is made about which conceptual instrument to use. Thus the United Nations went in for lists of human rights, not a utilitarianism of global welfare. Traditionally, the United States has operated chiefly with individual rights, supplemented by general welfare. Philosophers will recognize this as the perennial conflict of the right and the good, or duty and interest, or justice and welfare. But if the solutions I find in the several papers noted point a fruitful direction, we would do better to invoke a political model for analyzing the relation of private rights and public goods. There is a *separation of powers* in the realm of ideas too, analogous to that in our government, and with all the jockeying and contextual emphases that we find in political life. Thus there will be times when reckoning should be cast in terms of welfare, times when in terms of individual rights, but when to do which is a judgment of how best to further the idea of a kind of human life. It is precisely the working out of such an ideal, in close relation to what is going on in the world, that the humanities are asked to contribute.

If we turn back to that world, we find that the type of individualism dealt with above is cracking, not only from the pressure of large-scale organization, but from the growth of communal forms and intermediate groupings. Demands are increasingly being made in the name of groups—unions, blacks, ethnic groups, consumers, not to speak of experimental communes—so that confrontations with economic or governmental power are not just those of the private against the public. It is surprising that political and legal science has done so little with group rights in a rights-culture, when even Richard Price and Tom Paine could speak of the rights of a people to cashier its government, which is obviously not an individual right. In any case, we are obviously in the midst of a social revolution, guidedsby the movement to community of effort, if not yet focused on community of end. This is one front on the moral revolution I referred to in the last section.

> In Rosemary Park's "The Disestablished Humanities" and Robert Hanning's response, "A View from the Ivory Tower," there is an exchange on whether our cultural institutions beyond the academy can help give us a sense of community which was symbolized by the Parthenon in ancient Athens. Park suggests perhaps, Hanning says no since there is no sense of civic piety involved. Perhaps the secular moral equivalent of the sacred today would be the kind of self in which the communal is integrated within the self rather than confronting it from without— a view of the religious that John Dewey advocated in his *A Common Faith* (1934). Its achievement not only presupposes different social institutions but would be seriously complicated in our time since the City has long given way to the Nation and anticipatory strands of the global community are making inroads on the Nation.

Space and time do not permit me to enter here into the second major problem of private vs. public—the present status of private property. In the not-so-old days it was discussed in terms of individualist free enterprise vs. socialism. But the plurality of types of economic-political organization now defy these broad categories, though there are no doubt different forms of private and collective control of production and distribution for mankind to choose among. The underlying reality is that property has changed its forms rapidly in the twentieth century due to the rise of corporations and now multinational corporations and the recent rise of quasi property rights through social institutions of redistribution. Hence most of the arguments about the absolute rights of private property are meaningless today (though those concerning near-absolute power are not), and even social defenses of private property are reduced, as in effect Nisbet's attack on equalitarianism was, to threatening us with dire consequences to family and culture if we curb free enterprise. Obviously the study of the rationale of private property today requires the joint effort of three intellectual groups: first, anthropologists and economists to see what is really going on in global society as a whole; humanists to penetrate the folklore and to articulate human aspirations for a good life; and philosophers to devise new constructs for reformulating the issues realistically.

A brief word on I and Thou. This contrast to mine and thine comes (obviously) from Martin Buber's effort to focus on interpersonal relations as a distinct category. Intrasubjective feelings and institutional structures are, in his view, distracting escapes from the authentically interpersonal. Of course his own thought involves a religious background and a quasi-anarchist small-group political philosophy. But its tremendous influence came from the presentation of an alternative to the highly individualistic and the external total structures. Like George Herbert Mead's emphasis on interpersonal relations in the growth of selfhood, it helped break the monopoly of individual and group-social on the feelings of men. It restored on the intellectual level the plurality of categories of human relations (referred to as intermediate groupings above) that had been destroyed in the growth of individualism.

Finally, it should be stated (if it is not already clear) that this brief account is not directed against the social ideal of individuality and the development of individual powers: the task is rather, as Dewey stated it, to rescue it from the older individualism.

8. How To Be Practical though Humanistic

The humanities have gloried in the absence of practicality. Philosophy bakes no bread. But the philosopher, the poet, and the painter must still eat. And so endowment is necessary though no return other than their self-justifying work is to be expected.

On the other hand, there comes a time in the tangled affairs of men when even to ask a question is to make a significant practical contribution.

> Lionel Tiger said, at one of the conferences on education, that if good ideas may be slow to have effects, bad ideas get macro-development rapidly. Yet Bertrand Russell once argued that if stupidity has such serious effects in human life, there is no reason why intelligence cannot have equally serious effects.

Once, many years ago, I was to give a lecture in an Institute of Humanistic Studies for Executives, organized at the University of Pennsylvania for the Bell Telephone Company. I read the explanatory materials circulated by the administrators: junior executives, it was explained, have to solve problems, and this they can learn by apprenticeship on the job; but senior executives have to discover problems or think them up, and this requires a liberal education. (I do not know where the writer of this circular is now, but I hope he is employed by the National Endowment for the Humanities!)

Let me balance this with another story. Even more years ago, I attended a meeting on adult education for workers, presided over by Harry Carman, the well-known Columbia historian. After almost two hours of discussion, a union official representing one of the garment unions rose, pointed to her associate, and said solemnly, "What have we decided that will justify my having taken Mr. Shapiro out of the shop for two hours?" I must confess that since then, Mr. Shapiro's plight has been on my philosophical conscience (especially when a paper gets too long). In any case, I suspect that the truth about the practicality of the humanities lies somewhere between Ma Bell and Mr. Shapiro. But let us track it down, if possible.

Take, for example, the writing of utopias. It used to be a classic exercise in the humanities. Why has it gone out of fashion? I suppose one hypothesis would be that sensitive thinkers are pessimistic. In the time of the French Encyclopedists there could be excitement with the view that man is a machine because it meant he could recondition himself and was not bound by original sin. But now we seem to be too disillusioned by the machine to write utopias about it. And so we have had instead our *Brave New World* and our *1984*.

> Still, as James Clifford pointed out in the discussion of Elisabeth Hansot's paper, "Reflections on War, Utopias, and Temporary Systems," the eighteenth century also had its anti-utopian writings in Swift and Johnson. Hansot suggests that instead of literary utopias we have now the short-term quasi utopias of meetings, conferences, and commissions!

Perhaps there is a different explanation. Just as in science the time-span between invention and utilization has diminished remarkably, so in social affairs the movement of events overtakes too rapidly the stretch of the utopian imagination. Compare the modern organization of science with what Bacon envisaged in the *New Atlantis*. Even practical programs, such as Marx's list of next steps to be achieved (at the end of the *Communist Manifesto*), are long ago commonplaces in capitalist societies. Thus today instead of utopias we have the broad social programs of actually organizing philosophical-political movements. Why construct a utopia when one can advocate a reform or initiate a revolution? In such a perspective, utopias belong to the stage when possibilities of control have come into sight, when the will turns in a given direction, but the objective is not clarified sufficiently and the means have still to be discovered. Now what happens at an earlier stage, when there are no possibilities of control, and what will happen later, when the limits of control are still uncertain?

Viewed in the perspective of the growth of control in human life, the humanities have always had practical tasks. At the outer limit, where nothing could be done, they furnished attitudes of resignation or reconciliation or even the heroic stance, attuning emotion to face unavoidable suffering.

> The ancient Stoics generalized this function as a moral philosophy—controlling our impressions of things in the interest of peace of mind, since we could not ultimately control the things themselves. Greek tragedy, in Aristotle's analysis, served the cathartic functions. The Book of Job is dealing with the extremes

at the outer limits; only a modern like Archibald MacLeish could introduce the light touch (in his *J.B.*) of a momentary suspicion on God's part that maybe Job was forgiving Him!

But where control over human responses and human action is possible, even from ancient times, we see how eagerly and speedily the humanities move into the job of articulating attitudes relevant to policy.

> It is not just a question of the social import of the histories and the theater—e.g., Thucydides' possibly hoping to influence the Greeks by showing the corrupting character of class struggle, or Aristophanes and Euripides writing peace plays. Plato, as usual, shows it most clearly: physics is just a "likely story" and there seems little hope of controlling nature, but the *Republic* is a program for the total control of human nature. And the whole history of humanistic political theory embodies similar aspirations.

The insight conveyed in the great humanistic works, whether in myth or analysis, is the stabilization of an outlook that is always verging on action—where opportunities arise. In this sense, nothing is more practical than understanding. There is thus a continuity in the practicality of the humanities from points of minimal control to points of substantial control in human life. It is not simply a shift from inner to outer, and the development continues in the same way when science begins to bring the natural world into the scope of human powers. The myths of Daedalus in ancient times or the voyage to the moon in a Cyrano de Bergerac shift to the sketches of Leonardo and eventually to the airplane and the moon-shot; it is a progression from fantasy to hope, to aspiration, to plan, to achievement. This is not a discarding of the humanistic and its moral context: the landing on the moon had many of the aspects of theatrical spectacle and even of morality play. . . .

The practical responsibility of the humanities has always therefore been for the moral organization and the quality of human life. The dichotomy of a self-enclosed imagination responsible only to the quality of its fantasy and a self-enclosed science responsible only to its inner development has been an intellectual episode of a very short period in the Western world. No doubt its analysis reveals that it rested on its own myths of preestablished harmony or inevitable

progress or the like. The contemporary quest for the policy implications of the humanities is part of the wider effort to restore the older
integration of knowledge and value and practice in the new situation
of the vastly extended base of possible human control.

How close a basic humanistic question can come to the actual
problems of practical decision on life and death is strikingly
shown in Daniel Callahan's "Biomedical Progress and the Limits
of Human Health." We are used to the perennial question
whether life is worth living. It surfaces in ethical inquiries about
the nature of happiness and debates over pleasure and pain, in
the continuous struggle between optimism and pessimism, in the
Christian attitude to despair as the greatest of sins, even in psychological and psychiatric studies of suicide. Perhaps we are
tempted to dismiss it as one of the questions we shall always
have with us, and so relegate it to our spare time for philosophical reflection or poetic sentiment. Samuel Butler dismissed
it with the quip that it was a question for an embryo, not an
adult. I wonder whether he realized that so far from dismissing it
he was giving it a practical platform to stand on. Callahan's account of the progress of biomedical knowledge and technique
translates implicitly from whether life is worth living to what is a
life worth living, and explicitly from the latter to *what is a life
worth saving*. He notes the paradox that what began as an attempt to remedy limited birth defects has by its unintended consequences raised the question of evaluating the kind of life that
survival makes possible, with the result that fewer defective children may now be saved than before. Interestingly too he ex-
plores the way in which an overperfectionist definition of health
functions evaluatively to open up unrealistic demands in assuming a "right to health." His vigorous attack on individualism and
its consequences in hindering greater equalization of health care
may perhaps be seen as a demand for more organized planning,
but it has to be planning closely tied to basic answers to basic
value questions. In short, the problem of using technical knowledge and technical means is fast compelling humanistic answers
to humanistic questions—on pain of practical blundering.

Willard Gaylin's "The Technology of Life and Death" carries
similar lessons and shows even more strikingly that categories of
interpretation are critical in assessing a line of policy. Recent
redefinitions of "death" make it morally acceptable to pull the

plug on apparatus that continues vital functions. But suppose that instead of pulling the plug we instituted a bioemporium of "neomorts" and kept it going for use in medical training, transplants, experimenting and testing, harvesting medically essential bioproducts. A cost-benefit analysis might justify our doing so. Assuming our moral revulsion to these technologically possible procedures, Gaylin asks whether we would attribute our revulsion to the residue of habit, fear, and ignorance, and thereby risk our tenuous civility and human decency and erase the distinction between man and matter. In effect, he is posing a confrontation of the humanistic and the materialistic view of man. Yet surely this confrontation with its dire moral consequences does not follow merely from the incorporation of man wholly within the system of nature. It follows rather from identifying technological advance with an exploitative social attitude in which the cost-benefit analysis has not asked such vital questions as whether the individual consents to participate in the new institution. Simply to look at what goes on in the bioemporium does not tell its own story, for each item has its precursor in acts on the moral side of the ledger. Repeated blood contributions, sacrifice of an organ for another's dire need, submitting to medical experimentation for human benefit, willing one's body for medical research—all these are moral acts beyond the call of duty and some even approximate heroism. If to participate in the emporium were an act of autonomous decision made during one's lifetime, would not the acts of the emporium become allied with heroic categories rather than callous exploitation? Given a guarantee of no feeling, they might occasion less moral revulsion than the mutilations of war, traditionally accepted as moral sacrifices on behalf of country. Gaylin's account seems to make a metaphysical confrontation inevitable, while what is really required is a refined moral analysis and a refined moral sensitivity.

If we turn now to the future, the question of the extent and limits of control itself becomes central, and the new dangers and the new crises have tended to focus inquiry on mere survival. But the perennial question has not ceased to be the quality and moral organization of the life that is possible. Let us sample the two areas in which these essays face directly problems of policy for the future as enlightened by the humanities. One is education, where the practicality seems to lie in the strangely limited question of the fate of the humanities themselves. The other is technology and the attitudes it involves toward the prospects of mankind in the present world.

On the face of it, the central problems of education appear to be how to achieve established humanistic aims in an altered social and material milieu. But such a formulation is likely to ignore the real issues about aims that are in debate in the contemporary world. For this reason, inquiry has always to be critically focused on possible presuppositions in an analysis.

> For example, at one of the conferences John Silber, arguing against the slackening of standards, said that music schools are not troubled by the egalitarian thrust since one either plays the violin or one does not. But behind this apparent certitude lies the assumption that the present *function* of the music school is not to be questioned. For example, depending on the context, one might raise the question whether there ought not to be music schools where anyone could acquire some appreciation of the practice of music to the extent of his abilities; they would not thus be training places for only the highest performing excellence. Some colleges have the requirement of some practice in an art in connection with their required art course. It will be recalled that Aristotle in the *Politics* asks how much musical performance is required in education and answers it should be enough to give one a critical appreciation of performance. Contemporary aestheticians have often stressed the point that our appreciation tends to be that of the consumer reacting to surface features, rather than of one who understands the problems of creating and reacts to the total character of the art work. All these aspects of the problem can be made clear by a humanistic view of the issues. Of course there are questions of cost and type of institution for such purposes; but it is not unreasonable to hope that a community which builds resources for general gymnastic and athletic training might do so for general musical and cultural development.

Since the humanities have their primary home in the university, any consideration of their place becomes tied into the controversies over what the university is up to and where it should go in the contemporary world. Several different aspects have to be disentangled; for there are external pressures on the universities and there are sharp cultural and social changes, as well as internal conflicts about educational aims.

George Pierson's "The University and American Society" pinpoints the outside pressures by analyzing the different historical strata of American society whose demands at different times shaped the university. Since his ideal is the liberal and independent university (primarily sheltering teachers and scholars) with a humanistic emphasis, his standpoint is that of the humanities defining the university of their vision and seeking to control the flux of pressures sufficiently to maintain the desirable independence, for example by multiplying and diversifying sources of support. His confidence is breathtaking. It is almost as if he were saying that we humanists know what we want, let us work for it and maintain our ideal, let those who press for other aims and momentary relevance build their own institutions for these purposes, but the academy is ours.

Wm. Theodore de Bary, responding in "The University, Society, and the Critical Temper," does not believe we can "make a virtue of irrelevance." His eye is on the actual problems with students and the social milieu that actual universities are having: "There is a very close relationship between our learning or knowing function and what people around us are doing. We can ignore the implications of this, at our own peril." Looking at the American university in relation to its middle-class origins and seeing its unique character on a worldwide scene, he wonders whether the essential critical attitude directed to accepted institutions and values can avoid self-destruction for the universities without a deeper humanism.

Steven Marcus, in "Some Questions in General Education Today," studies with historical sensitivity both the inner conflict of science and the humanities and the relation of the humanities to middle-class culture. Tracing the interchange between T. H. Huxley and Matthew Arnold and the modern variants on Arnold's view of liberal education as the cultivation of the self, he concludes that "with the newly emerging context of the university as part of the system of production, the role of liberal or humanistic education becomes increasingly problematic." And the decline of liberal arts is entwined with the decomposition of bourgeois culture. In his concluding practical proposals Marcus analyzes the historical role of the high schools and suggests the university should consciously undertake reparative functions. He proposes several types of basic humanistic courses.

The net effect of such consideration is more a challenge than an unavoidable conclusion. There remains even at worst the possibility that a humanistic culture could attempt to humanize the professions and vocations. As suggested at the outset, a humanistic science would see itself as a great expression of the human imagination in a context of basic human purposes. And even very practical vocations could be carried on with an eye on fundamental values, as is seen in the current ferment in such fields as medicine, nursing, social work. The humanities hold back from facing the challenge, in part because they understand that it is the whole character of our society and culture that is involved, and this makes the task too gigantic for them alone.

> This seems to me to be the source of the underlying sadness in Marcus' view. It also gives us a different way of looking at Onora O'Neill's keen analysis, in "Some Inconsistent Educational Aims," of the incompatibility of happiness and success as educational objectives. Happiness, she points out, involves fulfilling a reasonable proportion of a person's desires while success involves not merely excellence in the chosen pursuit but preeminence or outdoing others. Hence success for all students as an aim is fostering ambitions sure to be disappointed. This is in effect much more than an exhibition of inconsistency. It is a critique, though in abstract terms, of a competitive culture whose economic institutions are unable to carry out what the ideology promises to its people.

In part also the humanities hold back from facing the contemporary challenge because they carry the heavy weight of their own traditional elitism. Rather than seeing them as the last gasp of bourgeois culture, a more hopeful view of the humanities would maintain that elitism is not of their essence but only a historical residue. After all, American thought has not been without its ideal of a rich democratic culture, expressed for example in Walt Whitman's *Democratic Vistas,* and the ideal of a higher education open to all the people has made considerable strides though too often negated in the very act of its expansion by a lack of faith in the capacities of the mass of human beings.

Rosemary Park's "The Disestablished Humanities" comes closest to a democratic vista for the humanities for several reasons. She is dealing with the humanities outside of the academy and precisely because they are there widely scattered, she is compelled to take an all-society perspective. Her focus is accordingly on educational *functions* not particular institutions. While the university often looks to itself as an institution and thus may limit its sights to what it can do for itself in the circumstances, Park's perspective would ask what it could do educationally through all and any institutions in the society. Again, Park relates what gets done and what is desirable to basic changes in the character of life and knowledge: for example, the fact that one used to be able to get the knowledge needed for a lifetime in the early years, but now it takes a whole lifetime; as well as the rapidity of growth and the extent of novelty in knowledge itself. There is also the greater flexibility and freedom in the extra-academic institutions. (Paul Goodman once pointed out that one does not have to present academic credentials to take a book out of the library nor be examined on it when he returns it.) Finally, Park evaluates progress in all corners of the society—in the expansion of book publishing, of college extension work, even of the quality of commercials on TV.

Robert Hanning's response, "A View from the Ivory Tower," is much more pessimistic, perhaps in part because he catches the impact of the current depression on the extra-academic cultural institutions. His lesson is their general vulnerability to manipulations of the profit motive and its vicissitudes in our society, and he hopes for more from the relation of the academy to the nonacademic cultural institutions.

Humanists certainly have ample ground to fear what happens to their fields when they are commercialized—in research grants as well as in television. But this is a problem of our whole life, affecting the sciences as well as the humanities. There is no answer but vigorous struggle, carried into the full arena of our culture, rather than a retreat behind a dubious Maginot line drawn within the university. There is ample room within for working at the reconciliation of the "two cultures" of science and the humanities, and there is ample room without for humanistic critique and creative development of a democratic culture.

On the question of attitudes to the prospects of mankind, the core lies obviously in technology and the changes it has brought. How intimately questions of value, attitude, and prediction of development are interwoven is clear from the influence that Ellul's *Technological Society* has had in evaluations of technology. It conveys the same mood for technological reckoning that Spengler's *Decline of the West* conveyed in its day for Western civilization. It is more philosophical in its scope than recent doomsday charts emanating from the Club of Rome, for the necessity it preaches concerns the decline of our values not merely our ability to achieve them. Ellul pictures technique becoming omnivoracious, bringing everything into line with large-scale technology and its corporatism, exalting the rationality of the means and enslaving ends. In the interpenetration of means and ends, we come eventually to hold only those debased ends that the machine allows, and so we are gladly enslaved for we get the material goods to which we then limit our desires. Ellul's position is, I think, open to criticism both on the conceptual side and the empirical side as well as for the neglect of alternatives all along the way. It is not so strong as to be faced only by a faith that mankind will not tolerate the outcome. We obviously cannot pursue the critique here. I raise it only to show how important is the humanistic task of a reckoning with technology and its works, and contributing to attitude and policy in facing the future.

In "Living with Scarcity," Roger Shinn seems to me to tap profoundly the problems involved for the humanities here. Considering the larger problems to be permanent not temporary crises, he sees that large-scale reorientations of understanding, attitude, and policy are required. His sober probe of attitudes to scarcity and the ways of dealing with it could serve as a model of what humanistic learning can do to clarify alternative policies and their grounds. I am tempted to suggest one extension—that the different alternatives be set in their concrete historical contexts. After all, scarcity in preindustrial and prescientific societies has a different character from that found in industrial and postindustrial society although the situations (and the attitudes) may be analogous. So too, given solutions, like slavery or war, which impose the belt-tightening on a part of society or on one society instead of another, are quite different in ancient slave systems or empires, and in modern class relations within countries and between developed and undeveloped countries. In part, these points are made in Shinn's treatment of justice. In general, he calls for a more realistic attitude than simply oscillation between an excessive confidence in technology and an excessive resignation. (The extremes are symbolized by Prometheus and Atlas.) Similarly, while the dynamic factors in social change are some kinds of compulsion, rather than moral reasons, the ethical does make some differences. The situation does not dictate of itself, but in its interaction with commitments, loyalties, and values.

The practical effect of a humanistic analysis may be as immediate today as the effect of scientific generalization. A good example is the meteoric rise of the ethics of *triage* in recent times. This, and variant "life-boat" ethics, see the inevitability of growing scarcity and argue that resources should be apportioned to those countries likely to make the best use of them. Slowly developing countries hitherto unsuccessful in controlling population growth would, if assisted, grow still more populous and more needy. They would bring all closer to doomsday. The slowly developing countries would thus be abandoned to mass starvation.

Such a speculation has the character of an antiutopia. Problems have been raised about its assumptions, particularly that it holds institutional forms constant in its extrapolations and therefore does not face the totality of issues which a really great crisis of global proportions requires. And there have been scientific objections too, that it has underestimated resources that are still available. But even if it is purely a speculative scenario, envisaging a possibility, it operates to distract us from our problems and responsibilities. It is not, however, pure speculation. It competes for national policy. Its effect on some of the world's poorest countries might very well be immediately disastrous. Not only do scenarios of the future today have practical import, but even the formulation of questions and judgment of priorities in facing questions have critical effects.

There is a rising scale in the guidance of social action which seems to vary with the degree of complexity and flux in the human field. In stable, relatively simple situations with constant values, we guide ourselves by rules. Increase the complexity and change, and we use not rules but broad principles and standards of evaluation. A further increase, and we appeal to methods. Still further, and our methods may not be enough; we appeal to virtues: face the problems courageously, carefully, deliberately, considering all relevant factors. Still further, even virtues do not help us, for the courageous man may be blundering in the dark. Then we ask for at least a general attitude to life as a whole; then we talk of authenticity and deciding for all mankind. As a last step, in the greatest of complexities and transformations, we can hope for nothing but wisdom. Perhaps the contributions of the humanities to public policy are to be found in what they can bring from their heritage and their continued exercise in the cultivation of wisdom. But wisdom is not a separate light shining from outside. It is rather a full sense of how, in the historical present, knowledge and value and the practical situation operate in their constant interaction, and what our directive values require in knowledge and for action.

Notes

1. Justice—Compensatory and Distributive

Herbert A. Deane

1. See Murray Kempton, "The Black Manifesto," *New York Review of Books,* 10 July 1969, where the text appears in full.

2. Michael Harrington and Arnold S. Kaufman, "Black Reparations—Two Views," *Dissent* 16 (1969): 317–20, esp. 317.

3. Hugo Adam Bedau, "Compensatory Justice and the Black Manifesto," *Monist* 56 (1972): 20–42.

4. For a useful survey of the various kinds of arguments in support of preferential policies, see James W. Nickel, "Preferential Policies in Hiring and Admissions: A Jurisprudential Approach," *Columbia Law Review* 75 (1975): 534–58, esp. 536–44. This entire issue of the *Review* is devoted to a discussion of the DeFunis case.

5. Bedau, "Compensatory Justice and the Black Manifesto," p. 37.

6. James W. Nickel, "Discrimination and Morally Relevant Characteristics," *Analysis* 32 (1972): 113–14, esp. 113.

7. James W. Nickel, "Should Reparations Be to Individuals or to Groups?" *Analysis* 34 (1974): 154–60, esp. 155 (my emphasis). See also his "Preferential Policies in Hiring and Admissions," pp. 550–53, 555–58.

8. For a discussion of this point, see Thomas Nagel, "Equal Treatment and Compensatory Discrimination," *Philosophy and Public Affairs* 2 (1973): 348–63.

9. Paul W. Taylor, "Reverse Discrimination and Compensatory Justice," *Analysis* 33 (1973): 177–82, esp. 181–82.

10. *Ibid.,* p. 181.

11. *Ibid.,* p. 182.

12. Michael D. Bayles, "Reparations to Wronged Groups," *Analysis* 33 (1973): 182–84, esp. 183.

13. Bayles, "Compensatory Reverse Discrimination in Hiring," *Social Theory and Practice* 2 (1973): 301–12, esp. 305–6.

14. Taylor, "Reverse Discrimination and Compensatory Justice," p. 1818.

15. Bayles, "Compensatory Reverse Discrimination in Hiring," p. 308.

2. Equality, Race, and Preferential Treatment

Paul A. Freund

1. For the references in this paragraph I am indebted to an unpublished study by J. R. Pole, of Cambridge University, on equality in America.

2. Carl Becker, *The Declaration of Independence: A Study in the History of Political Ideas* (New York: Harcourt, Brace, 1922), p. 278.

3. Cf. International Association of Machinists v. Street, 367 U.S. 740 (1961).

3. The Costs of Equality

Robert A. Nisbet

1. Lester Thurow, in *Public Interest,* no. 34 (Winter 1974), pp. 56–80.

2. *Ibid.,* pp. 91–101.

3. Bertrand de Jouvenel, *The Ethics of Redistribution* (Cambridge: At the University Press, 1951), pp. 37–38.

4. Friedrich von Hayek, *Constitution of Liberty* (Chicago: University of Chicago Press, 1960), p. 95.

5. Justice, Equality, and the Economic System

William S. Vickrey

1. Charles M. Tiebout, "A Pure Theory of Local Government Expenditures," *Journal of Political Economy* 64 (1956): 416–24.

2. Hans Neisser, "The Strategy of Expecting the Worse," *Social Research* 19 (1952): 346–63.

3. See my *Agenda for Progressive Taxation* (New York: Ronald Press, 1947).

4. Henry Simons, *Personal Income Taxation: The Definition of Income As a Problem of Fiscal Policy* (Chicago: University of Chicago Press, 1938).

7. Private Rights and the Public Good

Charles Frankel

1. Wesley Hohfeld, "Some Fundamental Legal Conceptions as Applied in Judicial Reasoning," *Fundamental Legal Conceptions as Applied in Judicial Reasoning and Other Legal Essays,* W. W. Cook, ed. (New Haven: Yale University Press, 1923), pp. 23–114, esp. 37; Glanville Williams, "The Concept of Legal Liberty." *Columbia Law Review* 56 (1956): 1129–50, esp. 1142–45.

2. Max Radin, "A Restatement of Hohfeld," *Harvard Law Review* 51 (1938): 1141–64, esp. 1149.

3. See Hans Kelsen, *General Theory of Law and State,* Anders Wedberg, trans. (Cambridge, Mass.: Harvard University Press, 1945), pp. 58–62; and "The Pure Theory of Law and Analytical Jurisprudence," *What Is Justice?* (Berkeley and Los Angeles: University of California Press, 1957), pp. 266–87.

4. Abrams v. U.S., 250 U.S. 616, 630 (1919).

10. Do Rocks Have Rights? Thoughts on Environmental Ethics

Roderick Nash

1. Douglas' 1972 opinion is printed, along with a useful analysis of the legal implications for environmental ethics, in Christopher D. Stone, *Should Trees Have Standing? Toward Legal Rights for Natural Objects* (Los Altos, Calif.: Kaufman, 1974), p. 73ff.

11. Living with Scarcity

Roger L. Shinn

1. E. F. Schumacher, "Small Is Beautiful: Toward a Theology of 'Enough,' " *Christian Century* 88 (1971): 900–2, esp. 900.

2. John Kenneth Galbraith, *The New Industrial State* (Boston: Houghton-Mifflin, 1967), pp. 164, 173.

3. I list a few of the major publications in chronological sequence: Kenneth Boulding, *Human Values on the Spaceship Earth* (New York: National Council of Churches, 1966), and "The Economics of the Coming Spaceship Earth," in *Environmental Quality in a Growing Economy,* Henry Jarrett, ed. (Baltimore: Johns Hopkins University Press, 1966); E. J. Mishan, *The Costs of Economic Growth* (New York: Praeger, 1967); Lynn White, Jr., "The Historical Roots of Our Ecological Crisis," *Science* 155 (1967): 1203–7; Garrett Hardin, "The Tragedy of the Commons," *Science* 162 (1968): 1243–48; Jay W. Forrester, *World Dynamics* (Cambridge, Mass.: Wright-Allen Press, 1971); E. F. Schumacher, "Small Is Beautiful: Toward a Theology of 'Enough,' " *Christian Century* 88 (1971): 900–2; Herman E. Daly, "The Steady-State Economy: Toward a Political Economy of Biophysical Equilibrium and Moral Growth," in *Toward a Steady-State Economy,* Herman E. Daly, ed. (San Francisco: Freeman, 1973), pp. 149–74; *Patient Earth,* John Harte and Robert H. Socolow, eds. (New York: Holt, Rinehart & Winston, 1971); Barry Commoner, *The Closing Circle* (New York: Knopf, 1971); Donella H. Meadows, Dennis L. Meadows, Jørgen Randers, and William W. Behrens III, *The Limits to Growth: A Report for the Club of Rome's Project on the Predicament of Mankind* (New York: Universe Books, 1972); *A Blueprint for Survival,* Edward Goldsmith et al., eds., being *Ecologist* 2 (January 1972) and published later that year as a book in Boston by Houghton-Mifflin; E. F. Schumacher, *Small Is Beautiful: Economics As If People Mattered* (New York: Harper & Row, 1973); *The No-Growth Society,* being *Daeda-*

lus: *Journal of the American Academy of Arts and Sciences* 102 (Fall 1973); Robert L. Heilbroner, *An Inquiry into the Human Prospect* (New York: Norton, 1974); Mihaijlo Mesarovic and Eduard Pestel, *Mankind at the Turning Point: The Second Report to the Club of Rome* (New York: Dutton, 1974).

The arguments among these authors, both on technical points and especially on values and goals, have been intense and sometimes bitter. Some of the writers may resent being included on the same list with others. Yet all of them, in contrast to the writers in the following note, have criticized in varying ways the possibility and desirability of conventional economic growth.

4. Much of the attack on the no-growth theme appears in critical articles and reviews discussing the books listed in note 3. But the countercase has now been stated in several books, including these: John Maddox, *The Doomsday Syndrome* (London: Macmillan, 1972); Peter Passell and Leonard Ross, *Retreat from Riches* (New York: Viking, 1973); *Thinking about the Future: A Critique of "The Limits to Growth,"* H. S. D. Cole et al., eds., for the Science Policy Research Unit of Sussex University (London: Chatto & Windus, 1973); Wilfred Beckerman, *In Defense of Economic Growth* (London: Jonathan Cape, 1974).

5. *Exploring New Ethics for Survival: The Voyage of the Spaceship Beagle* (New York: Viking, 1972), pp. 250–51. The same judgment is made by the authors of *The Limits to Growth,* p. 150.

6. Kermit Lansner, "This Is Your Life: A Crisis Guide," *Newsweek,* 25 February 1974, p. 35.

7. Jørgen Randers, "The Carrying Capacity of Our Global Environment—A Look at the Ethical Alternatives," paper presented to the Working Committee on Church and Society, World Council of Churches, Nemi, Italy, 20–26 June 1971, p. 9.

8. Margaret Mead, "A Dialogue: The Role of the Churches Now," in *To Love or to Perish: The Technological Crisis and the Churches,* J. Edward Carothers et al., eds. (New York: Friendship Press, 1972), p. 125.

9. Draft Report, Conference of the World Council of Churches on Science and Technology for Human Development, Bucharest, Romania, 24 June–2 July 1974.

12. The Technology of Life and Death

Willard Gaylin

1. William May, "Attitudes Toward the Newly Dead," *Hastings Center Studies* 1 (1973).

14. Technology and the Structuring of Cities

David P. Billington

1. This section is a summary and extension of my paper, "Structures and Machines: The Two Sides of Technology," *Soundings* 57 (1974): 275–88.

2. *The Ten Books of Architecture,* M. H. Morgan, trans. (1914; New York: Dover, 1960), p. 17. In his work, Vitruvius also considers structures and machines separately: "I have thought it not out of place, Emperor, since I have treated of buildings in the earlier books, to set forth and teach in this, which forms the final conclusion of my treatise, the principles which govern machines" (p. 282).

3. This table is made from values given in "Less Steel Per Sq. Ft.," *Modern Steel Construction* 12, no. 1 (1972): 8–9, esp. 9.

4. Fazlur Khan, "The John Hancock Tower," *Civil Engineering* 37 (October 1967): 38–42, esp. 39.

5. *Ibid.,* p. 40.

6. James S. Hornbeck, "Chicago's Multi-Use Giant," *Architectural Record* 141 (January 1967): 137–44.

7. Most of the discussion that follows derives from *The World Trade Center* (New York: Engineering News Record, 1972), a booklet of articles that appeared in the *Engineering News-Record* between 23 October 1958 and 23 December 1971.

8. Lester S. Feld, "Superstructure for 1,350-ft. World Trade Center," *Civil Engineering* 41 (June 1971): 66–70.

9. John Morris Dixon, "The Tall One," *Architectural Forum* 133 (July/August 1970): 37–45.

10. Kahn, "The John Hancock Tower," p. 38.

11. Lewis Mumford, *The Pentagon of Power* (New York: Harcourt, Brace, Jovanovich, 1970), p. 340.

12. Editorial, *Engineering News-Record* 172 (30 January 1964): 76.

13 *Ibid.*

17. The Lower Middle Class As a Historical Problem

Arno J. Mayer

1. Henri Pirenne, *A History of Europe: From the Invasions to the Sixteenth Century* (New York: University Books, 1956), pp. 210–42.

2. Karl Marx, *The Eighteenth Brumaire of Louis Bonaparte* (New York: International Publishers, 1935), p. 47; and *The Class Struggles in France, 1848–50* (New York: International Publishers, 1935), p. 89.

3. Friedrich Engels, *Germany: Revolution and Counter-Revolution,* in Friedrich Engels, *The German Revolutions,* Leonard Krieger, ed. (Chicago: University of Chicago Press, 1967), p. 128.

4. V. I. Lenin, *"Left-Wing" Communism: An Infantile Disorder* (New York: International Publishers, 1934).

5. Engels, *Germany: Revolution and Counter-Revolution,* pp. 232, 239–40; see also Marx, *The Eighteenth Brumaire,* p. 44.

6. Marx, *The Eighteenth Brumaire,* p. 22; *The Class Struggles in France,* p. 115.

7. For a definition of conservatism, reaction, and counterrevolution, see my *Dynamics of Counterrevolution in Europe, 1870–1956* (New York: Harper & Row, 1971), pp. 48–55 and chap. 3.

18. Reflections on War, Utopias, and Temporary Systems

Elisabeth Hansot

1. Iris Murdoch, "The Idea of Perfection," *Yale Review* 53 (1964): 344–45.

2. As Michael Holquist observes, in utopia man *is* his environment. See his article, "How to Play Utopia," in *Game, Play, Literature,* Jacques Ehrmann, ed. (Boston: Beacon Press, 1968), p. 115.

3. J. Glenn Gray, *The Warriors: Reflections on Men in Battle* (New York: Harper & Row, 1959), pp. 28ff.

4. Johan Huizinga, *Homo Ludens: A Study of the Play Element in Culture* (Boston: Beacon Press, 1950), p. 94.

5. Gray, *The Warriors,* pp. 225–26.

6. Michael Holquist believes it is a mistake to fault utopias for not exhibiting the psychological and social complexity of, say, a Balzac novel. According to Holquist, the utopist begins by postulating the best man; he then proceeds to construct the conditions which would best ensure the existence of such a man. For Holquist, the purpose of such constructs is play, the limitless combinations of ideas. The play element of utopia, its "let's pretend" quality, is surely one of that genre's chief delights, but to limit utopias to such playfulness does not do justice to the future-oriented author's discomfort with the present, his desire that utopia be a vehicle for change. See "How To Play Utopia," p. 113.

7. Anthony Storr, "Possible Substitutes for War," in *The Natural History of Aggression,* J. D. Carthy and F. J. Ebling, eds. (New York: Academic Press, 1964), pp. 137–44.

8. Huizinga, *Homo ludens,* p. 13.

9. Paul Weiss, *Sport: A Philosophic Inquiry* (Carbondale: Southern Illinois University Press, 1969), pp. 143, 137.

10. *Ibid.,* p. 153.

11. *Ibid.,* p. 141.

12. Matthew Miles, "On Temporary Systems," in *Innovation in Education,* Matthew B. Miles, ed. (New York: Bureau of Publications, Teachers College, Columbia University, 1964), pp. 437–92. The various points raised by Miles in this article are summarized in my next few paragraphs.

13. *Ibid.,* pp. 461, 463.

14. *Ibid.,* p. 465.

23. The Disestablished Humanities

Rosemary Park

1. Jerome Wiesner, "Science Advice for the White House," *Technology Review* 76 (January 1974): 17.

2. Irving Kristol, "When Virtue Loses All Her Loveliness," *Public Interest,* no. 21 (Fall 1970), p. 12.

3. Ivan Illich, *Deschooling Society* (New York: Harper & Row, 1972), p. 11.

4. Joe L. Spaeth and Andrew Greeley, *Recent Alumni and Higher Education* (New York: McGraw-Hill, 1970), p. 32.

5. "Dollars and Sense," *Moneysworth* 4 (3 March 1975): 2.

6. As quoted in "Publishing and 'The Knowledge Industry,' " *Publishers Weekly* 205 (20 May 1974): 46.

7. *The Bowker Annual of Library Book Trade Information* (New York: Bowker, 1974), p. 184.

8. *Ibid.,* p. 182.

9. "Dollars and Sense," p. 2.

10. "Publishing and 'The Knowledge Industry,' " p. 46.

24. A View from the Ivory Tower

Robert W. Hanning

1. Richard F. Kuhns, *The House, the City, and the Judge* (Indianapolis: Bobbs-Merrill, 1962), p. 29.

2. Daniel Bell, *The Reforming of General Education* (New York: Columbia University Press, 1966), p. 293.

Index

Abelard, 312
Abraham (and Isaac), 348
Academic Revolution, The, 297
Acropolis, 324
Action for Children's Television, 316
Addison, Joseph, 2-3
Adorno, Theodor W., 340, 341
Aeschylus, 324
Africa, North, political history of, 212
Akkad, 208
Alcoa Building (Pittsburgh), 203
Allende, Salvador, 140
Altruria, 251
America: anti-intellectualism in, 270-71; political and social goals of, 119; "problem-solving" in, 264-65
"America and the Future of Man," 318
Analysis, 14
Andromache, 346
Apollo, 360
Arabs, 213
Architectural Forum, 192
Arendt, Hannah: article, 103-8; discussed by Capouya, 111; discussed by Edel, 369
Aristophanes, 375
Aristotle, 15-17, 32, 96, 140, 337, 350-51, 368, 374, 378
Arnold, Matthew, 283, 284-90, 291, 293, 379

Assyria, ancient, 209
Aswan Dam, 319
Athena, 324
Athens, ancient, 110, 213-14, 309, 313, 314, 323-24, 325, 326, 371
Atlas, 151, 383
Augustine, 43
Authoritarian Personality, The, 340-41
Aztecs, 210, 212

Babylon, 209
Bacon, Sir Francis, 374
Bauhaus, 289
Bayles, Michael D., 20-21, 22
Becker, Carl, 27
Bedau, Hugo Adam, 14, 18, 22
Beecher, Henry Ward, 267
Bell, Daniel, 327
Bellamy, Edward, 249, 251
Bellow, Saul, 81-82
Bell Telephone Company, 373
Benedict, Ruth, 358
Bentham, Jeremy, 368
Bergson, Henri, 337, 345
Bernstein, Eduard, 230
Bertles, John F., 158
Bible, 3, 140, 252; Abraham and Isaac, 348; Job, 374-75; Matthew, 106; David, 208
Billington, David P.: article, 182-98;

Billington, David P. (*Continued*)
   discussed by Salvadori, 199-203;
   discussed by Edel, 361
Bill of Rights, 3
Bird in Flight, 200
Black, Hugo, 111
"Black Manifesto, The," 13-14, 18
Blacks, in America, 13-14, 18, 21,
   31-33
Boas, Franz, 345
Boulding, Kenneth, 151
Bozeman, Adda, 343, 362
Bradley, F. H., 291, 367
Brancusi, Constantin, 200
Brandeis, Louis, 99, 106, 112
Brave New World, 374
Bronowski, Jacob, 336
Brown University, 268
Bruni, Leonardo, 2
Buber, Martin, 110, 372
Bukharin, Nikolai, 221
Burckhardt, Jacob, 301-2
Bush, Vannevar, 308
Butler, Samuel, 376
Byron, Lord, 80

Callahan, Daniel: article, 170-81; dis-
   cussed by Edel, 376
Calley, William, 125
Call Girls, The, 8-9
Capital, 346
Capouya, Emile: article, 109-19; dis-
   cussed by Edel, 359
Cardozo, Benjamin, 26
Carman, Harry, 373
Carnegie Commission on Higher
   Education, reports, 315
Casa Italiana of Columbia University,
   328
Central Intelligence Agency (CIA),
   140

Change in the Village, 76-77
Chapman, John Jay, 113, 115-16
Charles V, 208
Chautauqua, 312
Chicago School of Architecture, 191
Childe, V. Gordon, 350
China: administrative structures in,
   213; education in, 279; Han and
   Sung Dynasties of, 209, 210;
   Mao's, 47; Northwest frontier of,
   207
Christ, Jesus, 84, 359-60
Christianity, 148, 151
Christians, early, and private rights,
   106
Church of England, 267
Cicero, 2
City University of New York, 113-14
Civil Engineering, 192
Civil War (U.S.), 126
Clark, John Bates, 59-60
Clement of Alexandria, 359-60
Clifford, James L., 374
Club of Rome, 382
Coleridge, Samuel Taylor, 287
Columbia College Alumni Associa-
   tion, 327
Columbia Reports, 274, 277-78
Columbia University, 4, 288; Casa
   Italiana of, 328; as King's College,
   267
Columbia University Law School,
   164, 165
Common Faith, A, 371
Common Market, 215
Communist Manifesto, The, 234-35,
   374
Conferences on the Humanities and
   Public Policy Issues, history of, 4-5
Confessions (Augustine), 43
Confessions (Rousseau), 42

Confucius, 47
Congress, definition of "humanities," 336
Constitution (U.S.), 111; First Amendment of, 104; Fourteenth Amendment of, 27
Constitution of Liberty, 49
Cornell, Ezra, 269
Cornell University, 269
Crosland, Anthony, 43
Culture: A Critical Review of Concepts and Definitions, 358
Culture and Anarchy, 286
Culture and Commitment, 351-52
Cupid, 329-30

Daedalus, 375
Daly, Herman, 151
Dante, 348
Darwin, Charles, 353
Darwinian controversy (Tennessee), 310
David, 208
da Vinci, Leonardo, 375
Deane, Herbert A.: article, 13-25; discussed by Edel, 370
de Bary, Wm. Theodore: article, 277-280; discussed by Edel, 379; on general education, 274, 277-78
de Bergerac, Cyrano, 375
Declaration of Independence (U.S.), 3, 6-7, 26-27
Decline of the West, 382
DeFunis v. Odegaard, 14
de Jouvenel, Bertrand, 40, 53
Delaunay, Jules, 200
Democratic Vistas, 380
DeMott, Benjamin: article, 72-84; discussed by Edel, 359, 364-65
Department of Health, Education and Welfare (HEW), 45, 272

Dewey, John, 274, 348, 355, 369, 371
Discourse on Political Economy, 46-47
Discourse on the Arts and Sciences, 42
Discourses in America, 284
Dostoyevski, Fyodor, 346
Douglas, William O., 128
Down's Syndrome, 166, 173, 178

Edel, Abraham, 6, 8; article, 335-84
Egypt, ancient, 208
Eiffel, Alexandre, 183, 184
Eiffel Tower, 199-200
Einstein, Albert, 9
"Einstein Letter," 9, 10
Eiseley, Loren, 128
Elective System, 268
Eliot, Charles W., 268, 269
Eliot, T. S., 290-93, 345, 351
Ellul, Jacques, 143, 382
Emerson, Ralph Waldo, 81, 344
Empire State Building, 188, 194, 197
Engels, Friedrich, 226, 235
England, linguistic sense in, 363
English Revolution, 225
Epicureans, 338
Epicurus, 2
Epstein, Jacob, 337
Erikson, Erik H., 342
Essays (Montaigne), 43
Ethics of Redistribution, The, 40
Euripides, 375
Eumenides, 324
Europe: former cultural hegemony of, 290-93; medieval, lower middle class in, 222-24, 226; modern, national units of, 207-8; secondary education in, 297-98
Evolutionary Socialism, 230

Exorcist, The, 322
Expo '74 (Spokane, Washington), 133

Faulkner, William, 43
Flaubert, Gustave, 227
Ford Foundation, 66
Ford Motor Company, 65-66
Foreman, James, 13
Frankel, Charles, 112; article, 87-102; discussed by Arendt, 103-8; discussed by Edel, 369
Franklin, Benjamin, 3, 267
Freeland, 249
French Revolution, 215, 225
Frenkel-Brunswik, Else, 340
Freud, Sigmund, 353
Freund, Paul: article, 26-33; discussed by Edel, 370
Friedman, Milton, 342-43
Fromm, Erich, 341-42
"Function of Criticism at the Present Time, The," 290
Future of Socialism, The, 43

Galbraith, John Kenneth, 138
Gans, Herbert J., 34, 35, 44; article, 50-58; discussed by Edel, 364-66
Gaylin, Willard: article, 152-69; discussed by Edel, 376-77
Geertz, Clifford, 358
Geiger, Theodor, 230
General Education in a Free Society, 288n
Germania, 213
Germany: pre-WWI, lower middle class in, 229-30; Nazi, anti-Semitism in, 340-41
G.I. Bill of Rights, 270
Gilman, D. C., 269
Gladstone, William, 26

Glass, Bentley, 178
Godkin, E. L., 26
Goethe, Johann Wolfgang von, 289
Golden Rule, 27, 125
Goodman, Paul, 381
Gramsci, Antonio, 238-39
Gray, Glenn, 252-53, 255
Great Chain of Being, 27
Greece, ancient, 107, 221-22
Green Revolution, 139
Gross National Product (GNP), of U.S., 146-47, 270-71

Hallowell, Irving, 364
Hamlet, 29, 320
Hanning, Robert W.: article, 321-30; discussed by Edel, 371, 381
Hansot, Elisabeth: article, 246-59; discussed by Edel, 343-44, 374
Hardin, Garrett, 143
Harper, William Rainey, 269
Harrington, Michael, 14
Harvard Medical School, Ad Hoc Committee of, on brain death, 155
Harvard University, 270, 288n
Harvard Values Studies, 363-64
Hayek, Friedrich von, 49, 83, 365
Hector, 346
Hegel, Georg Wilhelm Friedrich, 291, 353
Heilbroner, Robert, 121, 151
Hemingway, Ernest, 80
Henkin, Louis, 355-56
Herodotus, 214
Hertzka, Theodor, 249
Herzog, 81-82
Himalaya mountains, 207
Hiroshima, 153, 273
Hitler, Adolf, 125, 215, 273
Hittite Empire, 208
Hobbes, Thomas, 27, 88

Hohfeld, Wesley, 88n, 89
Holmes, Oliver Wendell, 43, 97
Homer, 76, 101, 346
Homo Ludens, 253, 254
Howells, William Dean, 248n, 249, 251
How Natives Think, 345
Hugins, Walter, 47
Huizinga, Johan, 253, 254, 256
Hulme, T. E., 337
Husserl, Edmund, 337, 358
Hutcheson, Francis, 350
Huxley, T. H., 283-84, 287, 289-90, 297, 379

Iliad, 345
Illich, Ivan, 313, 323
Incas, 212
Independents, of English Revolution, 225
India: effect of petroleum shortage on, 147; northern frontier of, 207; political structures in, 209-10
Inequality (Jencks), 34, 35, 37-39, 51, 82-83
Internal Revenue Service (IRS), 45
Interpretation of Cultures, 358
Interstate Commerce Commission (ICC), 45
Introduction to Metaphysics (Bergson), 337
Irnerius, 312
Isocrates, 2, 309
Italian-Americans, in New York City, 328
Italy, futurism in, 337; thirteenth-century city states of, 214

Jacobins, of French Revolution, 225
James, William, 344, 355
Jaspers, Karl, 348

Jaws, 322, 323
J.B., 375
Jefferson, Thomas, 3, 27, 78, 313
Jencks, Christopher, 34, 35, 36, 37-39, 44, 46, 50, 51, 53, 82-83, 297
Jews: and anti-Semitism, 340-41; in America, 18; in Germany, 31; and Hitler, 125
Job, Book of, 374-75
John Hancock Tower, 187, 188, 188-91, 200; compared to World Trade Center, 191-97, 198, 200-2
Johns Hopkins University, 269
Johnson, Samuel, 3, 374
Jonas, Hans, 161-62
Jordan, David Starr, 269

Kant, Immanuel, 314, 352-53
Kelsen, Hans, 88n
Keyfitz, Nathan, 37-39
Khaldun, Ibn, 212
Khan, Fazlur, 188-90, 191
Kierkegaard, Søren, 348, 350
King's College (Columbia University), 267
Kissinger, Henry A., 137
Kittridge, George Lyman, 309
Kluckhohn, Clyde, 358, 363-64
Koestler, Arthur, 8
Kristol, Irving, 313
Kroeber, Alfred Louis, 358
Kuhns, Richard, 324

Lansner, Kermit, 143
Lappe, Marc, 159
Laws (Plato), 339
Le Corbusier, 199
Lederer, Emil, 230
"Left-Wing" Communism: An Infantile Disorder, 230-32

Lenin, V. I., 230-32
Leonardo da Vinci, 375
Leopold, Aldo, 120, 122, 124, 125-26, 126-28, 131, 132, 133, 134, 350
Levelers, of English Revolution, 225
Levinson, Daniel J., 340
Lévy-Bruhl, Lucien, 345
Lewis, Oscar, 76
Life of Addison, 3
Life of Reason (Santayana), 360
Limits to Growth, The, 148
Lindsay, John V., 146
Lipset, Seymour, 47
"Literature and Science," 284-90
Livy, 213
Lodge, Henry Cabot, 80
Looking Backward, 250
Los Angeles, Calif., 312
Lowenthal, John, 8

Machlup, Fritz, 316
MacIver, R. M., 357
MacLeish, Archibald, 375
McNeill, William H.: article, 207-19; discussed by Edel, 344
Mailer, Norman, 80
Mandingo, 322
Man Makes Himself, 350
Marcus, Steven, 279; article, 281-302; discussed by de Bary, 280; discussed by Edel, 366, 379-80, 380
Maritain, Jacques, 367-68
Marx, Karl, 43, 226, 235, 295-96, 346, 348, 374
Matthew, Gospel According to, 106
May, William, 161, 162
Mayer, Arno J.: article, 220-45; discussed by Edel, 354-55, 362
Mead, George Herbert, 372

Mead, Margaret, 151, 351-52, 354, 358
Medicaid, 140, 145
Medicare, 140, 145
Melville, Herman, 7
Meno, 283, 284
Merchant of Venice, The, 91-92
Mesopotamia, ancient, 210
Metropolitan Museum of Art (New York City), 327-28, 329
Mexico, "Imperial Age" of, 210
Middle Ages: education in, 310, 323; lower middle class in, 222-25, 226; religion in, 287
Middle America, 232-33, 275
Middle East, history of, 208, 212
Miles, Matthew, 255-57
Mill, John Stuart, 93-95, 99, 337, 345
Mind of Primitive Man, 345
Montaigne, 43
Montesquieu, 95-96
Moore, G. E., 337
More, Thomas, 43, 247-49, 251, 255, 258
More Equality (Gans), 34, 50-51, 53
Morrill Act of 1862, 310
Mossadegh, Muhammed, 140
Mumford, Lewis, 194, 197, 203
Murdoch, Iris, 249
My Lai (Vietnam), 125

Napoleonic empire, 215
Nash, Roderick: article, 120-34; discussed by Edel, 350-51, 360
National Endowment for the Humanities (NEH), 4, 315, 316
National Institutes of Health (NIH), 272
National Science Foundation (NSF), 314-15
Neisser, Hans, 61

Nervi, Pier Luigi, 200, 202
New Atlantis, 374
Newton, Isaac, 352
New York City: financial crisis in, 328; public transportation in, 146
New York Public Library (New York City), 316
New York Times, 137
Nichomachean Ethics, 15-16, 96
Nickel, James W., 19-21
Niebuhr, Reinhold, 348
Nietzsche, Friedrich, 253, 258, 345, 350, 368
1984 (Orwell), 374
Nineteenth Century, 284
Nisbet, Robert A.: article, 34-49; discussed by Gans, 50-58; discussed by DeMott, 83; discussed by Edel, 364-66, 372
Notes from Underground, 346

Oakeshott, Michael, 345
Odysseus, 125
Odyssey, 125
Oedipus, myth of, 341-42
Oedipus at Colonus, 342
O'Neill, Onora: article, 303-7; discussed by Edel, 366, 380
"On Fortune and Misfortune in History," 301-2
On Liberty (Mill), 93-95
Open University, 268
"Open University," 318
Oresteia, 324
Orestes, 324
Organization of Petroleum Exporting Countries (OPEC), 147
Opus Posthumous (Stevens), 79
Ostrogoths, 213
Ottoman Empire, 208

Paine, Tom, 371
Pan Am Building (New York City), 197
Pareto, Vilfredo, 347
Park, Rosemary: article, 308-20; discussed by Hanning, 321-30; discussed by Edel, 371, 381
Parsons, Talcott, 357-58
Parthenon, 314, 324, 371
Péguy, Charles Pierre, 72, 84
Peloponnesian War, 214
Pentagon, 45, 55
People's Colleges, 312
Pericles, 214, 309
Persia, 209
Peru, "Imperial Age" of, 210
Philosophical Essays (Jonas), 161-62
Philosophy and Public Affairs, 14
Pierson, George W., 1; article, 263-76; discussed by deBary, 277-80; discussed by Edel, 379
Pinkham, Lydia, 318
Pirenne, Henri, 222-23, 226
Place Victoria (Montreal), 202
Plato, 2, 43, 46, 110, 140, 247-49, 251, 255, 258, 284, 288, 336-37, 339, 375
Plotinus, 360
Politics (Aristotle), 15, 16, 378
Port of New York Authority, and World Trade Center, 192-94, 197
"Portrait of the Antisemite" (Sartre), 341
Price, Richard, 371
Program of General Education in the Humanities, Columbia University, 274, 277-78, 327
Progressive Education Association, 274-75
Prometheus, 151, 383
Proposed Roads to Freedom, 342-43

Public and Its Problem, The, 369
Puritans, American, 266, 267, 310
Pythagoras, 16

Radin, Max, 89
Rainwater, Lee, 52
Randers, Jørgen, 148-49
Randolph, Philip, 24
Rawls, John, 24, 34, 35-36, 37, 38,
    44, 46, 50, 53, 60, 61-62, 68,
    69-70, 365
Redfield, Robert, 364
Reflections on the Jewish Question
    (Sartre), 341
Reflections on World History (Burck-
    hardt), 302
Reforming of General Education, The
    (Bell), 327
Renaissance, 311; education in, 310,
    323
Republic, The (Plato), 43, 247-49,
    251, 284, 288, 339, 375
Reston, James, 137
Revolts of 1848, 225-26
Richards, I. A., 6
Riesman, David, 297
Robinson Crusoe, 109
Rockefeller, John D., 269
Rome, ancient, 45, 107-8, 213,
    221-22
Roosevelt, Franklin D., 9
Roth, Emery, and Sons, 194
Rousseau, Jean-Jacques, 42, 44-45,
    46-47, 52
Russell, Bertrand, 342-43, 373
Rustin, Bayard, 24

Salvadori, Mario G.: article, 199-203;
    discussed by Edel, 361
Samuelson, Paul, 61
Sand County Almanac and Sketches

Here and There, A, 120, 122, 124,
    126-28, 131, 132
Sanford, R. Nevitt, 340
Sansculottes, of French Revolution,
    225
Santayana, George, 343, 345, 348,
    360
Sartre, Jean-Paul, 341, 348
Schiller, Friedrich von, 318
Schopenhauer, Arthur, 338
Schweitzer, Albert, 127, 131
Science: The Endless Frontier, 308
"Science and Culture" (Huxley),
    283-84, 287, 289-90, 297
Sears, Roebuck, & Co. Building
    (Chicago), 188
Seattle, Chief, 134
Sermon on the Mount, 151
Seven Against Thebes (Sophocles),
    342
*Sex and Temperament in Three
    Primitive Societies,* 354
Shakespeare, William, 29, 320, 348
Shinn, Roger L.: article, 137-51; dis-
    cussed by Edel, 382-83
Shumacher, E. F., 138
Silber, John, 378
Simons, Henry, 66
Singer, Milton, 364
Sirica, John J., 78
"Six Significant Landscapes"
    (Stevens), 30-31
Skilling, Helle, Christiansen, Robert-
    son (Seattle), 197
Skinner, B. F., 248n
Social Security, 140
Social Theory and Practice, 14
Socrates, 2, 337, 338-39
Solovief, Nikolai Borissovitch, 9
Solzhenitsyn, Aleksandr, 107
Sombart, Werner, 230

Spectator (Addison and Steele), 3, 10

Speer, Albert, 215

Spengler, Oswald, 337, 382

Spirit of the Laws, The (Montesquieu), 95-96

Sputnik, 270

Stalin, Josef, 107

Stanford, Leland, 269

Steele, Richard, 3

Stevens, Wallace, 30-31, 79-80

Stiles, Ezra, 268

Stoics, 27, 140, 304, 338, 374

Sturt, George B., 76-77, 78

Sullivan, Louis, 191

Sumer, 208

"Sunday Morning" (Stevens), 79-80

Sun God, 210

Supreme Court, U.S.: on abortion, 174, 356; on *DeFunis* case, 14; on desegregation of schools, 97; on right to privacy, 7; on World Trade Center, 194

Swift, Jonathan, 374

Tacitus, 213

Taoism, 150

Tawney, R. H., 364

Taylor, Paul W., 20-21, 22

Tay-Sachs disease, 166, 178

Tchambouli, males, 354

Technological Society, The, 143, 382

Terence, 319

Tertullian, 106

Thrasymachus, 339

Theory of Justice, A (Rawls), 34, 35-36, 60

Thucydides, 214, 375

Thurow, Lester, 37

Tiebout, Charles M., 61

Tiger, Lionel, 373

Titian, 329-30

Tocqueville, Alexis de, 6, 43, 45, 54

Tolman, R. C., 357

Tours, battle of, 351

Toynbee, Arnold, 150

"Tradition and the Individual Talent" (Eliot), 290-93

"Tragedy of the Commons, The" (Hardin), 143

Trilling, Lionel, 279

Truman, Harry S., 1

Turin Exhibition Hall, 200

Tylor, Edward Burnett, 357

Union of Soviet Socialist Republics (USSR): post-WWII, 215; present-day government, 107; under Stalin, 107

United Nations (UN), 370

United States: deaths in, 161; lower middle class in, 232-33; post-WWII, 215; secondary education in, 296-98

United States Forest Service, 133

University Extension, 312

University of California, 299; at Berkeley, 299; at Los Angeles, 299, 312

University of Chicago, 269

University of Pennsylvania, 267

University Seminars, Columbia University, 4

Uriah the Hittite, 208

Utopia (More), 43, 247-49, 251

Valery, Paul, 26

Vandals, 213

Venus, 329-30

Venus and Musician, 329-30

Vérités Fondamentales, Les, 72, 84

Vickrey, William S.: article, 59-71; discussed by Edel, 364-65

Vico, Giambattista, 353
Vietnam, 273
Vitruvius, 187

Walden II, 340
Wanderer and His Shadow, The, 253
War and Peace (television version),
  317
Warriors, The, 252-53
Washington Post, 315
Waste Land, The, 351
Watergate, 2, 273, 348
Wayland, Francis, 268
Weber, Max, 346, 364
Weiss, Paul, 254
Wells, H. G., 249, 252
Wernicke, Johann, 230
What Money Buys, 52
White, Andrew D., 269
White, Jr., Lynn, 151
Whitehead, Alfred North, 351
Whitman, Walt, 320, 380

Wiesner, Jerome, 308
Williams, Glanville, 89
Willowbrook, 158
World Council of Churches (WCC),
  148-49, 151
World Health Organization (WHO),
  174
World Trade Center (New York City),
  187, 188, 191-97, 200, 212; com-
  pared to John Hancock Tower,
  191-97, 198, 200-2
World War I, 215, 229
World War II, 215

Yale University, 268, 310
Yamasaki, Minoru, and Associates
  (Detroit), 194, 197

Zeus, 84, 338-39
Zola, Émile, 227
Zoroastrianism, 209